FINITE ELEMENT ANALYSIS

LICENSE, DISCLAIMER OF LIABILITY, AND LIMITED WARRANTY

By purchasing or using this book (the "Work"), you agree that this license grants permission to use the contents contained herein, but does not give you the right of ownership to any of the textual content in the book or ownership to any of the information or products contained in it. *This license does not permit uploading of the Work onto the Internet or on a network (of any kind) without the written consent of the Publisher.* Duplication or dissemination of any text, code, simulations, images, etc. contained herein is limited to and subject to licensing terms for the respective products, and permission must be obtained from the Publisher or the owner of the content, etc., in order to reproduce or network any portion of the textual material (in any media) that is contained in the Work.

MERCURY LEARNING AND INFORMATION ("MLI" or "the Publisher") and anyone involved in the creation, writing, or production of the companion disc, accompanying algorithms, code, or computer programs ("the software"), and any accompanying Web site or software of the Work, cannot and do not warrant the performance or results that might be obtained by using the contents of the Work. The author, developers, and the Publisher have used their best efforts to insure the accuracy and functionality of the textual material and/or programs contained in this package; we, however, make no warranty of any kind, express or implied, regarding the performance of these contents or programs. The Work is sold "as is" without warranty (except for defective materials used in manufacturing the book or due to faulty workmanship).

The author, developers, and the publisher of any accompanying content, and anyone involved in the composition, production, and manufacturing of this work will not be liable for damages of any kind arising out of the use of (or the inability to use) the algorithms, source code, computer programs, or textual material contained in this publication. This includes, but is not limited to, loss of revenue or profit, or other incidental, physical, or consequential damages arising out of the use of this Work.

Images of ANSYS menus, dialog boxes and plots are copyright of ANSYS Incorporation, United States of America and have been used with prior consent. Commercial software name, company name, other product trademarks, registered trademark logos are the properties of the ANSYS Incorporation, U.S.A.

The sole remedy in the event of a claim of any kind is expressly limited to replacement of the book, and only at the discretion of the Publisher. The use of "implied warranty" and certain "exclusions" vary from state to state, and might not apply to the purchaser of this product.

Finite Element Analysis
A Primer

By
S. M. Musa
A. V. Kulkarni
V. K. Havanur

Mercury Learning and Information
Dulles, Virginia
Boston, Massachusetts
New Delhi

Copyright ©2014 by MERCURY LEARNING AND INFORMATION LLC. All rights reserved.
Reprinted and revised with permission.

Original title and copyright: *A Primer on Finite Element Analysis.* Copyright © 2011 by Laxmi Publications Pvt. Ltd. All rights reserved. ISBN: 978-93-81159-10-1

This publication, portions of it, or any accompanying software may not be reproduced in any way, stored in a retrieval system of any type, or transmitted by any means, media, electronic display or mechanical display, including, but not limited to, photocopy, recording, Internet postings, or scanning, without prior permission in writing from the publisher.

Publisher: David Pallai
MERCURY LEARNING AND INFORMATION
22841 Quicksilver Drive
Dulles, VA 20166
info@merclearning.com
www.merclearning.com
1-800-758-3756

This book is printed on acid-free paper.

S. M. Musa, A. V. Kulkarni, V. K. Havanur. *FINITE ELEMENT ANALYSIS: A Primer.*
ISBN: 978-1-938549-34-2

The publisher recognizes and respects all marks used by companies, manufacturers, and developers as a means to distinguish their products. All brand names and product names mentioned in this book are trademarks or service marks of their respective companies. Any omission or misuse (of any kind) of service marks or trademarks, etc. is not an attempt to infringe on the property of others.

Images of ANSYS menus, dialog boxes and plots are copyright of ANSYS Incorporation, United States of America and have been used with prior consent. Commercial software name, company name, other product trademarks, registered trademark logos are the properties of the ANSYS Incorporation, U.S.A.

Library of Congress Control Number: 2013944461
131415321

Our titles are available for adoption, license, or bulk purchase by institutions, corporations, etc. For additional information, please contact the Customer Service Dept. at 1-800-758-3756 (toll free).

The sole obligation of MERCURY LEARNING AND INFORMATION to the purchaser is to replace the disc, based on defective materials or faulty workmanship, but not based on the operation or functionality of the product.

*Dedicated to my late father, Mahmoud,
my mother, Fatmeh, and my wife, Lama.*

Contents

Preface — xiii
Acknowledgments — xv

Chapter 1: Mathematical Preliminaries — 1
    1.1 Introduction — 1
    1.2 Matrix Definition — 1
    1.3 Types of Matrices — 3
    1.4 Addition or Subtraction of Matrices — 6
    1.5 Multiplication of a Matrix by Scalar — 8
    1.6 Multiplication of a Matrix by Another Matrix — 8
    1.7 Rules of Matrix Multiplications — 9
    1.8 Transpose of a Matrix Multiplication — 12
    1.9 Trace of a Matrix — 14
    1.10 Differentiation of a Matrix — 14
    1.11 Integration of a Matrix — 15
    1.12 Equality of Matrices — 15
    1.13 Determinant of a Matrix — 16
    1.14 Direct Methods for Linear Systems — 18
    1.15 Gaussian Elimination Method — 19
    1.16 Cramer's Rule — 21
    1.17 Inverse of a Matrix — 24
    1.18 Vector Analysis — 27
    1.19 Eigenvalues and Eigenvectors — 34
    1.20 Using MATLAB — 36
    Problems — 44
    References — 45

Chapter 2: Introduction to Finite Element Method — 47
    2.1 Introduction — 47
    2.2 Methods of Solving Engineering Problems — 47
        2.2.1 Experimental Method — 48
        2.2.2 Analytical Method — 48
        2.2.3 Numerical Method — 48
    2.3 Procedure of Finite Element Analysis (Related to Structural Problems) — 48
    2.4 Methods of Prescribing Boundary Conditions — 50
        2.4.1 Elimination Method — 50
        2.4.2 Penalty Method — 51
        2.4.3 Multipoint Constrains Method — 51
    2.5 Practical Applications of Finite Element Analysis — 51
    2.6 Finite Element Analysis Software Package — 52
    2.7 Finite Element Analysis for Structure — 52
    2.8 Types of Elements — 53
    2.9 Direct Method for Linear Spring — 55
    Problems — 56
    References — 57

Chapter 3: Finite Element Analysis of Axially Loaded Members — 59
    3.1 Introduction — 59
        3.1.1 Two-Node Bar Element — 62
        3.1.2 Three-Node Bar Element — 64
    3.2 Bars of Constant Cross-Section Area — 65
    3.3 Bars of Varying Cross-Section Area — 97
    3.4 Stepped Bar — 117
    Problems — 141
    References — 144

Chapter 4: Finite Element Analysis Trusses — 145
    4.1 Introduction — 145
    4.2 Truss — 145
    Problems — 171
    References — 174

Chapter 5: Finite Element Analysis of Beams — 175
    5.1 Introduction — 175
    5.2 Simply Supported Beams — 176

	5.3 Cantilever Beams	202
	Problems	232
	References	235

Chapter 6: Stress Analysis of a Rectangular Plate with a Circular Hole — 237
- 6.1 Introduction — 237
- 6.2 A Rectangular Plate with a Circular Hole — 238
- Problems — 251
- References — 255

Chapter 7: Thermal Analysis — 257
- 7.1 Introduction — 257
- 7.2 Procedure of Finite Element Analysis (Related to Thermal Problems) — 258
- 7.3 One-Dimensional Heat Conduction — 258
- 7.4 Two-Dimensional Problem with Conduction and with Convection Boundary Conditions — 285
- Problems — 287
- References — 289

Chapter 8: Fluid Flow Analysis — 291
- 8.1 Introduction — 291
- 8.2 Procedure of Finite Element Analysis (Related to Fluid Flow Problems) — 292
- 8.3 Potential Flow Over a Cylinder — 293
- 8.4 Potential Flow Around an Airfoil — 296
- Problems — 302
- References — 304

Chapter 9: Dynamic Analysis — 305
- 9.1 Introduction — 305
- 9.2 Procedure of Finite Element Analysis (Related to Dynamic Problems) — 306
- 9.3 Fixed-Fixed Beam for Natural Frequency Determination — 307
- 9.4 Transverse Vibrations of a Cantilever Beam — 316
- 9.5 Fixed-Fixed Beam Subjected to Forcing Function — 323
- 9.6 Axial Vibrations of a Bar — 335
- 9.7 Bar Subjected to Forcing Function — 345
- Problems — 347
- References — 349

Chapter 10: Engineering Electromagnetics Analysis   351
    10.1 Introduction to Electromagnetics   351
    10.2 Maxwell's Equations and Continuity Equation   351
           10.2.1 Maxwell's Equations and Continuity Equation in Differential Form   352
           10.2.2 Maxwell's Equations and Continuity Equation in Integral Form   353
           10.2.3 Divergence and Stokes Theorems   353
           10.2.4 Maxwell's Equations and Continuity Equation in Quasi-Statics Case   354
           10.2.5 Maxwell's Equations and Continuity Equation in Statics Case   354
           10.2.6 Maxwell's Equations and Continuity Equation in Source-Free Regions of Space Case   354
           10.2.7 Maxwell's Equations and Continuity Equation in Time-Harmonic Fields Case   355
    10.3 Lorentz Force Law and Continuity Equation   356
    10.4 Constitutive Relations   357
    10.5 Potential Equations   360
    10.6 Boundary Conditions   362
    10.7 Laws for Static Fields in Unbounded Regions   364
           10.7.1 Coulomb's Law and Field Intensity   364
           10.7.2 Bio-Savart's Law and Field Intensity   365
    10.8 Electromagnetic Energy and Power Flow   365
    10.9 Loss in Medium   369
    10.10 Skin Depth   370
    10.11 Poisson's and Laplace's Equations   371
    10.12 Wave Equations   372
    10.13 Electromagnetic Analysis   373
           10.13.1 One-Dimensional Elements   374
                  10.13.1.1 The Approach to FEM Standard Steps Procedure   374
                  10.13.1.2 Application to Poisson's Equation in One-Dimension   377
                  10.13.1.3 Natural Coordinates in One-Dimension   381
           10.13.2 Two-Dimensional Elements   382
                  10.13.2.1 Applications of FEM to Electrostatic Problems   382
                          10.13.2.1.1 Solution of Laplace's Equation $\nabla^2 V = 0$ with FEM   382

		10.13.2.1.2	Solution of Passion's Equation $\nabla^2 V = -\frac{\rho_v}{\varepsilon}$ with FEM	394
		10.13.2.1.3	Solution of Wave's Equation $\nabla^2 \Phi + k^2 \Phi = g$ with FEM	398

    10.14 Automatic Mesh Generation   401
              10.14.1 Rectangular Domains   401
              10.14.2 Arbitrary Domains   402
    10.15 Higher Order Elements   405
              10.15.1 Pascal Triangle   405
              10.15.2 Local Coordinates   406
              10.15.3 Shape Functions   409
              10.15.4 Fundamental Matrices   411
    10.16 Three-Dimensional Element   415
    10.17 Finite Element Methods for External Problems   420
              10.17.1 Infinite Element Method   420
              10.17.2 Boundary Element Method   422
              10.17.3 Absorbing Boundary Conditions   422
    10.18 Modeling and Simulation of Shielded Microstrip Lines with COMSOL Multiphysics   423
              10.18.1 Rectangular Cross-Section Transmission Line   426
              10.18.2 Square Cross-Section Transmission Line   427
              10.18.3 Rectangular Line with Diamondwise Structure   428
              10.18.4 A Single-Strip Shielded Transmission Line   429
    10.19 Multistrip Transmission Lines   432
              10.19.1 Double-strip Shielded Transmission Line   433
              10.19.2 Three-strip Line   435
              10.19.3 Six-strip Line   437
              10.19.4 Eight-strip Line   439
    10.20 Solenoid Actuator Analysis with ANSYS   441
    Problems   459
    References   463

Appendix A ANSYS   (On the companion disc)
Appendix B MATLAB   (On the companion disc)
Appendix C Color Figures   (On the companion disc)

Index   555

PREFACE

Today, the finite element method (FEM) has become a common tool for solving engineering problems in industries for the obvious reasons of its versatility and affordability. To expose an undergraduate student in engineering to this powerful method, most of the universities have included this subject in the undergraduate curriculum. This book contains materials applied to mechanical engineering, civil engineering, electrical engineering, and physics. It is written primarily for students and educators as a simple introduction to the practice of FEM analysis in engineering and physics. This book contains many 1D and 2D problems solved by the analytical method, by FEM using hand calculations, and by using ANSYS 11 academic teaching software and COMSOL. Results of all the methods have been compared. This book compromises 10 chapters and 3 appendices.

Chapter 1 contains mathematical preliminaries needed for understanding the chapters of the book. Chapter 2 provides a brief introduction to FEA, a theoretical background, and its applications. Chapter 3 contains the linear static analysis of bars of a constant cross-section, tapered cross-section, and stepped bar. In each section, a different variety of exercise problems is given. Chapter 4 contains the linear static analysis of trusses. Trusses problems are also selected in such a way that each problem has different boundary conditions to apply. Chapter 5 provides the linear static analysis of simply supported and cantilever beams. In Chapters 3 to 5, all the problems are considered as one dimensional in nature. Indeed, stress analysis of a rectangular plate with a circular hole is covered in Chapter 6. In this chapter, emphasis is given on the concept of exploiting symmetric geometry and symmetric loading conditions. Also, stress and deformation plots are given. Chapter 7 introduces the thermal analysis of cylinders and plates. Here both one dimensional and two dimensional problems are considered. Chapter 8 contains the problems of potential flow distribution over a cylinder and over an airfoil. Chapter 9 provides the dynamic analysis (modal and transient analysis) of bars and beams. Chapter 10 provides the engineering electromagnetics analysis. The chapter gives an overview of electromagnetics theory and provides the finite element method analysis toward the electromagnetics, some models are demonstrated using the COMSOL multiphysics application and also ANSYS.

The appendices are located on the companion disc in the back of the book. Appendix A contains the introduction to Classic ANSYS and ANSYS Workbench. Appendix B contains an overview of computational MATLAB. Appendix C contains the color figures in the book.

Sarhan M. Musa
Houston, Texas

Acknowledgments

Thank you to Brain Gaskin and James Gaskin for their wonderful hearts and for being great American neighbors. It is my pleasure to acknowledge the outstanding help and support of the team at Mercury Learning and Information in preparing this book, especially from David Pallai. My gratitude to Dr. John Burghduff and Professor Mary Jane Ferguson for their support, understanding, and for being great friends. I would also like to thank Dr. Kendall Harris, my college dean, for his constant support. Finally, the book would never have seen the light of day if not for the constant support, love, and patience of my family.

Chapter 1
MATHEMATICAL PRELIMINARIES

1.1 INTRODUCTION

This chapter introduces matrix and vector algebra that is essential in the formulation and solution of finite element problems. Finite element analysis procedures are most commonly described using matrix and vector notations. These procedures eventually lead to the solutions of a large set of simultaneous equations. This chapter will be a good help in understanding the remaining chapters of the book.

1.2 MATRIX DEFINITION

A *matrix* is an array of numbers or mathematical terms arranged in rows (horizontal lines) and columns (vertical lines). The numbers, or mathematical terms, in the matrix are called the *elements of the matrix*. We denote the matrix through this book, by a **boldface-letter**, a letter in brackets [ ], or a letter in braces {}. We sometimes use {} for a column matrix. Otherwise, we define the symbols of the matrices.

Example 1.1

The following are matrices.

$$\mathbf{A} = \begin{bmatrix} 0 & 1 \\ 3 & \pi \end{bmatrix}, \quad [B] = \begin{bmatrix} \sin\theta & 0 & 0 \\ 0 & \cos\theta & 0 \\ 0 & \tan\theta & 0 \end{bmatrix},$$

$$\{C\} = \begin{bmatrix} \int_0^3 x\,dx \\ \int_4^{11} y\,dy \end{bmatrix}, \quad [D] = \begin{bmatrix} \dfrac{\partial f(x,y)}{\partial x} & \dfrac{\partial f(x,y)}{\partial y} \end{bmatrix}, \quad \mathbf{E} = [e]$$

The size (dimension or order) of the matrices varies and is described by the number of rows (m) and the number of columns (n). Therefore, we write the size of a matrix as $m \times n$ (m by n). The sizes of the matrices in Example 1.1 are 2×2, 3×3, 2×1, 1×2, and 1×1, respectively.

We use a_{ij} to denote the element that occurs in row i and column j of matrix $\mathbf{A}$. In general, matrix $\mathbf{A}$ can be written

$$\mathbf{A} = [A] = \begin{bmatrix} a_{11} & a_{12} & \cdots & a_{1j} & \cdots & a_{1n} \\ a_{21} & a_{22} & \cdots & a_{2j} & \cdots & a_{2n} \\ \vdots & \vdots & & \vdots & & \vdots \\ a_{i1} & a_{i2} & \cdots & a_{ij} & \cdots & a_{in} \\ \vdots & \vdots & & \vdots & & \vdots \\ a_{m1} & a_{m2} & \cdots & a_{mj} & \cdots & a_{mn} \end{bmatrix} \quad (1.1)$$

Example 1.2

Location of an element in a matrix.

$$\text{Let } \mathbf{A} = \begin{bmatrix} a_{11} & a_{12} & a_{13} \\ a_{21} & a_{22} & a_{23} \\ a_{31} & a_{32} & a_{33} \end{bmatrix}$$

Find (a) size of the matrix $\mathbf{A}$
(b) location of elements a_{11}, a_{12}, a_{32}, and a_{33}

MATHEMATICAL PRELIMINARIES

Solution

(a) Size of the matrix **A** is 3×3
(b) a_{11} is element a at row 1 and column 1
a_{12} is element a at row 1 and column 2
a_{32} is element a at row 3 and column 2
a_{33} is element a at row 3 and column 3

Note that, two matrices are equal if they have the same size and their corresponding elements in the two matrices are equal. For example,

let, $[A] = \begin{bmatrix} 1 & 3 & 7 \end{bmatrix}, [B] = \begin{bmatrix} \pi & 0 \\ 1 & e \end{bmatrix}, [C] = \begin{bmatrix} e & 0 \\ 1 & \pi \end{bmatrix}$, then $[A] \neq [B]$ since $[A]$ and $[B]$

are not the same size. Also, $[B] \neq [C]$ since the corresponding elements are not all equal.

1.3 TYPES OF MATRICES

The types of matrices are based on the number of rows (m) and the number of columns (n) in addition to the nature of elements and the way the elements are arranged in the m atrix.

(a) **Rectangular matrix** is a matrix of different number of rows and columns, that is, $m \neq n$. For example, the matrix

$$[X] = \begin{bmatrix} 1 & 2 \\ -3 & 5 \\ 7 & 0 \end{bmatrix}, \text{ is rectangular matrix.}$$

(b) **Square matrix** is a matrix of equal number of rows and columns, that is, $m = n$. For example, the matrix

$$[K] = \begin{bmatrix} k_1 & k_2 \\ k_3 & k_4 \end{bmatrix}, \text{ is square matrix.}$$

(c) **Row matrix** is a matrix that has one row and has more than one column, that is, $m = 1$ and $n > 1$. For example, the matrix

$$[F] = \begin{bmatrix} x & y & z \end{bmatrix}, \text{ is row matrix.}$$

(d) **Column matrix** is a matrix that has one column and has more than one row, that is, $n = 1$ and $m > 1$. For example, the matrix

$$\mathbf{N} = \{N\} = \begin{Bmatrix} 0 \\ 2 \\ 4 \end{Bmatrix}, \text{ is column matrix.}$$

(e) **Scalar matrix** is a matrix that has the number of columns and the number of rows equal to 1, that is, $m = 1$ and $n = 1$. For example, the matrix

$[M] = [7]$, is a scalar matrix; we can write it as 7 without bracket.

(f) **Null matrix** is a matrix whose elements are all zero. For example, the matrix

$$\begin{bmatrix} 0 & 0 & 0 \\ 0 & 0 & 0 \end{bmatrix}, \text{ is a null matrix.}$$

(g) **Diagonal matrix** is a square matrix that has zero elements everywhere except on its main diagonal. That is, for diagonal matrix $a_{ij} = 0$ when $i \neq j$ and not all are zero for a_{ii} when $i = j$. For example, the matrix

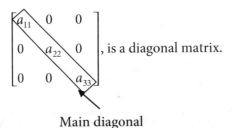

Main diagonal

Main diagonal elements have equal row and column subscripts. The main diagonal runs from the upper-left corner to the lower-right corner. The main diagonal of the matrix here is a_{11}, a_{22}, and a_{33}.

(h) **Identity (unit) matrix** $[I]$ or **I**, is a diagonal matrix whose main diagonal elements are equal to unity (1's) for any square matrix. That is, if the elements of an identity matrix are denoted as e_{ij}, then

$$e_{ij} = \begin{cases} 1, & i = j \\ 0, & i \neq j \end{cases}. \tag{1.2}$$

For example, the matrix

$$[I] = \begin{bmatrix} 1 & 0 & 0 & 0 \\ 0 & 1 & 0 & 0 \\ 0 & 0 & 1 & 0 \\ 0 & 0 & 0 & 1 \end{bmatrix}, \text{ is an identity matrix.}$$

(i) **Banded matrix** is a square matrix that has a band of nonzero elements parallel to its main diagonal. For example, the matrix

$$\begin{bmatrix} a_{11} & a_{12} & 0 & 0 & 0 \\ a_{21} & a_{22} & a_{23} & 0 & 0 \\ 0 & a_{32} & a_{33} & a_{34} & 0 \\ 0 & 0 & a_{43} & a_{44} & a_{45} \\ 0 & 0 & 0 & a_{54} & a_{55} \end{bmatrix}, \text{ is a banded matrix.}$$

(j) **Symmetric matrix** is a square matrix whose elements satisfy the condition $a_{ij} = a_{ji}$ for $i \neq j$. For example, the matrix

$$\begin{bmatrix} a_{11} & 5 & 8 \\ 5 & a_{22} & 2 \\ 8 & 2 & a_{33} \end{bmatrix}, \text{ is a symmetric matrix.}$$

(k) **Anti-symmetric (Skew-symmetric) matrix** is a square matrix whose elements $a_{ij} = -a_{ji}$ for $i \neq j$, and $a_{ii} = 0$. For example, the matrix

$$\begin{bmatrix} 0 & 3 & -7 \\ -3 & 0 & 2 \\ 7 & -2 & 0 \end{bmatrix}, \text{ is an anti-symmetric matrix.}$$

(l) **Triangular matrix** is a square matrix whose elements on one side of the main diagonal are all zero. There are two types of triangular matrices; first, an *upper triangular matrix* whose elements below the main diagonal are zero, that is, $a_{ij} = 0$ for $i > j$; second, a *lower triangular matrix* whose elements above the main diagonal are all zero, that is $a_{ij} = 0$ for $i < j$. For example, the matrix

$$\begin{bmatrix} a_{11} & a_{12} & a_{13} \\ 0 & a_{22} & a_{23} \\ 0 & 0 & a_{33} \end{bmatrix}, \text{ is an upper triangular matrix.}$$

While, the matrix

$$\begin{bmatrix} a_{11} & 0 & 0 \\ a_{21} & a_{22} & 0 \\ a_{31} & a_{32} & a_{33} \end{bmatrix}, \text{ is a lower triangular matrix.}$$

(m) **Partitioned matrix (Super-matrix)** is a matrix that can be divided into smaller arrays (*sub-matrices*) by horizontal and vertical lines, that is, the elements of the partitioned matrix are matrices. For example, the matrix

$$\begin{bmatrix} a_{11} & a_{12} & \vdots & a_{13} \\ a_{21} & a_{22} & \vdots & a_{23} \\ a_{31} & a_{32} & \vdots & a_{33} \end{bmatrix} = \begin{bmatrix} \mathbf{A} & \mathbf{B} \\ \mathbf{C} & \mathbf{D} \end{bmatrix}, \text{ is partitioned matrix with four smaller matrices,}$$

where

$\mathbf{A} = \begin{bmatrix} a_{11} & a_{12} \end{bmatrix}$, $\mathbf{B} = \begin{bmatrix} a_{13} \end{bmatrix}$, $\mathbf{C} = \begin{bmatrix} a_{21} & a_{22} \\ a_{31} & a_{32} \end{bmatrix}$, and $\mathbf{D} = \begin{bmatrix} a_{23} \\ a_{33} \end{bmatrix}$. For example, the matrix

$$\begin{bmatrix} 0 & 1 & 5 \\ 8 & 3 & 4 \\ 6 & 2 & 9 \end{bmatrix} = \begin{bmatrix} \mathbf{A} & \mathbf{B} \\ \mathbf{C} & \mathbf{D} \end{bmatrix}, \text{ is a partitioned matrix,}$$

where $\mathbf{A} = \begin{bmatrix} 0 \\ 8 \end{bmatrix}$, $\mathbf{B} = \begin{bmatrix} 1 & 5 \\ 3 & 4 \end{bmatrix}$, $\mathbf{C} = \begin{bmatrix} 6 \end{bmatrix}$, and $\mathbf{D} = \begin{bmatrix} 2 & 9 \end{bmatrix}$.

1.4 ADDITION OR SUBTRACTION OF MATRICES

Addition and subtraction of matrices can only be performed for matrices of the same size, that is, the matrices must have same number of rows and columns. The addition is accomplished by adding corresponding elements of each matrix. For example, for addition of two matrices **A** and **B**, can give **C** matrix, that is, $\mathbf{C} = \mathbf{A} + \mathbf{B}$ implies that $c_{ij} = a_{ij} + b_{ij}$. Where c_{ij}, a_{ij}, and b_{ij} are typical elements of the **C**, **A**, and **B** matrices, respectively.

Now, the subtraction of matrices is accomplished by subtracting corresponding elements of each matrix. For example, for subtraction of two matrices **A** and **B**, can give you **C** matrix, that is, $\mathbf{C} = \mathbf{A} - \mathbf{B}$ implies that $c_{ij} = a_{ij} - b_{ij}$. Note that, both **A** and **B** matrices are the same size, $m \times n$, then the resulting matrix **C** is also of size $m \times n$.

For example, let $[A] = \begin{bmatrix} 1 & 2 \\ 5 & 7 \end{bmatrix}$ and $[B] = \begin{bmatrix} 0 & 6 \\ 9 & 12 \end{bmatrix}$, then

$$[A]+[B] = \begin{bmatrix} 1 & 2 \\ 5 & 7 \end{bmatrix} + \begin{bmatrix} 0 & 6 \\ 9 & 12 \end{bmatrix} = \begin{bmatrix} 1+0 & 2+6 \\ 5+9 & 7+12 \end{bmatrix} = \begin{bmatrix} 1 & 8 \\ 14 & 19 \end{bmatrix}, \text{ and}$$

$$[A]-[B] = \begin{bmatrix} 1 & 2 \\ 5 & 7 \end{bmatrix} - \begin{bmatrix} 0 & 6 \\ 9 & 12 \end{bmatrix} = \begin{bmatrix} 1-0 & 2-6 \\ 5-9 & 7-12 \end{bmatrix} = \begin{bmatrix} 1 & -4 \\ -4 & -5 \end{bmatrix}.$$

Matrices addition and subtraction are associative; that is

$$\begin{aligned} \mathbf{A}+\mathbf{B}+\mathbf{C} &= (\mathbf{A}+\mathbf{B})+\mathbf{C} = \mathbf{A}+(\mathbf{B}+\mathbf{C}) \\ \mathbf{A}+\mathbf{B}-\mathbf{C} &= (\mathbf{A}+\mathbf{B})-\mathbf{C} = \mathbf{A}+(\mathbf{B}-\mathbf{C}) \end{aligned} \qquad (1.3)$$

For example,

let $[A] = \begin{bmatrix} 1 & 3 \\ 7 & 8 \end{bmatrix}$, $[B] = \begin{bmatrix} 2 & 5 \\ 3 & 1 \end{bmatrix}$, and $[C] = \begin{bmatrix} 9 & 8 \\ 4 & 6 \end{bmatrix}$.

Then, $([A]+[B])+[C] = \begin{bmatrix} 1+2 & 3+5 \\ 7+3 & 8+1 \end{bmatrix} + \begin{bmatrix} 9 & 8 \\ 4 & 6 \end{bmatrix} = \begin{bmatrix} 12 & 16 \\ 14 & 15 \end{bmatrix}$

$[A]+([B]+[C]) = \begin{bmatrix} 1 & 3 \\ 7 & 8 \end{bmatrix} + \begin{bmatrix} 2+9 & 5+8 \\ 3+4 & 1+6 \end{bmatrix} = \begin{bmatrix} 12 & 16 \\ 14 & 15 \end{bmatrix}.$

Therefore, $(A+B) + C = A + (B + C)$.
Matrices addition and subtraction are commutative; that is

$$\begin{aligned} \mathbf{A}+\mathbf{B} &= \mathbf{B}+\mathbf{A} \\ \mathbf{A}-\mathbf{B} &= -\mathbf{B}+\mathbf{A} \end{aligned} \qquad (1.4)$$

For example,

let $[A] = \begin{bmatrix} 6 & 5 \\ 2 & 1 \end{bmatrix}$ and $[B] = \begin{bmatrix} 3 & 2 \\ 1 & 5 \end{bmatrix}$, then $[A]+[B] = \begin{bmatrix} 6 & 5 \\ 2 & 1 \end{bmatrix} + \begin{bmatrix} 3 & 2 \\ 1 & 5 \end{bmatrix} = \begin{bmatrix} 9 & 7 \\ 3 & 6 \end{bmatrix}$,

and $[B]+[A] = \begin{bmatrix} 3 & 2 \\ 1 & 5 \end{bmatrix} + \begin{bmatrix} 6 & 5 \\ 2 & 1 \end{bmatrix} = \begin{bmatrix} 9 & 7 \\ 3 & 6 \end{bmatrix}$, therefore, $[A] + [B] = [B] + [A]$.

1.5 MULTIPLICATION OF A MATRIX BY SCALAR

A matrix is multiplied by a scalar, c, by multiplying each element of the matrix by this scalar. That is, the multiplication of a matrix $[A]$ by a scalar c is defined as

$$c[A] = [ca_{ij}]. \qquad (1.5)$$

The scalar multiplication is commutative.
For example,

$$\text{Let } [A] = \begin{bmatrix} -3 & 1 \\ 4 & 2 \end{bmatrix}, \text{ then } 5[A] = \begin{bmatrix} -15 & 5 \\ 20 & 10 \end{bmatrix}.$$

1.6 MULTIPLICATION OF A MATRIX BY ANOTHER MATRIX

The product of two matrices is $\mathbf{C} = \mathbf{AB}$, if and only if, the number of columns in $\mathbf{A}$ is equal to the number of rows in $\mathbf{B}$. The product of matrix $\mathbf{A}$ of size $m \times n$ and matrix $\mathbf{B}$ of size $n \times r$, the result in matrix $\mathbf{C}$ has size $m \times r$.

$$\text{That is, } [A]_{m \times n}[B]_{n \times r} = [C]_{m \times r}, \qquad (1.6)$$

must be equal

$$\text{and } c_{ij} = \sum_{k=1}^{n} a_{ik} b_{kj}, \qquad (1.7)$$

where, the (ij)th component of matrix $\mathbf{C}$ is obtained by taking the dot product $C_{ij} = (i\text{th row of } \mathbf{A}) \cdot (j\text{th column of } \mathbf{B})$.

That is, to find the element in row i and column j of $[A][B]$, you need to single out row i from $[A]$ and column j from $[B]$, then multiply the corresponding elements from the row and column together and add up the resulting products.

For example,

$$\text{let } [A]_{2 \times 2} = \begin{bmatrix} 1 & 3 \\ 7 & 8 \end{bmatrix} \text{ and } [B]_{2 \times 3} = \begin{bmatrix} 3 & 2 & 4 \\ -1 & 0 & 6 \end{bmatrix}, \text{ then}$$

MATHEMATICAL PRELIMINARIES

$$[A][B] = \begin{bmatrix} 1 & 3 \\ 7 & 8 \end{bmatrix} \begin{bmatrix} 3 & 2 & 4 \\ -1 & 0 & 6 \end{bmatrix} = \begin{bmatrix} 1\times 3 + 3\times(-1) & 1\times 2 + 3\times 0 & 1\times 4 + 3\times 6 \\ 7\times 3 + 8\times(-1) & 7\times 2 + 8\times 0 & 7\times 4 + 8\times 6 \end{bmatrix}$$

$$= \begin{bmatrix} 0 & 2 & 22 \\ 13 & 14 & 76 \end{bmatrix}$$

Size of $[A][B] = 2 \times 3$.

1.7 RULES OF MATRIX MULTIPLICATIONS

Matrix multiplication is associative; that is

$$\mathbf{ABC} = (\mathbf{AB})\mathbf{C} = \mathbf{A}(\mathbf{BC}). \tag{1.8}$$

Matrix multiplication is distributive; that is

$$\mathbf{A}(\mathbf{B}+\mathbf{C}) = \mathbf{AB} + \mathbf{AC} \tag{1.9}$$

or

$$(\mathbf{A}+\mathbf{B})\mathbf{C} = \mathbf{AC} + \mathbf{BC}. \tag{1.10}$$

Matrix multiplication is not commutative; that is

$$\mathbf{AB} \neq \mathbf{BA}. \tag{1.11}$$

A square matrix multiplied by its identity matrix is equal to same matrix; that is

$$\mathbf{AI} = \mathbf{IA} = \mathbf{A}. \tag{1.12}$$

A square matrix can be raised to an integer power n; that is

$$\mathbf{A}^n = \overbrace{\mathbf{A}\,\mathbf{A}\ldots\mathbf{A}}^{n}. \tag{1.13}$$

A same square matrix multiplication with different integer power n and m can be given as

$$\mathbf{A}^n \mathbf{A}^m = \mathbf{A}^{n+m} \text{ and } \left(\mathbf{A}^n\right)^m = \mathbf{A}^{nm}. \tag{1.14}$$

Transpose of product of matrices rule is given as

$$(\mathbf{AB})^T = \left(\mathbf{B}^T \mathbf{A}^T\right), (\mathbf{ABC})^T = \mathbf{C}^T \mathbf{B}^T \mathbf{A}^T. \tag{1.15}$$

Example 1.3

Given matrices

$$\{A\} = \begin{Bmatrix} 2 \\ 4 \\ 1 \\ 3 \end{Bmatrix}, [B] = \begin{bmatrix} 6 & 1 & 2 & -1 \\ 4 & -3 & 5 & 9 \\ 8 & -2 & 6 & 7 \\ 0 & 7 & -8 & 3 \end{bmatrix}, [C] = \begin{bmatrix} 4 & 0 & -3 & 2 \\ 1 & 8 & 4 & -4 \\ 5 & 3 & -2 & 6 \\ 9 & -1 & 0 & 7 \end{bmatrix},$$

$$[D] = \begin{bmatrix} -1 & 1 & 0 \\ 2 & 3 & -1 \\ 4 & 0 & 5 \end{bmatrix}$$

Find the following:
a. $[B]+[C]$
b. $[B]-[C]$
c. $5\{A\}$
d. $[B]\{A\}$
e. $[D]^2$
f. show that $[D][I] = [I][D] = [D]$

Solution

$$\text{a. } [B]+[C] = \begin{bmatrix} 6 & 1 & 2 & -1 \\ 4 & -3 & 5 & 9 \\ 8 & -2 & 6 & 7 \\ 0 & 7 & -8 & 3 \end{bmatrix} + \begin{bmatrix} 4 & 0 & -3 & 2 \\ 1 & 8 & 4 & -4 \\ 5 & 3 & -2 & 6 \\ 9 & -1 & 0 & 7 \end{bmatrix} = \begin{bmatrix} (6+4) & (1+0) & (2-3) & (-1+2) \\ (4+1) & (-3+8) & (5+4) & (9-4) \\ (8+5) & (-2+3) & (6-2) & (7+6) \\ (0+9) & (7-1) & (-8+0) & (3+7) \end{bmatrix}$$

$$= \begin{bmatrix} 10 & 1 & -1 & 1 \\ 5 & 5 & 9 & 5 \\ 13 & 1 & 4 & 13 \\ 9 & 6 & -8 & 10 \end{bmatrix}$$

b. $[B]-[C] = \begin{bmatrix} 6 & 1 & 2 & -1 \\ 4 & -3 & 5 & 9 \\ 8 & -2 & 6 & 7 \\ 0 & 7 & -8 & 3 \end{bmatrix} - \begin{bmatrix} 4 & 0 & -3 & 2 \\ 1 & 8 & 4 & -4 \\ 5 & 3 & -2 & 6 \\ 9 & -1 & 0 & 7 \end{bmatrix} = \begin{bmatrix} (6-4) & (1-0) & (2+3) & (-1-2) \\ (4-1) & (-3-8) & (5-4) & (9+4) \\ (8-5) & (-2-3) & (6+2) & (7-6) \\ (0-9) & (7+1) & (-8-0) & (3-7) \end{bmatrix}$

$= \begin{bmatrix} 2 & 1 & 5 & -3 \\ 3 & -11 & 1 & 13 \\ 3 & -5 & 8 & 1 \\ -9 & 8 & -8 & -4 \end{bmatrix}$

c. $5\{A\} = 5 \begin{Bmatrix} 2 \\ 4 \\ 1 \\ 3 \end{Bmatrix} = \begin{Bmatrix} 10 \\ 20 \\ 5 \\ 15 \end{Bmatrix}$

d. $[B]\{A\} = \begin{bmatrix} 6 & 1 & 2 & -1 \\ 4 & -3 & 5 & 9 \\ 8 & -2 & 6 & 7 \\ 0 & 7 & -8 & 3 \end{bmatrix} \begin{Bmatrix} 2 \\ 4 \\ 1 \\ 3 \end{Bmatrix} = \begin{Bmatrix} (6 \times 2)+(1 \times 4)+(2 \times 1)+(-1 \times 3) \\ (4 \times 2)+(-3 \times 4)+(5 \times 1)+(9 \times 3) \\ (8 \times 2)+(-2 \times 4)+(6 \times 1)+(7 \times 3) \\ (0 \times 2)+(7 \times 4)+(-8 \times 1)+(3 \times 3) \end{Bmatrix} = \begin{Bmatrix} 13 \\ 28 \\ 35 \\ 29 \end{Bmatrix}$

e. $[D]^2 = [D][D] = \begin{bmatrix} -1 & 1 & 0 \\ 2 & 3 & -1 \\ 4 & 0 & 5 \end{bmatrix} \begin{bmatrix} -1 & 1 & 0 \\ 2 & 3 & -1 \\ 4 & 0 & 5 \end{bmatrix}$

$= \begin{bmatrix} (-1 \times -1)+(1 \times 2)+(0 \times 4) & (-1 \times 1)+(1 \times 3)+(0 \times 0) & (-1 \times 0)+(1 \times -1)+(0 \times 5) \\ (2 \times -1)+(3 \times 2)+(-1 \times 4) & (2 \times 1)+(3 \times 3)+(-1 \times 0) & (2 \times 0)+(3 \times -1)+(-1 \times 5) \\ (4 \times -1)+(0 \times 2)+(5 \times 4) & (4 \times 1)+(0 \times 3)+(5 \times 0) & (4 \times 0)+(0 \times -1)+(5 \times 5) \end{bmatrix}$

$= \begin{bmatrix} 3 & 2 & -1 \\ 0 & 10 & -8 \\ 16 & 4 & 25 \end{bmatrix}$

f. $[D][I] = \begin{bmatrix} -1 & 1 & 0 \\ 2 & 3 & -1 \\ 4 & 0 & 5 \end{bmatrix} \begin{bmatrix} 1 & 0 & 0 \\ 0 & 1 & 0 \\ 0 & 0 & 1 \end{bmatrix}$

$= \begin{bmatrix} (-1 \times 1)+(1 \times 0)+(0 \times 0) & (-1 \times 0)+(1 \times 1)+(0 \times 0) & (-1 \times 0)+(1 \times 0)+(0 \times 1) \\ (2 \times 1)+(3 \times 0)+(-1 \times 0) & (2 \times 0)+(3 \times 1)+(-1 \times 0) & (2 \times 0)+(3 \times 0)+(-1 \times 1) \\ (4 \times 1)+(0 \times 0)+(5 \times 0) & (4 \times 0)+(0 \times 1)+(5 \times 0) & (4 \times 0)+(0 \times 0)+(5 \times 1) \end{bmatrix}$

$= \begin{bmatrix} -1 & 1 & 0 \\ 2 & 3 & -1 \\ 4 & 0 & 5 \end{bmatrix} = [D]$

and

$[I][D] = \begin{bmatrix} 1 & 0 & 0 \\ 0 & 1 & 0 \\ 0 & 0 & 1 \end{bmatrix} \begin{bmatrix} -1 & 1 & 0 \\ 2 & 3 & -1 \\ 4 & 0 & 5 \end{bmatrix}$

$= \begin{bmatrix} (1 \times -1)+(0 \times 2)+(0 \times 4) & (1 \times 1)+(0 \times 3)+(0 \times 0) & (1 \times 0)+(0 \times -1)+(0 \times 5) \\ (0 \times -1)+(1 \times 2)+(0 \times 4) & (0 \times 1)+(1 \times 3)+(0 \times 0) & (0 \times 0)+(1 \times -1)+(0 \times 5) \\ (0 \times -1)+(0 \times 2)+(1 \times 4) & (0 \times 1)+(0 \times 3)+(1 \times 0) & (0 \times 0)+(0 \times -1)+(1 \times 5) \end{bmatrix}$

$= \begin{bmatrix} -1 & 1 & 0 \\ 2 & 3 & -1 \\ 4 & 0 & 5 \end{bmatrix} = [D]$

1.8 TRANSPOSE OF A MATRIX MULTIPLICATION

The *transpose* of a matrix $\mathbf{A} = [a_{ij}]$ is denoted as $\mathbf{A}^T = [a_{ji}]$. It is obtained by interchanging the rows and columns in matrix $\mathbf{A}$. Thus, if a matrix $\mathbf{A}$ is of order $m \times n$, then $\mathbf{A}^T$ will be of order $n \times m$.

For example,

$$\text{let } [A]_{2 \times 3} = \begin{bmatrix} 0 & 1 & 3 \\ -1 & 2 & 5 \end{bmatrix}, \text{ then } [A]^T_{3 \times 2} = \begin{bmatrix} 0 & -1 \\ 1 & 2 \\ 3 & 5 \end{bmatrix}.$$

Note that it is valid that, $(\mathbf{AB})^T = \mathbf{B}^T\mathbf{A}^T$, $(\mathbf{A}+\mathbf{B})^T = \mathbf{A}^T+\mathbf{B}^T$, $(c\mathbf{B})^T = c\mathbf{B}^T$, and $(\mathbf{A}^T)^T = \mathbf{A}$. Also note, if $\mathbf{A}^T = \mathbf{A}$, then $\mathbf{A}$ is a symmetric matrix.

Example 1.4

Consider that matrix $[A] = \begin{bmatrix} 1 & 2 \\ 3 & 4 \end{bmatrix}$ and $[B] = \begin{bmatrix} -1 & 0 & -3 \\ -4 & -2 & 5 \end{bmatrix}$.

Show that $([A][B])^T = [B]^T[A]^T$.

Solution

$$([A][B]) = \begin{bmatrix} 1 & 2 \\ 3 & 4 \end{bmatrix}\begin{bmatrix} -1 & 0 & -3 \\ -4 & -2 & 5 \end{bmatrix} = \begin{bmatrix} (1\times-1)+(2\times-4) & (1\times0)+(2\times-2) & (1\times-3)+(2\times5) \\ (3\times-1)+(4\times-4) & (3\times0)+(4\times-2) & (3\times-3)+(4\times5) \end{bmatrix}$$

$$= \begin{bmatrix} -9 & -4 & 7 \\ -19 & -8 & 11 \end{bmatrix}$$

$$([A][B])^T = \begin{bmatrix} -9 & -19 \\ -4 & -8 \\ 7 & 11 \end{bmatrix}$$

$$[A]^T = \begin{bmatrix} 1 & 3 \\ 2 & 4 \end{bmatrix}, [B]^T = \begin{bmatrix} -1 & -4 \\ 0 & -2 \\ -3 & 5 \end{bmatrix}$$

$$[B]^T[A]^T = \begin{bmatrix} -1 & -4 \\ 0 & -2 \\ -3 & 5 \end{bmatrix}\begin{bmatrix} 1 & 3 \\ 2 & 4 \end{bmatrix} = \begin{bmatrix} (-1\times1)+(-4\times2) & (-1\times3)+(-4\times4) \\ (0\times1)+(-2\times2) & (0\times3)+(-2\times4) \\ (-3\times1)+(5\times2) & (-3\times3)+(5\times4) \end{bmatrix}$$

$$= \begin{bmatrix} -9 & -19 \\ -4 & -8 \\ 7 & 11 \end{bmatrix}$$

Therefore, $([A][B])^T = [B]^T[A]^T$.

1.9 TRACE OF A MATRIX

A *trace* of a matrix **A**, tr(**A**), is a square matrix and is defined to be the sum of the elements on the main diagonal of matrix **A**.

For example, let, $[A] = \begin{bmatrix} 3 & 5 & 8 \\ 5 & 7 & 2 \\ 8 & 2 & -1 \end{bmatrix}$, then tr(**A**) = 3 + 7 + (−1) = 9.

Example 1.5

Consider that matrix

$$[A] = \begin{bmatrix} 5 & 8 \\ 7 & 6 \end{bmatrix}.$$

Find the tr(A).

Solution

$$\text{tr}(A) = 5 + 6 = 11.$$

1.10 DIFFERENTIATION OF A MATRIX

Differentiation of a matrix is differentiation of every element of the matrix separately. For example, if the elements of the matrix **A** are a function of t, then

$$\frac{d\mathbf{A}}{dt} = \left[\frac{da_{ij}}{dt}\right]. \tag{1.16}$$

Example 1.6

Consider the matrix $[A] = \begin{bmatrix} 3x^5 & x^2 \\ 7x & 6 \end{bmatrix}$, find the derivative $\frac{d[A]}{dx}$.

Solution

$$\frac{d[A]}{dx} = \begin{bmatrix} 15x^4 & 2x \\ 7 & 0 \end{bmatrix}$$

1.11 INTEGRATION OF A MATRIX

Integration of a matrix is integration of every element of the matrix separately. For example, if the elements of the matrix **A** are a function of t, then

$$\int \mathbf{A}\, dt = \left[\int a_{ij} dt\right]. \qquad (1.17)$$

Example 1.7

Consider the matrix $[A] = \begin{bmatrix} 4x^3 & 2 \\ 8x & 1 \end{bmatrix}$, find the derivative $\int [A]\, dx$.

Solution

$$\int [A]\, dx = \begin{bmatrix} x^4 & 2x \\ 4x^2 & x \end{bmatrix}$$

1.12 EQUALITY OF MATRICES

Two matrices are equal if they have the same sizes and their corresponding elements are equal.

Example 1.8

Let $\mathbf{A} = \begin{bmatrix} 1 & -4 \\ 5 & 3 \end{bmatrix}$ and $\mathbf{B} = \begin{bmatrix} 2x & w \\ z-2 & k+1 \end{bmatrix}$.

If the matrices **A** and **B** are equal, find the value of $x, w, z,$ and k.

Solution

$1 = 2x \longrightarrow x = \frac{1}{2}$
$w = -4$
$z - 2 = 5 \longrightarrow z = 7$
$3 = k + 1 \longrightarrow k = 2$

1.13 DETERMINANT OF A MATRIX

The determinant of a square matrix **A** is a scalar number denoted by $|A|$ or det $[A]$. It is the sum of the products $(-1)^{i+j} a_{ij} M_{ij}$, where a_{ij} are the elements along any one row or column and M_{ij} are the deleted elements of ith row and jth column from the matrix $[A]$.

For example, the value of the determinant of matrix $[A]$ is a and can be obtained by expanding along the first row as:

$$a = \begin{vmatrix} a_{11} & a_{12} & a_{13} & \cdots & a_{1n} \\ a_{21} & a_{22} & a_{23} & \cdots & a_{2n} \\ a_{31} & a_{32} & a_{33} & \cdots & a_{3n} \\ \vdots & \vdots & \vdots & \cdots & \vdots \\ a_{n1} & a_{n2} & a_{n3} & \cdots & a_{nn} \end{vmatrix} \quad (1.18)$$

$$= a_{11} M_{11} - a_{12} M_{12} + a_{13} M_{13} + \ldots + (-1)^{n+1} a_{1n} M_{1n}$$

where the minor M_{ij} is a $(n-1) \times (n-1)$ determinant of the matrix formed by removing the ith row and jth column.

Also, the value a can be obtained by expanding along the first column as:

$$a = a_{11} M_{11} - a_{21} M_{21} + a_{31} M_{31} + \ldots + (-1)^{n+1} a_{n1} M_{n1}. \quad (1.19)$$

Now, the value of a second-order determinant of (2×2) matrix is calculated by

$$a = \det \begin{bmatrix} a_{11} & a_{12} \\ a_{21} & a_{22} \end{bmatrix} = \begin{vmatrix} a_{11} & a_{12} \\ a_{21} & a_{22} \end{vmatrix} = a_{11} a_{22} - a_{12} a_{21}. \quad (1.20)$$

The value of a third-order determinate of (3×3) matrix is calculated by

$$a = \det \begin{bmatrix} a_{11} & a_{12} & a_{13} \\ a_{21} & a_{22} & a_{23} \\ a_{31} & a_{32} & a_{33} \end{bmatrix} = \begin{vmatrix} a_{11} & a_{12} & a_{13} \\ a_{21} & a_{22} & a_{23} \\ a_{31} & a_{32} & a_{33} \end{vmatrix}$$

$$= a_{11}(-1)^2 \begin{vmatrix} a_{22} & a_{23} \\ a_{32} & a_{33} \end{vmatrix} + a_{12}(-1)^3 \begin{vmatrix} a_{21} & a_{23} \\ a_{31} & a_{33} \end{vmatrix} + a_{13}(-1)^4 \begin{vmatrix} a_{21} & a_{22} \\ a_{31} & a_{32} \end{vmatrix}$$

$$= a_{11} \begin{vmatrix} a_{22} & a_{23} \\ a_{32} & a_{33} \end{vmatrix} - a_{12} \begin{vmatrix} a_{21} & a_{23} \\ a_{31} & a_{33} \end{vmatrix} + a_{13} \begin{vmatrix} a_{21} & a_{22} \\ a_{31} & a_{32} \end{vmatrix}$$

$$= a_{11}(a_{22} a_{33} - a_{23} a_{32}) - a_{12}(a_{21} a_{33} - a_{23} a_{31}) + a_{13}(a_{21} a_{32} - a_{22} a_{31}). \quad (1.21)$$

Example 1.9

Find the value, a, of the following determinants:

a. $\begin{vmatrix} 2 & 3 \\ -1 & 4 \end{vmatrix}$

b. $\begin{vmatrix} 1 & 3 & 4 \\ -2 & -1 & 2 \\ 5 & -4 & 6 \end{vmatrix}$

Solution

a. $a = \begin{vmatrix} 2 & 3 \\ -1 & 4 \end{vmatrix} = (2 \times 4) - (3 \times -1) = 8 + 3 = 11$

b. $a = \begin{vmatrix} 1 & 3 & 4 \\ -2 & -1 & 2 \\ 5 & -4 & 6 \end{vmatrix} = 1 \times \begin{vmatrix} -1 & 2 \\ -4 & 6 \end{vmatrix} - 3 \times \begin{vmatrix} -2 & 2 \\ 5 & 6 \end{vmatrix} + 4 \times \begin{vmatrix} -2 & -1 \\ 5 & -4 \end{vmatrix}$

$= 1(-6+8) - 3(-12-10) + 4(8+5) = 2 + 66 + 52 = 120$

An alternative method of obtaining the determinant of a (3×3) matrix is by using the sign rule of each term that is determined by the first row in the diagram as follows:

$\begin{vmatrix} + & - & + \\ - & + & - \\ + & - & + \end{vmatrix}$, or by repeating the first two rows and multiplying the terms diagonally as follows:

$a = \begin{vmatrix} a_{11} & a_{12} & a_{13} \\ a_{21} & a_{22} & a_{23} \\ a_{31} & a_{32} & a_{33} \\ a_{11} & a_{12} & a_{13} \\ a_{21} & a_{22} & a_{23} \end{vmatrix}$

$= a_{11}a_{22}a_{33} + a_{21}a_{32}a_{13} + a_{31}a_{12}a_{23} - a_{13}a_{22}a_{31} - a_{23}a_{32}a_{11} - a_{33}a_{12}a_{21}.$

1.14 DIRECT METHODS FOR LINEAR SYSTEMS

Many engineering problems in finite element analysis will result in a set of simultaneous equations represented by $[A]\{X\} = \{B\}$.

For a set of simultaneous equations having the form

$$\begin{aligned}
a_{11}x_1 + a_{12}x_2 + a_{13}x_3 + \ldots + a_{1n}x_n &= b_1 \\
a_{21}x_1 + a_{22}x_2 + a_{23}x_3 + \ldots + a_{2n}x_n &= b_2 \\
a_{31}x_1 + a_{32}x_2 + a_{33}x_3 + \ldots + a_{3n}x_n &= b_3 \\
&\vdots \\
a_{n1}x_1 + a_{n2}x_2 + a_{n3}x_3 + \ldots + a_{nn}x_n &= b_n
\end{aligned} \quad (1.22)$$

where there are n unknown $x_1, x_2, x_3, \ldots, x_n$ to be determined. These equations can be written in matrix form as

$$\begin{bmatrix} a_{11} & a_{12} & a_{13} & \ldots & a_{1n} \\ a_{21} & a_{22} & a_{23} & \ldots & a_{2n} \\ a_{31} & a_{32} & a_{33} & \ldots & a_{3n} \\ \vdots & \vdots & \vdots & & \vdots \\ a_{n1} & a_{n2} & a_{n3} & \ldots & a_{nn} \end{bmatrix} \begin{bmatrix} x_1 \\ x_2 \\ x_3 \\ \vdots \\ x_n \end{bmatrix} = \begin{bmatrix} b_1 \\ b_2 \\ b_3 \\ \vdots \\ b_n \end{bmatrix}.$$

This matrix equation can be written in a compact form as

$$\mathbf{AX} = \mathbf{B}, \quad (1.23)$$

where $\mathbf{A}$ is a square matrix with order $n \times n$, while $\mathbf{X}$ and $\mathbf{B}$ are column matrices defined as

$$\mathbf{A} = \begin{bmatrix} a_{11} & a_{12} & a_{13} & \ldots & a_{1n} \\ a_{21} & a_{22} & a_{23} & \ldots & a_{2n} \\ a_{31} & a_{32} & a_{33} & \ldots & a_{3n} \\ \vdots & \vdots & \vdots & & \vdots \\ a_{n1} & a_{n2} & a_{n3} & \ldots & a_{nn} \end{bmatrix}, \quad \mathbf{X} = \begin{bmatrix} x_1 \\ x_2 \\ x_3 \\ \vdots \\ x_n \end{bmatrix}, \quad \mathbf{B} = \begin{bmatrix} b_1 \\ b_2 \\ b_3 \\ \vdots \\ b_n \end{bmatrix}$$

There are several methods for solving a set of simultaneous equations such as by substitution, Gaussian elimination, Cramer's rule, matrix inversion, and numerical analysis.

1.15 GAUSSIAN ELIMINATION METHOD

In the *argument matrix* of a system, the variables of each equation must be on the left side of the equal sign (vertical line) and the constants on the right side. For example, the *argument matrix* of the system

$$2x_1 - 3x_2 = -5$$
$$x_1 - 4x_2 = 8$$

is $\begin{bmatrix} 2 & -3 & | & -5 \\ 1 & -4 & | & 8 \end{bmatrix}$.

The *argument matrix* is used in Gaussian elimination method. The Gaussian elimination method is summarized by the following steps:

1. Write the system of equations in the *argument matrix* form.
2. Perform *elementary row operations* to get zeros below the main diagonal.
   a. interchange any two rows
   b. replace a row by a nonzero multiply of that row
   c. replace a row by the sum of that row and a constant nonzero multiple of some other row
3. Use back substitution to find the solution of the system.

We demonstrate the Gaussian elimination method in Example 1.10.

Example 1.10

Solve the linear system using the Gaussian elimination method.

$$x_2 + x_3 - 2 = 0$$
$$2x_1 + 3x_3 - 5 = 0$$
$$x_1 + x_2 + x_3 - 3 = 0$$

Solution

We use R_i to represent the ith row. Write the argument matrix of the system as:

$$\begin{bmatrix} 0 & 1 & 1 & | & 2 \\ 2 & 0 & 3 & | & 5 \\ 1 & 1 & 1 & | & 3 \end{bmatrix}.$$

Interchange R_1 and R_2, this gives: $\begin{bmatrix} 2 & 0 & 3 & | & 5 \\ 0 & 1 & 1 & | & 2 \\ 1 & 1 & 1 & | & 3 \end{bmatrix}$.

$\frac{1}{2}R_1$, this gives: $\begin{bmatrix} 1 & 0 & \frac{3}{2} & | & \frac{5}{2} \\ 0 & 1 & 1 & | & 2 \\ 1 & 1 & 1 & | & 3 \end{bmatrix}$.

$-R_1 + R_3$, this gives: $\begin{bmatrix} 1 & 0 & \frac{3}{2} & | & \frac{5}{2} \\ 0 & 1 & 1 & | & 2 \\ 0 & 1 & -\frac{1}{2} & | & \frac{1}{2} \end{bmatrix}$.

$-R_2 + R_3$, this gives: $\begin{bmatrix} 1 & 0 & \frac{3}{2} & | & \frac{5}{2} \\ 0 & 1 & 1 & | & 2 \\ 0 & 1 & -\frac{3}{2} & | & -\frac{3}{2} \end{bmatrix}$.

$-\frac{2}{3}R_3$, this gives: $\begin{bmatrix} 1 & 0 & \frac{3}{2} & | & \frac{5}{2} \\ 0 & 1 & 1 & | & 2 \\ 0 & 1 & 1 & | & 1 \end{bmatrix}$.

R_3 gives $x_3 = 1$, substitute the value of x_3 in R_2 and R_3, this gives $x_2 = 1$, and $x_1 = 1$, respectively.

1.16 CRAMER'S RULE

Cramer's rule can be used to solve the simultaneous equations for $x_1, x_2, x_3, \ldots, x_n$ as

$$x_1 = \frac{a_1}{a}, x_2 = \frac{a_2}{a}, x_3 = \frac{a_3}{a}, \ldots, x_n = \frac{a_n}{a} \qquad (1.24)$$

where the α's are the determinations expressed as

$$a = \begin{bmatrix} a_{11} & a_{12} & a_{13} & \cdots & a_{1n} \\ a_{21} & a_{22} & a_{23} & \cdots & a_{2n} \\ a_{31} & a_{32} & a_{33} & \cdots & a_{3n} \\ \cdot & \cdot & \cdot & \cdot & \cdot \\ \cdot & \cdot & \cdot & & \cdot \\ \cdot & \cdot & \cdot & & \cdot \\ a_{n1} & a_{n2} & a_{n3} & \cdots & a_{nn} \end{bmatrix}, a_1 = \begin{bmatrix} b_1 & a_{12} & a_{13} & \cdots & a_{1n} \\ b_2 & a_{22} & a_{23} & \cdots & a_{2n} \\ b_3 & a_{32} & a_{33} & \cdots & a_{3n} \\ \cdot & \cdot & \cdot & & \cdot \\ \cdot & \cdot & \cdot & & \cdot \\ \cdot & \cdot & \cdot & & \cdot \\ b_n & a_{n2} & a_{n3} & \cdots & a_{nn} \end{bmatrix}, a_2 = \begin{bmatrix} a_{11} & b_1 & a_{13} & \cdots & a_{1n} \\ a_{21} & b_2 & a_{23} & \cdots & a_{2n} \\ a_{31} & b_3 & a_{33} & \cdots & a_{3n} \\ \cdot & \cdot & \cdot & & \cdot \\ \cdot & \cdot & \cdot & & \cdot \\ \cdot & \cdot & \cdot & & \cdot \\ a_{n1} & b_n & a_{n3} & \cdots & a_{nn} \end{bmatrix},$$

$$a_3 = \begin{bmatrix} a_{11} & a_{12} & b_1 & \cdots & a_{1n} \\ a_{21} & a_{22} & b_2 & \cdots & a_{2n} \\ a_{31} & a_{32} & b_3 & \cdots & a_{3n} \\ \cdot & \cdot & \cdot & & \cdot \\ \cdot & \cdot & \cdot & & \cdot \\ \cdot & \cdot & \cdot & & \cdot \\ a_{n1} & a_{n2} & b_n & \cdots & a_{nn} \end{bmatrix}, \ldots, a_n = \begin{bmatrix} a_{11} & a_{12} & a_{13} & \cdots & b_1 \\ a_{21} & a_{22} & a_{23} & \cdots & b_2 \\ a_{31} & a_{32} & a_{33} & \cdots & b_3 \\ \cdot & \cdot & \cdot & & \cdot \\ \cdot & \cdot & \cdot & & \cdot \\ \cdot & \cdot & \cdot & & \cdot \\ a_{n1} & a_{n2} & a_{n3} & \cdots & b_n \end{bmatrix}.$$

(1.25)

It is worth noting that a is the determinant of matrix **A** and a_n is the determinant of the matrix formed by replacing the nth column of **A** by **B**. Also, Cramer's rule applies only when $a \neq 0$, but when $a = 0$, the set of questions has no unique solution, because the equations are linearly dependent.

Summary of Cramer's Rule

1. Form the coefficient matrix of **A** and column matrix **B**.
2. Compute the determinant of matrix of **A**. If $\det[A] = 0$, then the system has no solution; otherwise, go to the next step.
3. Compute the determinant of the new matrix $[A_i]$, by replacing the ith matrix with the column vector **B**.
4. Repeat Step 3 for $i = 1, 2, \ldots, n$.
5. Solve for the unknown variable X_i using

$$X_i = \frac{|A_i|}{|A|}, \text{ for } i = 1, 2, \ldots, n. \qquad (1.26)$$

Example 1.11

Solve the simultaneous equations

$$2x_1 - 5x_2 = 13, \quad 5x_1 + 3x_2 = -14$$

Solution

The matrix form of the given equations is

$$\begin{bmatrix} 2 & -5 \\ 5 & 3 \end{bmatrix} \begin{bmatrix} x_1 \\ x_2 \end{bmatrix} = \begin{bmatrix} 13 \\ -14 \end{bmatrix}.$$

The determinants are calculated as

$$a = \begin{vmatrix} 2 & -5 \\ 5 & 3 \end{vmatrix} = (2 \times 3) - (-5 \times 5) = 6 + 25 = 31$$

$$a_1 = \begin{vmatrix} 13 & -5 \\ -14 & 3 \end{vmatrix} = (13 \times 3) - (-5 \times -14) = 39 - 70 = -31$$

$$a_2 = \begin{vmatrix} 2 & 13 \\ 5 & -14 \end{vmatrix} = (2 \times -14) - (13 \times 5) = -28 - 65 = -93$$

Thus,

$$x_1 = \frac{a_1}{a} = \frac{-31}{31} = -1, \quad x_2 = \frac{a_2}{a} = \frac{-93}{31} = -3$$

Example 1.12

Solve the simultaneous equations

$$10x_1 - 3x_2 - 4x_3 = 15, \quad 2x_1 + 5x_2 - 2x_3 = 0, \quad -2x_1 + x_2 + 6x_3 = 0,$$

Solution

In matrix form, the given set of equations becomes

$$\begin{bmatrix} 10 & -3 & -4 \\ 2 & 5 & -2 \\ -2 & 1 & 6 \end{bmatrix} \begin{bmatrix} x_1 \\ x_2 \\ x_3 \end{bmatrix} = \begin{bmatrix} 15 \\ 0 \\ 0 \end{bmatrix}.$$

The determinants are calculated as

$$a = \begin{vmatrix} 10 & -3 & -4 \\ 2 & 5 & -2 \\ -2 & 1 & 6 \end{vmatrix} = 10[(5 \times 6) - (-2 \times 1)] - (-3)[(2 \times 6) - (-2 \times -2)] + (-4)[(2 \times 1) - (5 \times -2)]$$

$$= 320 + 24 - 48 = 296$$

$$a_1 = \begin{vmatrix} 15 & -3 & -4 \\ 0 & 5 & -2 \\ 0 & 1 & 6 \end{vmatrix} = 15[(5 \times 6) - (-2 \times 1)] - (-3)[(0 \times 6) - (-2 \times 0)] + (-4)[(0 \times 1) - (5 \times 0)]$$

$$= 480 + 0 - 0 = 480$$

$$a_2 = \begin{vmatrix} 10 & 15 & -4 \\ 2 & 0 & -2 \\ -2 & 0 & 6 \end{vmatrix} = 10[(0 \times 6) - (-2 \times 0)] - (15)[(2 \times 6) - (-2 \times -2)] + (-4)[(2 \times 0) - (0 \times -2)]$$

$$= 0 - 120 - 0 = -120$$

$$a_3 = \begin{vmatrix} 10 & -3 & 15 \\ 2 & 5 & 0 \\ -2 & 1 & 0 \end{vmatrix} = 10[(5 \times 0) - (0 \times 1)] - (-3)[(2 \times 0) - (0 \times -2)] + (15)[(2 \times 1) - (5 \times -2)]$$

$$= 0 - 0 + 180 = 180$$

Thus,

$$x_1 = \frac{a_1}{a} = \frac{480}{296} = 1.62, \quad x_2 = \frac{a_2}{a} = \frac{-120}{296} = -0.41, \quad x_3 = \frac{a_3}{a} = \frac{180}{296} = 0.61$$

1.17 INVERSE OF A MATRIX

Matrix inversion is used in many applications including the linear system of equations.

For the matrix equation $\mathbf{AX} = \mathbf{B}$, we can invert $\mathbf{A}$ to obtain $\mathbf{X}$, that is,

$$\mathbf{X} = \mathbf{A}^{-1}\mathbf{B} \tag{1.27}$$

where $\mathbf{A}^{-1}$ is the inverse matrix of $\mathbf{A}$. The inverse matrix satisfies

$$\mathbf{A}\mathbf{A}^{-1} = \mathbf{A}^{-1}\mathbf{A} = \mathbf{I} \tag{1.28}$$

where

$$\mathbf{A}^{-1} = \frac{\mathrm{Adj}\,[A]}{|\mathbf{A}|}, \tag{1.29}$$

where $\mathrm{Adj}\,[A]$ is the adjoint of matrix $\mathbf{A}$. The $\mathrm{Adj}\,[A]$ is the transpose of the cofactors of matrix $\mathbf{A}$. For example, let the $n \times n$ matrix $\mathbf{A}$ be presented as

$$\mathbf{A} = \begin{bmatrix} a_{11} & a_{12} & \cdots & a_{1n} \\ a_{21} & a_{22} & \cdots & a_{2n} \\ \vdots & \vdots & & \vdots \\ a_{n1} & a_{n2} & \cdots & a_{nn} \end{bmatrix}.$$

The cofactors of the matrix $\mathbf{A}$ are written in matrix $\mathbf{F}$ as

$$\mathbf{F} = \mathrm{cof}\,[A] = \begin{bmatrix} f_{11} & f_{12} & \cdots & f_{1n} \\ f_{21} & f_{22} & \cdots & f_{2n} \\ \vdots & \vdots & & \vdots \\ f_{n1} & f_{n2} & \cdots & f_{nn} \end{bmatrix} \tag{1.30}$$

where f_{ij} is the product of $(-1)^{i+j}$ and the determinant of the $(n-1)\times(n-1)$ submatrix is obtained by removing the ith row and jth column from matrix $\mathbf{A}$.

For instance, by removing the first row and the first column of matrix **A**, we find the cofactor f_{11} as

$$(-1)^2 f_{11} = \begin{vmatrix} a_{22} & a_{23} & \cdots & a_{2n} \\ a_{32} & a_{33} & \cdots & a_{3n} \\ \vdots & \vdots & \vdots & \vdots \\ a_{n2} & a_{n3} & \cdots & a_{nn} \end{vmatrix}. \qquad (1.31)$$

Now the adjoint of matrix **A** can be obtained as

$$\text{Adj}[A] = [F]^T = \begin{bmatrix} f_{11} & f_{12} & \cdots & f_{1n} \\ f_{21} & f_{22} & \cdots & f_{2n} \\ \vdots & \vdots & \vdots & \vdots \\ f_{n1} & f_{n2} & \cdots & f_{nn} \end{bmatrix}^T. \qquad (1.32)$$

So, the inverse of **A** matrix can be written as

$$\mathbf{A}^{-1} = \frac{[F]^T}{|\mathbf{A}|}. \qquad (1.33)$$

A matrix that possesses an inverse is called *invertible matrix (nonsingular matrix)*. A matrix without an inverse is called a *non-invertible matrix (singular matrix)*.

Consider a 2×2 matrix, if

$$\mathbf{A} = \begin{bmatrix} a & b \\ c & d \end{bmatrix}, \text{ and } ad - bc \neq 0, \text{ then}$$

$$\mathbf{A}^{-1} = \frac{1}{|\mathbf{A}|}\begin{bmatrix} d & -b \\ -c & a \end{bmatrix} = \frac{1}{ad-bc}\begin{bmatrix} d & -b \\ -c & a \end{bmatrix} = \begin{bmatrix} \dfrac{d}{ad-bc} & \dfrac{-b}{ad-bc} \\ \dfrac{-c}{ad-bc} & \dfrac{a}{ad-bc} \end{bmatrix}. \qquad (1.34)$$

The inverse of product of matrices rule can be presented as

$$(\mathbf{AB})^{-1} = (\mathbf{B}^{-1}\mathbf{A}^{-1}), (\mathbf{ABC})^{-1} = \mathbf{C}^{-1}\mathbf{B}^{-1}\mathbf{A}^{-1}. \tag{1.35}$$

Example 1.13

Let matrix, $\mathbf{A} = \begin{bmatrix} 3 & 2 \\ 2 & 1 \end{bmatrix}$, find its inverse matrix, $\mathbf{A}^{-1}$.

Solution

$$\mathbf{A}^{-1} = \frac{1}{|\mathbf{A}|}\begin{bmatrix} 1 & -2 \\ -2 & 3 \end{bmatrix} = \frac{1}{3-4}\begin{bmatrix} 1 & -2 \\ -2 & 3 \end{bmatrix} = \begin{bmatrix} -1 & 2 \\ 2 & -3 \end{bmatrix}$$

Example 1.14

Let matrix, $\mathbf{A} = \begin{bmatrix} 2 & 1 \\ -1 & 1 \end{bmatrix}$, find its inverse matrix, $\mathbf{A}^{-1}$.

Solution

Using the concept of equation (1.27), we get

$$\mathbf{A}^{-1} = \begin{bmatrix} a & b \\ c & d \end{bmatrix}, \text{ then } \mathbf{A}\mathbf{A}^{-1} = \begin{bmatrix} 2 & 1 \\ -1 & 1 \end{bmatrix}\begin{bmatrix} a & b \\ c & d \end{bmatrix} = \begin{bmatrix} 1 & 0 \\ 0 & 1 \end{bmatrix}$$

$2a + c = 1 \longrightarrow 3a = 1 \longrightarrow a = 1/3$
$2b + d = 0 \longrightarrow 3b = -1 \longrightarrow b = -1/3$
$-a + c = 0 \longrightarrow a = c = 1/3$
$-b + d = 1 \longrightarrow d = 1 + b = 2/3$

Therefore,

$$\mathbf{A}^{-1} = \begin{bmatrix} 1/3 & -1/3 \\ 1/3 & 2/3 \end{bmatrix}.$$

1.18 VECTOR ANALYSIS

A *vector* is a special case of a matrix with just one row or one column. A vector is a quantity (mathematical or physical) that has both magnitude and direction. Examples of vectors are force, momentum, acceleration, velocity, electric field intensity, and displacement. A *scalar* is a quantity that has only magnitude. Examples of scalars are mass, time, length, volume, distance, temperature, and electric potential.

A vector **A** has both magnitude and direction. A vector **A** in Cartesian (rectangular) coordinates can be written as (A_x, A_y, A_z) where $A_x, A_y,$ and A_z are components of vector **A** in the x, y, and z directions, respectively. The magnitude of vector **A** is a scalar written as $|\mathbf{A}|$ or A and given as

$$|\mathbf{A}| = \sqrt{A_x^2 + A_y^2 + A_z^2}. \tag{1.36}$$

A unit vector $\mathbf{a}_A$ along vector **A** is defined as a vector whose magnitude is unity (i.e., 1) and its direction is along vector **A**, that is,

$$\mathbf{A} = |\mathbf{A}|\mathbf{a}_A, \tag{1.37}$$

thus, $(A_x, A_y, A_z) = A_x\mathbf{a}_x + A_y\mathbf{a}_y + A_z\mathbf{a}_z$ (1.38)

$$\text{and } \mathbf{a}_A = \frac{\mathbf{A}}{|\mathbf{A}|} = \frac{A_x\mathbf{a}_x + A_y\mathbf{a}_y + A_z\mathbf{a}_z}{\sqrt{A_x^2 + A_y^2 + A_z^2}}. \tag{1.39}$$

Figure 1.1 (a) illustrates the components of vector **A** and Figure 1.1 (b) shows the unit vectors.

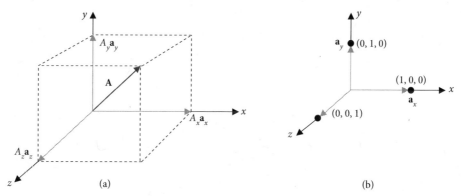

Figure 1.1. (a) Components of vector A (b) Unit vectors.

(a) Vectors equality

Two vectors are equal if they are the same type (row or column) and their corresponding elements are equal to each other.

(b) Vector addition and subtraction

Two vectors can be added or subtracted only if they are of the same type, (i.e., both row vectors or both column vectors) and they are of the same number of components (elements).

Two vectors $\mathbf{A} = (A_x, A_y, A_z)$ and $\mathbf{B} = (B_x, B_y, B_z)$ can be added together to give another vector $\mathbf{C}$, that is,

$$\mathbf{C} = \mathbf{A} + \mathbf{B} \tag{1.40}$$

$$\mathbf{C} = (A_x + B_x)\mathbf{a}_x + (A_y + B_y)\mathbf{a}_y + (A_z + B_z)\mathbf{a}_z. \tag{1.41}$$

Vector subtraction is similarly presented as

$$\mathbf{D} = \mathbf{A} - \mathbf{B} = \mathbf{A} + (-\mathbf{B}) \tag{1.42}$$

$$\mathbf{D} = (A_x - B_x)\mathbf{a}_x + (A_y - B_y)\mathbf{a}_y + (A_z - B_z)\mathbf{a}_z. \tag{1.43}$$

(c) Multiplication of a scalar by a vector

When a vector is multiplied by a scalar, each element is manipulated by the scalar. Let, vector $\mathbf{A} = (A_x, A_y, A_z)$ and scalar k, then

$$k\mathbf{A} = (kA_x, kA_y, kA_z). \tag{1.44}$$

There are three basic laws of algebra for given vectors $\mathbf{A}$, $\mathbf{B}$, and $\mathbf{C}$ when k and l are scalars, summarized in Table 1.1.

Table 1.1 Three Basic Laws of Vector Algebra

Law	Addition	Multiplication
Commutative	$\mathbf{A} + \mathbf{B} = \mathbf{B} + \mathbf{A}$	$k\mathbf{A} = \mathbf{A}k$
Associative	$\mathbf{A} + (\mathbf{B} + \mathbf{C}) = (\mathbf{A} + \mathbf{B}) + \mathbf{C}$	$k(l\mathbf{A}) = (kl)\mathbf{A}$
Distributive	$k(\mathbf{A} + \mathbf{B}) = k\mathbf{B} + k\mathbf{A}$	

(d) Vector multiplication

There are two types of vector multiplication:

1. Scalar (dot) product, $\mathbf{A} \cdot \mathbf{B}$
2. Vector (cross) product, $\mathbf{A} \times \mathbf{B}$

1. The dot product of two vectors $\mathbf{A} = (A_x, A_y, A_z)$ and $\mathbf{B} = (B_x, B_y, B_z)$ written as $\mathbf{A} \cdot \mathbf{B}$ is defined as

$$\mathbf{A} \cdot \mathbf{B} = |\mathbf{A}||\mathbf{B}| = \cos\theta_{AB}, \qquad (1.45)$$

where θ_{AB} is the smallest angle between vectors $\mathbf{A}$ and $\mathbf{B}$. Also, the dot product is defined as,

$$\mathbf{A} \cdot \mathbf{B} = A_x B_x + A_y B_y + A_z B_z. \qquad (1.46)$$

It is worth it to know that, two vectors $\mathbf{A}$ and $\mathbf{B}$ are perpendicular (orthogonal) if and only if $\mathbf{A} \cdot \mathbf{B} = 0$. Also, two vectors $\mathbf{A}$ and $\mathbf{B}$ are parallel if and only if $\mathbf{B} = k\mathbf{A}$. For vectors $\mathbf{A}, \mathbf{B}, \mathbf{C}$ and k scalar, the following prosperities dot product hold:

(a) $\mathbf{A} \cdot \mathbf{B} = \mathbf{B} \cdot \mathbf{A}$ \hfill (1.47)

(b) $\mathbf{A} \cdot (\mathbf{B} + \mathbf{C}) = \mathbf{A} \cdot \mathbf{B} + \mathbf{A} \cdot \mathbf{C}$ \hfill (1.48)

(c) $\mathbf{A} \cdot \mathbf{A} = |\mathbf{A}|^2 = A^2$ \hfill (1.49)

(d) $k(\mathbf{A} \cdot \mathbf{B}) = (k\mathbf{A}) \cdot \mathbf{B} = \mathbf{A} \cdot (k\mathbf{B})$ \hfill (1.50)

(e) $\mathbf{a}_x \cdot \mathbf{a}_y = \mathbf{a}_y \cdot \mathbf{a}_z = \mathbf{a}_z \cdot \mathbf{a}_x = 0$
$\mathbf{a}_x \cdot \mathbf{a}_x = \mathbf{a}_y \cdot \mathbf{a}_y = \mathbf{a}_z \cdot \mathbf{a}_z = 1$ \hfill (1.51)

2. The cross product of two vectors $\mathbf{A} = (A_x, A_y, A_z)$ and $\mathbf{B} = (B_x, B_y, B_z)$ written as $\mathbf{A} \times \mathbf{B}$, is defined as

$$\mathbf{A} \times \mathbf{B} = |\mathbf{A}||\mathbf{B}| \sin\theta_{AB} \mathbf{a}_n, \qquad (1.52)$$

where $\mathbf{a}_n$ is a unit vector normal to the plane containing vectors $\mathbf{A}$ and $\mathbf{B}$. The direction of $\mathbf{a}_n$ is taken as the direction of the right thumb when the fingers of the right hand rotate from vector $\mathbf{A}$ to vector $\mathbf{B}$ as shown in Figure 1.2.

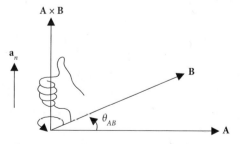

Figure 1.2. Right-hand rule for the direction of $\mathbf{A} \times \mathbf{B}$ and $\mathbf{a}_n$.

Also, the cross product is defined as,

$$\mathbf{A} \times \mathbf{B} = \begin{vmatrix} \mathbf{a}_x & \mathbf{a}_y & \mathbf{a}_z \\ A_x & A_y & A_z \\ B_x & B_y & B_z \end{vmatrix} = (A_y B_z - A_z B_y)\mathbf{a}_x - (A_x B_z - A_z B_x)\mathbf{a}_y + (A_x B_y - A_y B_x)\mathbf{a}_z.$$

(1.53)

Because of the direction requirement of the cross product, the commutative law does not apply to the cross product. Instead,

$$\mathbf{A} \times \mathbf{B} = -\mathbf{B} \times \mathbf{A}.$$

(1.54)

Example 1.15

Given $\mathbf{A} = (3, -2, 5)$ and $\mathbf{A} = (2, 4, -6)$, find
a. $\mathbf{A} + \mathbf{B}$
b. $\mathbf{A} - \mathbf{B}$
c. $|\mathbf{A} - \mathbf{B}|$

Solution

a. $\mathbf{A} + \mathbf{B} = (3+2)\mathbf{a}_x + (-2+4)\mathbf{a}_y + (5-6)\mathbf{a}_z$
$\mathbf{A} + \mathbf{B} = 5\mathbf{a}_x + 2\mathbf{a}_y - \mathbf{a}_z = (5, 2, -1)$

b. $\mathbf{A} - \mathbf{B} = (3-2)\mathbf{a}_x + (-2-4)\mathbf{a}_y + (5+6)\mathbf{a}_z$
$\mathbf{A} - \mathbf{B} = \mathbf{a}_x - 6\mathbf{a}_y + 11\mathbf{a}_z = (1, -6, 11)$

c. $|\mathbf{A} - \mathbf{B}| = \sqrt{(3-2)^2 + (-2-4)^2 + (5+6)^2}$
$|\mathbf{A} - \mathbf{B}| = \sqrt{1 + 36 + 121} = \sqrt{158} = 12.57$

Example 1.16

Given $\mathbf{A} = 3\mathbf{a}_x + 2\mathbf{a}_y - \mathbf{a}_z$ and $\mathbf{B} = \mathbf{a}_x + \mathbf{a}_y$, find $\mathbf{A} \cdot \mathbf{B}$ and $\mathbf{A} \times \mathbf{B}$.

Solution

$\mathbf{A} \cdot \mathbf{B} = (3)(1) + (2)(1) + (-1)(0) = 5$

$$\mathbf{A} \times \mathbf{B} = \begin{vmatrix} \mathbf{a}_x & \mathbf{a}_y & \mathbf{a}_z \\ 3 & 2 & -1 \\ 1 & 1 & 0 \end{vmatrix} = (2 \times 0 - (-1) \times 1)\mathbf{a}_x - (3 \times 0 - (-1) \times 1)\mathbf{a}_y + (3 \times 1 - 2 \times 1)\mathbf{a}_z$$

$\mathbf{A} \times \mathbf{B} = \mathbf{a}_x - \mathbf{a}_y + \mathbf{a}_z$

(e) The Del (∇) operator

The *Del* (∇) operator is a vector differential operator and known as *gradient* operator.

We obtain ∇ in Cartesian coordinates (x,y,z) as,

$$\nabla = \mathbf{a}_x \frac{\partial}{\partial x} + \mathbf{a}_y \frac{\partial}{\partial y} + \mathbf{a}_z \frac{\partial}{\partial z}. \tag{1.55}$$

We obtain ∇ in cylindrical coordinates (ρ,ϕ,z) as,

$$\nabla = \mathbf{a}_\rho \frac{\partial}{\partial \rho} + \mathbf{a}_\phi \frac{1}{\rho} \frac{\partial}{\partial \phi} + \mathbf{a}_z \frac{\partial}{\partial z}. \tag{1.56}$$

We obtain ∇ in spherical coordinates (r,θ,ϕ) as,

$$\nabla = \mathbf{a}_r \frac{\partial}{\partial r} + \mathbf{a}_\theta \frac{1}{r} \frac{\partial}{\partial \theta} + \mathbf{a}_\phi \frac{1}{r\sin\theta} \frac{\partial}{\partial \phi}. \tag{1.57}$$

The Del (∇) operator is useful in defining the following operations on a scalar or a vector:

1. ∇A is the *gradient* of a scalar A (the result of this operation is a vector)

   (a) For Cartesian coordinates, $\nabla A = \mathbf{a}_x \dfrac{\partial A}{\partial x} + \mathbf{a}_y \dfrac{\partial A}{\partial y} + \mathbf{a}_z \dfrac{\partial A}{\partial z}.$ (1.58)

   (b) For cylindrical coordinates, $\nabla A = \mathbf{a}_\rho \dfrac{\partial A}{\partial \rho} + \mathbf{a}_\phi \dfrac{1}{\rho} \dfrac{\partial A}{\partial \phi} + \mathbf{a}_z \dfrac{\partial A}{\partial z}.$ (1.59)

   (c) For spherical coordinates, $\nabla A = \mathbf{a}_r \dfrac{\partial A}{\partial r} + \mathbf{a}_\theta \dfrac{1}{r} \dfrac{\partial A}{\partial \theta} + \mathbf{a}_\phi \dfrac{1}{r\sin\theta} \dfrac{\partial A}{\partial \phi}.$ (1.60)

   Considering A and B are scalars and n is an integer, the following formulas are true on gradient:

   - $\nabla(A+B) = \nabla A + \nabla B$ (1.61)
   - $\nabla(AB) = A\nabla B + B\nabla A$ (1.62)
   - $\nabla\left(\dfrac{A}{B}\right) = \dfrac{B\nabla A - A\nabla B}{B^2}$ (1.63)
   - $\nabla A^n = nA^{n-1}\nabla A$ (1.64)

2. $\nabla \cdot \mathbf{A}$ is the *divergence* of a vector $\mathbf{A}$ (the result of this operation is a scalar)

   (a) For Cartesian coordinates, $\nabla \cdot \mathbf{A} = \dfrac{\partial A_x}{\partial x} + \dfrac{\partial A_y}{\partial y} + \dfrac{\partial A_z}{\partial z}.$ \hfill (1.65)

   (b) For cylindrical coordinates, $\nabla \cdot \mathbf{A} = \dfrac{1}{\rho}\dfrac{\partial}{\partial \rho}(\rho A_\rho) + \dfrac{1}{\rho}\dfrac{\partial A_\phi}{\partial \phi} + \dfrac{\partial A_z}{\partial z}.$ \hfill (1.66)

   (c) For spherical coordinates,
   $$\nabla \cdot \mathbf{A} = \dfrac{1}{r^2}\dfrac{\partial}{\partial r}(r^2 A_r) + \dfrac{1}{r \sin\theta}\dfrac{\partial}{\partial \theta}(A_\theta \sin\theta) + \dfrac{1}{r \sin\theta}\dfrac{\partial A_\phi}{\partial \phi}. \qquad (1.67)$$

   Considering $\mathbf{A}$ and $\mathbf{B}$ are vectors and k is a scalar, the following formulas are true on divergence of a vector:

   - $\nabla \cdot (\mathbf{A} + \mathbf{B}) = \nabla \cdot \mathbf{A} + \nabla \cdot \mathbf{B}$ \hfill (1.68)
   - $\nabla \cdot (k\mathbf{A}) = k\nabla \cdot \mathbf{A} + \mathbf{A} \cdot \nabla k$ \hfill (1.69)

3. $\nabla \times \mathbf{A}$ is the *curl* of a vector $\mathbf{A}$ (the result of this operation is a vector)

   (a) For Cartesian coordinates, $\nabla \times \mathbf{A} = \begin{vmatrix} \mathbf{a}_x & \mathbf{a}_y & \mathbf{a}_z \\ \dfrac{\partial}{\partial x} & \dfrac{\partial}{\partial y} & \dfrac{\partial}{\partial z} \\ A_x & A_y & A_z \end{vmatrix}$ \hfill (1.70)

   or
   $$\nabla \times \mathbf{A} = \left(\dfrac{\partial A_z}{\partial y} - \dfrac{\partial A_y}{\partial z}\right)\mathbf{a}_x + \left(\dfrac{\partial A_x}{\partial z} - \dfrac{\partial A_z}{\partial x}\right)\mathbf{a}_y + \left(\dfrac{\partial A_y}{\partial x} - \dfrac{\partial A_x}{\partial y}\right)\mathbf{a}_z. \qquad (1.71)$$

   (b) For cylindrical coordinates, $\nabla \times \mathbf{A} = \dfrac{1}{\rho}\begin{vmatrix} \mathbf{a}_\rho & \rho\mathbf{a}_\phi & \mathbf{a}_z \\ \dfrac{\partial}{\partial \rho} & \dfrac{\partial}{\partial \phi} & \dfrac{\partial}{\partial z} \\ A_\rho & \rho A_\phi & A_z \end{vmatrix}$ \hfill (1.72)

   or
   $$\nabla \times \mathbf{A} = \left(\dfrac{1}{\rho}\dfrac{\partial A_z}{\partial \phi} - \dfrac{\partial A_\phi}{\partial z}\right)\mathbf{a}_\rho + \left(\dfrac{\partial A_\rho}{\partial z} - \dfrac{\partial A_z}{\partial \rho}\right)\mathbf{a}_\phi + \dfrac{1}{\rho}\left(\dfrac{\partial(\rho A_\phi)}{\partial \rho} - \dfrac{\partial A_\rho}{\partial \phi}\right)\mathbf{a}_z. \qquad (1.73)$$

(c) For spherical coordinates, $\nabla \times \mathbf{A} = \dfrac{1}{r^2 \sin\theta} \begin{vmatrix} \mathbf{a}_r & r\mathbf{a}_\theta & r\sin\theta\, \mathbf{a}_\phi \\ \dfrac{\partial}{\partial r} & \dfrac{\partial}{\partial \theta} & \dfrac{\partial}{\partial \phi} \\ A_r & rA_\theta & r\sin\theta A_\phi \end{vmatrix}$ (1.74)

or

$$\nabla \times \mathbf{A} = \frac{1}{r\sin\theta}\left(\frac{\partial(A_\phi \sin\theta)}{\partial \theta} - \frac{\partial A_\theta}{\partial \phi}\right)\mathbf{a}_r + \frac{1}{r}\left(\frac{1}{\sin\theta}\frac{\partial A_r}{\partial \phi} - \frac{\partial(rA_\phi)}{\partial r}\right)\mathbf{a}_\theta \quad (1.75)$$
$$+ \frac{1}{r}\left(\frac{\partial(rA_\theta)}{\partial r} - \frac{\partial A_r}{\partial \theta}\right)\mathbf{a}_\phi.$$

Considering **A** and **B** are vectors and k is a scalar, the following formulas are true on curl of a vector:

- $\nabla \times (\mathbf{A} + \mathbf{B}) = \nabla \times \mathbf{A} + \nabla \times \mathbf{B}$ (1.76)
- $\nabla \times (\mathbf{A} \times \mathbf{B}) = \mathbf{A}(\nabla \cdot \mathbf{B}) - \mathbf{B}(\nabla \cdot \mathbf{A}) + (\mathbf{B} \cdot \nabla)\mathbf{A} - (\mathbf{A} \cdot \nabla)\mathbf{B}$ (1.77)
- $\nabla \times (k\mathbf{A}) = k\nabla \times \mathbf{A} + \nabla k \times \mathbf{A}$ (1.78)
- $\nabla \cdot (\nabla \times \mathbf{A}) = 0$ (1.79)
- $\nabla \times \nabla k = 0$ (1.80)

4. $\nabla^2 A$-Laplacian of a scalar A (the result of this operation is a scalar)

(a) For Cartesian coordinates, $\nabla^2 A = \dfrac{\partial^2 A}{\partial x^2} + \dfrac{\partial^2 A}{\partial y^2} + \dfrac{\partial^2 A}{\partial z^2}.$ (1.81)

(b) For cylindrical coordinates, $\nabla^2 A = \dfrac{1}{\rho}\dfrac{\partial}{\partial \rho}\left(\rho\dfrac{\partial A}{\partial \rho}\right) + \dfrac{1}{\rho^2}\dfrac{\partial^2 A}{\partial \phi^2} + \dfrac{\partial^2 A}{\partial z^2}.$ (1.82)

(c) For spherical coordinates,

$$\nabla^2 A = \frac{1}{r^2}\frac{\partial}{\partial r}\left(r^2 \frac{\partial A}{\partial r}\right) + \frac{1}{r^2 \sin\theta}\frac{\partial}{\partial \theta}\left(\sin\theta \frac{\partial A}{\partial \theta}\right) + \frac{1}{r^2 \sin^2\theta}\frac{\partial^2 A}{\partial \phi^2}. \quad (1.83)$$

The Laplacian of a vector **A**, can be defined as

$$\nabla^2 \mathbf{A} = \nabla(\nabla \cdot \mathbf{A}) - \nabla \times \nabla \times \mathbf{A}. \quad (1.84)$$

Example 1.17

Find the gradient of the scalar field $A = e^{-z}\sin 3x \cosh y$.

Solution

$\nabla A = 3e^{-z}\cos 3x \cosh y \mathbf{a}_x + e^{-z}\sin 3x \sinh y \mathbf{a}_y - e^{-z}\sin 3x \cosh y \mathbf{a}_z$

1.19 EIGENVALUES AND EIGENVECTORS

Eigenvalues problems arises from many branches of engineering especially in analysis of vibration of elastic structures and electrical systems.

The eigenvalue problem is presented in linear equations in the form

$$[A]\cdot\{X\} - \lambda\{X\} = \{0\}. \tag{1.85}$$

Where $[A]$ is a square matrix; λ is a scalar and called eigenvalue of matrix $[A]$; $\{X\}$ is eigenvector of matrix $[A]$ corresponding to λ.

To find the eigenvalues of a square matrix $[A]$ we rewrite the equation (1.55) as

$$[A]\{X\} = \lambda[I]\{X\} \tag{1.86}$$

or

$$[\lambda\mathbf{I} - \mathbf{A}]\cdot\{X\} = \{0\}. \tag{1.87}$$

There must be a nonzero solution of equation (1.87) in order for λ to be an eigenvalue. However, equation (1.87) can have a nonzero solution if and only if

$$|\lambda\mathbf{I} - \mathbf{A}| = 0. \tag{1.88}$$

Equation (1.88) is called the *characteristic equation* of matrix $[A]$ and the scalars satisfy the equation (1.88) are the eigenvalues of matrix $[A]$. If matrix $[A]$ has the form

$$\mathbf{A} = \begin{bmatrix} a_{11} & a_{12} & a_{13} & \cdots & a_{1n} \\ a_{21} & a_{22} & a_{23} & \cdots & a_{2n} \\ a_{31} & a_{32} & a_{33} & \cdots & a_{3n} \\ \cdot & \cdot & \cdot & \cdot & \cdot \\ \cdot & \cdot & \cdot & & \cdot \\ \cdot & \cdot & \cdot & & \cdot \\ a_{n1} & a_{n2} & a_{n3} & \cdots & a_{nn} \end{bmatrix}, \text{ then equation (1.88) can be written as}$$

$$\begin{vmatrix} a_{11}-\lambda & a_{12} & a_{13} & \cdots & a_{1n} \\ a_{21} & a_{22}-\lambda & a_{23} & \cdots & a_{2n} \\ a_{31} & a_{32} & a_{33}-\lambda & \cdots & a_{3n} \\ \vdots & \vdots & \vdots & & \vdots \\ a_{n1} & a_{n2} & a_{n3} & \cdots & a_{nn}-\lambda \end{vmatrix} = 0. \tag{1.89}$$

The equation (1.89) can be expanded to a polynomial equation in λ as

$$\lambda^n + c_1 \lambda^{n-1} + \ldots + c_{n-1}\lambda + c_n = 0. \tag{1.90}$$

Thus, the nth degree polynomial is

$$|\lambda \mathbf{I} - \mathbf{A}| = \lambda^n + c_1 \lambda^{n-1} + \ldots + c_{n-1}\lambda + c_n. \tag{1.91}$$

Equation (1.91) is called a *characteristic polynomial* of $n \times n$ matrix $[A]$. Indeed, the nth roots of the polynomial equation are the nth eigenvalues of matrix $[A]$. The solutions of equation (1.87) with the eigenvalues substituted on the equation are called *eigenvectors*.

Example 1.18

Find the eigenvalues and eigenvectors of the 2×2 matrix $\mathbf{A} = \begin{bmatrix} 6 & -3 \\ -4 & 5 \end{bmatrix}$

Solution

Since

$$[\lambda \mathbf{I} - \mathbf{A}] = \lambda \begin{bmatrix} 1 & 0 \\ 0 & 1 \end{bmatrix} - \begin{bmatrix} 6 & -3 \\ -4 & 5 \end{bmatrix} = \begin{bmatrix} \lambda-6 & 3 \\ 4 & \lambda-5 \end{bmatrix},$$

the characteristic polynomial of matrix $[A]$ is

$$|\lambda \mathbf{I} - \mathbf{A}| = \begin{vmatrix} \lambda-6 & 3 \\ 4 & \lambda-5 \end{vmatrix} = (\lambda-6)(\lambda-5) - (3 \times 4) = \lambda^2 - 11\lambda + 18.$$

and the characteristic equation of matrix $[A]$ is

$$\lambda^2 - 11\lambda + 18 = 0.$$

The solutions of this equation are $\lambda_1 = 2$ and $\lambda_2 = 9$; these values are the eigenvalues of matrix $[A]$.

The eigenvectors for each of the above eigenvalues are calculated using equation (1.87).

For $\lambda_1 = 2$, we obtain

$$\begin{bmatrix} 2-6 & 3 \\ 4 & 2-5 \end{bmatrix} \begin{Bmatrix} x_1 \\ x_2 \end{Bmatrix} = \begin{Bmatrix} 0 \\ 0 \end{Bmatrix}.$$

The above equation yields to two simultaneous equations for x_1 and x_2, as follows:

$$-4x_1 + 3x_2 = 0 \text{ gives } x_1 = \frac{3}{4} x_2$$

$$4x_1 - 3x_2 = 0 \text{ gives } x_1 = \frac{3}{4} x_2.$$

Thus, choosing $x_2 = 4$, we obtain the eigenvector $\mathbf{x}_2 = k \begin{Bmatrix} 3 \\ 4 \end{Bmatrix}$, where k is an arbitrary constant.

For $\lambda_2 = 9$, we obtain

$$\begin{bmatrix} 9-6 & 3 \\ 4 & 9-5 \end{bmatrix} \begin{Bmatrix} x_1 \\ x_2 \end{Bmatrix} = \begin{Bmatrix} 0 \\ 0 \end{Bmatrix}.$$

The above equation yields to two simultaneous equations for x_1 and x_2, as follows:

$$3x_1 + 3x_2 = 0 \text{ gives } x_1 = -x_2$$

$$4x_1 + 4x_2 = 0 \text{ gives } x_1 = -x_2.$$

Thus, choosing $x_1 = -1$, we obtain the eigenvector $\mathbf{x}_1 = k \begin{Bmatrix} -1 \\ 1 \end{Bmatrix}$, where k is an arbitrary constant.

1.20 USING MATLAB

MATLAB is a numerical computation and simulation tool that uses matrices and vectors. Also, MATLAB enables users to solve wide analytical problems. The majority of engineering systems are presented by matrix and vector equations. Therefore, MATLAB becomes essential to reduce the computational workload.

All MATLAB commands or expressions are entered in the command window at the MATLAB prompt "≫". To execute a command or statement, we must press *return* or *enter* at the end. If the command does not fit on one line, we can continue the command on the next line by typing three consecutive periods (…) at the end of the first line. A semicolon (;) at the end of a command suppresses the screen output and the command is carried out. Typing anything following a % is considered as comment, except when the % appears in a quote enclosed character string or certain I/O format statements. Comment statements are not executable. To get help on a topic (such as matrix), you can type the command *helpmatrix*. Here, we introduce basic ideas of matrices and vectors operations. For more details, see Appendix B.

Elements of a matrix are enclosed in brackets and they are row-wise. The consecutive elements of a row are separated by a comma or a space and are entered in rows separated by a space or a comma, and the rows are separated by semicolons (;) or carriage returns (enter).

A vector is entered in the MATLAB environment the same way as a matrix. For example, matrix **A**,

$$\mathbf{A} = \begin{bmatrix} 1 & 0 \\ 3 & 2 \end{bmatrix}, \text{ is typed in MATLAB as}$$

```
>> A = [1 0; 3 2]
A =
    1   0
    3   2
```

The basic scalar operations are shown in Table 1.2. In addition to operating on mathematical scalar, MATLAB allows us to work easily with vectors and matrices. Arithmetic operations can apply to matrices and Table 1.3 shows extra common operations that can be implemented to matrices and vectors.

Table 1.2 **MATLAB Common Arithmetic Operators**

Operators Symbols	Descriptions
+	addition
−	subtraction
*	multiplication
/	right division (means $\frac{a}{b}$)
\	left division (means $\frac{b}{a}$)
^	exponentiation (raising to a power)
'	converting to complex conjugate transpose
( )	specify evaluation order

Table 1.3 **Matrix Operations**

Operations	Descriptions
A'	Transpose of matrix A
det (A)	Determinant of matrix A
inv (A)	Inverse of matrix A
eig (A)	Eigenvalues of matrix A
diag (A)	Diagonal elements of matrix A
rank (A)	Rank of matrix A
cond (A)	Condition number of matrix A
eye (n)	The $n \times n$ identity matrix (1's on the main diagonal)
eye (m, n)	The $m \times n$ identity matrix (1's on the main diagonal)
trace (A)	Summation of diagonal elements of matrix A
zeros (m, n)	The $m \times n$ matrix consisting of all zeros
ones (m, n)	The $m \times n$ matrix consisting of all ones
rand (m, n)	The $m \times n$ matrix consisting of random numbers
randn (m, n)	The $m \times n$ matrix consisting of normally distributed numbers
diag (A)	Extraction of the diagonal matrix A as vector
diag (A,1)	Extracting of first upper off-diagonal vector of matrix A
diag (**u**)	Generating of a diagonal matrix with a vector **u** on the diagonal
expm (A)	Exponential of matrix A
ln (A)	LU decomposition of matrix A
svd (A)	Singular value decomposition of matrix A
qr (A)	QR decomposition of matrix A
min (A)	Minimum of vector A
max (A)	Maximum of vector A
sum (A)	Sum of elements of vector A
std (A)	Standard deviation of the data collection of vector A
sort (A)	Sort the elements of vector A
mean (A)	Means value of vector A
triu (A)	Upper triangular of matrix A
triu (A, I)	Upper triangular with zero diagonals of matrix A
tril (A)	lower triangular of matrix A
tril (A, I)	lower triangular with zero diagonals of matrix A

Example 1.19

Given the following matrices:

$$[A] = \begin{bmatrix} 1 & 2 & 3 \\ 4 & 5 & 6 \\ 7 & 8 & 9 \end{bmatrix}, [B] = \begin{bmatrix} 0 & -1 & 0 \\ 2 & -3 & 1 \\ 4 & -5 & 3 \end{bmatrix}, \text{ and } [C] = \begin{Bmatrix} 2 \\ 0 \\ 4 \end{Bmatrix}$$

Use MATLAB to perform the following operations:

a. $[A] + [B]$
b. $[A] - [B]$
c. $5[B]$
d. $[A][B]$
e. $[A][C]$
f. $[A]^2$
g. $[A]^T$
h. $[B]^{-1}$
i. tr (**A**)
j. |**B**|

Solution

a. $[A] + [B]$

```
>> A=[1 2 3;4 5 6;7 8 9];
>> B=[0 -1 0;2 -3 1;4 -5 3];
>> A+B
ans =
    1    1    3
    6    2    7
   11    3   12
```

b. $[A] - [B]$

```
>> A=[1 2 3;4 5 6;7 8 9];
>> B=[0 -1 0;2 -3 1;4 -5 3];
>> A-B
ans =
    1    3    3
    2    8    5
    3   13    6
```

c. $5[B]$

```
>> B=[0 -1 0;2 -3 1;4 -5 3];
>> 5*B
ans =
    0   -5    0
   10  -15    5
   20  -25   15
```

d. $[A][B]$

```
>> A=[1 2 3;4 5 6;7 8 9];
>> B=[0 -1 0;2 -3 1;4 -5 3];
>> A*B
ans =
   16  -22   11
   34  -49   23
   52  -76   35
```

e. $[A][C]$

```
>> A=[1 2 3;4 5 6;7 8 9];
>> C=[2;0;4];
>> A*C
ans =
   14
   32
   50
```

f. $[A]^2$

```
>> A=[1 2 3;4 5 6;7 8 9];
>> A^2
ans =
   30   36   42
   66   81   96
  102  126  150
```

Mathematical Preliminaries

g. $[A]^T$

```
>> A=[1 2 3;4 5 6;7 8 9];
>> A'
ans =
    1   4   7
    2   5   8
    3   6   9
```

h. $[B]^{-1}$

```
>> B=[0 -1 0;2 -3 1;4 -5 3];
>> inv(B)
ans =
   -2.0000   1.5000  -0.5000
   -1.0000        0        0
    1.0000  -2.0000   1.0000
```

i. tr (**A**)

```
>> A=[1 2 3;4 5 6;7 8 9];
>> trace(A)
ans =
    15
```

j. |**B**|

```
>> B=[0 -1 0;2 -3 1;4 -5 3];
>> det(B)
ans =
    2
```

Example 1.20

Solve the following system of three equations:

$$5x + y + 2z = 6$$
$$-x + 4y + z = 7$$
$$x - 2y - z = -3$$

using the following methods:

a. the matrix inverse
b. Gaussian elimination
c. Reverse Row Echelon Function

Solution

a. Since we know $A^{-1}A = 1$, we can find the solution of the system of linear equations $AX = B$ by using $X = A^{-1}B$.

Now, we write the system of equations by using the following matrices:

$$A = \begin{bmatrix} 5 & 1 & 2 \\ -1 & 4 & 1 \\ 1 & -2 & -1 \end{bmatrix}, \quad X = \begin{bmatrix} x \\ y \\ z \end{bmatrix}, \quad B = \begin{bmatrix} 6 \\ 7 \\ -3 \end{bmatrix}$$

```
>> A = [5 1 2; –1 4 1; 1 –2 –1];
>> B = [6; 7; –3];
>> X = inv(A)*B
X =
    0.8571
    2.0000
   –0.1429
```

Generally, using the matrix inverse to solve linear systems of equations should be avoided due to the excessive round-off errors.

b. We use the left division operator in MATLAB $X = A\backslash B$ to solve linear systems of equations using Gaussian elimination.

```
>> A = [5 1 2; –1 4 1; 1 –2 –1];
>> B = [6; 7; –3];
>> X = A\B
X =
    0.8571
    2.0000
   –0.1429
```

c. The reduced row echelon function use, *rref*, to solve the system of linear equations. The *rref* function requires an expanded matrix as input, representing the coefficients and results. The last column in the output array represents the solution of equations.

Mathematical Preliminaries

```
>> A = [5 1 2; -1 4 1; 1 -2 -1];
>> B = [6; 7; -3];
>> C = [A,B];
>> rref(C)
ans =
    1.0000        0        0   0.8571
         0   1.0000        0   2.0000
         0        0   1.0000  -0.1429
```

Example 1.21

Solve the following set of equations using the Cramer's rule:

$$5x_1 + x_3 + 2x_4 = 3$$
$$x_1 + x_2 + 3x_3 + x_4 = 5$$
$$x_1 + x_2 + 2x_4 = 1$$
$$x_1 + x_2 + x_3 + x_4 = -1$$

Solution

```
>> A = [5 0 1 2;1 1 3 1; 1 1 0 2;1 1 1 1];
>> B = [3;5;1;-1];
>> A1 = [B A(:,[2:4])];
>> A2 = [A(:,1) B A(:,[3:4])];
>> A3 = [A(:,[1:2]) B A(:,4)];
>> A4 = [A(:,[1:3]) B];
>> x1 = det(A1)/det(A)
x1 =
    -2
>> x2 = det(A2)/det(A)
x2 =
    -7
>> x3 = det(A3)/det(A)
x3 =
     3
>> x4 = det(A4)/det(A)
x4 =
     5
```

PROBLEMS

1. Identify the size and the type of the given matrices. Identify if the matrix is a square, column, diagonal, row, identity, banded, symmetric, or triangular.

    a. $\begin{Bmatrix} x_1 \\ y_1 \\ z_1 \\ t \end{Bmatrix}$
    b. $\begin{bmatrix} 7 & 5 & 3 & 1 \end{bmatrix}$
    c. $\begin{bmatrix} -1 & 0 & 1 \\ 2 & 6 & 4 \\ 7 & 5 & 2 \end{bmatrix}$
    d. $\begin{bmatrix} 1 & 0 \\ 0 & 1 \end{bmatrix}$
    e. $\begin{bmatrix} 2 & 0 \\ 0 & 4 \end{bmatrix}$

    f. $\begin{bmatrix} 1 & 3 & 0 \\ 5 & 6 & 4 \\ 2 & 0 & 7 \end{bmatrix}$
    g. $\begin{bmatrix} 1 & b & c & d \\ 0 & 1 & e & f \\ 0 & 0 & 1 & a \\ 0 & 0 & 0 & 1 \end{bmatrix}$
    h. $\begin{bmatrix} 2 & 4 & 0 & 0 & 0 \\ 3 & 9 & -1 & 0 & 0 \\ 0 & 4 & 8 & 2 & 0 \\ 0 & 0 & 6 & 7 & 3 \\ 0 & 0 & 0 & 1 & 5 \end{bmatrix}$
    i. $\begin{bmatrix} a & 0 & 0 & 0 \\ 0 & b & 0 & 0 \\ 0 & 0 & c & 0 \\ 0 & 0 & 0 & d \end{bmatrix}$

2. Given the matrices $\mathbf{A} = \begin{bmatrix} 2 & 1 & 6 \\ 0 & 3 & 5 \\ 1 & -7 & 4 \end{bmatrix}$, $\mathbf{B} = \begin{bmatrix} 5 & 2 & 4 \\ 3 & 1 & 6 \\ 0 & -2 & 1 \end{bmatrix}$, and $\mathbf{C} = \begin{Bmatrix} 3 \\ 2 \\ 1 \end{Bmatrix}$ find

    a. A+B
    b. A−B
    c. 4A
    d. AB
    e. A{C}
    f. A^2
    g. IA
    h. AI

3. Given the matrices $\mathbf{A} = \begin{bmatrix} 1 & 8 & 3 \\ 5 & 3 & 1 \\ 0 & -3 & 4 \end{bmatrix}$, $\mathbf{B} = \begin{bmatrix} 2 & 3 & 0 \\ 1 & 5 & -6 \\ 0 & 4 & 7 \end{bmatrix}$, find the following:

    a. A^T
    b. B^T
    c. $|A|$
    d. $|A|^{-1}$
    e. $|B|^{-1}$

MATHEMATICAL PRELIMINARIES

4. What is the 3×3 null matrix and the 5×5 identity matrix?
5. Express the following systems of equations in matrix form $\mathbf{AX} = \mathbf{B}$.
   a. $3x_1 + 2x_2 = 10, \quad 3x_1 + 4x_2 = -8$
   b. $2x_1 + 3x_2 + 5x_3 - 20, \quad x_1 + 3x_2 - 5x_3 = 0, \quad 2x_1 - 3x_2 - 4x_3 = 0,$
6. Solve the system using the Gaussian elimination method.

$$x_1 + x_2 + 2x_3 = 8$$
$$-x_1 - 2x_2 + 3x_3 = 1$$
$$3x_1 - 7x_2 + 4x_3 = 10$$

7. Solve the simultaneous equations using Cramer's rule.

$$2x_1 + 3x_2 = 8$$
$$3x_1 + 4x_2 - 5x_3 = 2$$
$$x_1 - x_2 + 2x_3 = 1$$

8. Show that $\mathbf{A} \cdot \mathbf{B} = A_x B_x + A_y B_y + A_z B_z$, know that $\mathbf{a}_x \cdot \mathbf{a}_x = \mathbf{a}_y \cdot \mathbf{a}_y = \mathbf{a}_z \cdot \mathbf{a}_z = 1$.
9. Given $\mathbf{A} = 2\mathbf{a}_x + 4\mathbf{a}_y - 5\mathbf{a}_z$ and $\mathbf{B} = 3\mathbf{a}_x - \mathbf{a}_y + \mathbf{a}_z$, find $\mathbf{A} \cdot \mathbf{B}$ and $\mathbf{A} \times \mathbf{B}$.
10. Show that if $\mathbf{A} = 5\mathbf{a}_x - 4\mathbf{a}_y - \mathbf{a}_z$ and $\mathbf{B} = \mathbf{a}_x + 2\mathbf{a}_y + 2\mathbf{a}_z$, then they are perpendicular or not.
11. Determine the gradient of the scalar fields $A = x^3 y + xyz$.
12. Find the eigenvalues and eigenvectors of the 2×2 matrix $\mathbf{A} = \begin{bmatrix} 3 & 4 \\ 2 & 7 \end{bmatrix}$.
13. Solve problem 2 using MATLAB.
14. Solve problem 3 using MATLAB.

REFERENCES

1. R. Butt, "Applied Linear Algebra and Optimization using MATLAB," Mercury Learning and Information, 2011.
2. H. Anton, "Elementary Linear Algebra, 6th Edition," John Wiley and Sons, INC., 1991.
3. S. Nakamura, "Applied Numerical Methods with Software," Prentice-Hall, 1991.
4. B. Kolman, "Introductory Linear Algebra with Applications, 6th Edition," John, Prentice Hall, 1997.

Chapter 2: Introduction to Finite Element Method

2.1 INTRODUCTION

Finite element analysis method is a numerical procedure that applies to many areas in real-world engineering problems, including structural/stress analysis, fluid flow analysis, heat transfer analysis, and electromagnetics analysis. Indeed, finite element has several advantages and features such as the capability of solving complicated and complex geometries, flexibility, strong mathematical foundation, and high-order approximation. Therefore, finite element analysis (FEA) has become an important method in the design and modeling of a physical event in many engineering disciplines. The actual component in the FEA method is placed by a simplified model that is identified by a finite number of *elements* connected at common points called *nodes*, with an assumed response of each element to applied loads, and then evaluating the unknown field variable (displacement) at these nodes.

2.2 METHODS OF SOLVING ENGINEERING PROBLEMS

There are 3 common methods to solve any engineering problem:

1. Experimental method
2. Analytical method
3. Numerical method

2.2.1 Experimental Method

This method involves actual measurement of the system response. This method is time consuming and needs expensive set up. This method is applicable only if physical prototype is available. The results obtained by this method cannot be believed blindly and a minimum of 3 to 5 prototypes must be tested. Examples of this method are strain photo elasticity, heat transfer for a gas turbine engine, static and dynamic response for aircraft and spacecraft, amount of water which is lost for groundwater seepage, etc.

2.2.2 Analytical Method

This is a classic approach. This method gives closed form solutions. The results obtained with this method are accurate within the assumptions made. This method is applicable only for solving problems of simple geometry and loading, like cantilever and simply supported beams, etc. Analytical methods produce exact solutions of the problem. Examples of this method are integral solutions (such as Laplace and Fourier transforms), conformal mapping, perturbation methods, separation of variables, and series expansion.

2.2.3 Numerical Method

This approximate method is resorted to when analytical method fails. This method is applicable to real-life problems of a complex nature. Results obtained by this method cannot be believed blindly and must be carefully assessed against experience and the judgment of the analyst. Examples of this method are finite element method, finite difference method, moment method, etc.

2.3 PROCEDURE OF FINITE ELEMENT ANALYSIS (RELATED TO STRUCTURAL PROBLEMS)

Step (i). Discretization of the structure
This first step involves dividing the structure or domain of the problem into small divisions or elements. The analyst has to decide about the type, size, and arrangement of the elements.

Step (ii). Selection of a proper interpolation (or displacement) model
A simple polynomial equation (linear/quadratic/cubic) describing the variation of state variable (e.g., displacement) within an element is assumed. This model generally is the interpolation/shape function type. Certain conditions are to be satisfied by this model so that the results are meaningful and converging.

Step (iii). Derivation of element stiffness matrices and load vectors
Response of an element to the loads can be represented by element equation of the form

$$[k]\{q\} = \{Q\} \qquad (2.1)$$

where, $[k]$ = Element stiffness matrix,
$\{q\}$ = Element response matrix or element nodal displacement vector, or nodal degree of freedom,
$\{Q\}$ = Element load matrix or element nodal load vector.

From the assumed displacement model, the element properties, namely stiffness matrix and the load vector are derived. Element stiffness matrix $[k]$ is a characteristic property of the element and depends on geometry as well as material. There are 3 approaches for deriving element equations. They are

(a) Direct approach,

(b) Variational approach,

(c) Weighted residual approach.

(a) **Direct approach:** In this method, direct physical reasoning is used to establish the element properties (stiffness matrices and load vectors) in terms of pertinent variables. Although this approach is limited to simple types of elements, it helps to understand the physical interpretation of the finite element method.

(b) **Variational approach:** This approach can be adopted when the variational theorem (extremum principle) that governs the physics of the problem is available. This method involves minimizing a scalar quantity known as functional that is typical of the problem at hand (e.g., potential energy in stress analysis problems).

(c) **Weighted residual approach:** This approach is more general in the sense that it is applicable to all situations where the governing differential equation of the problem is available. This method involves minimizing error resulting from substituting trial solution in to the differential equation.

Step (iv). Assembling of element equations to obtain the global equations
Element equations obtained in *Step (iii)* are assembled to form global equations in the form of

$$[K]\{r\} = \{R\} \qquad (2.2)$$

where, [K] is the global stiffness matrix,

{r} is the vector of global nodal displacements, and

{R} is the global load vector of nodal forces for the complete structure.

Equation (2.2) describes that the behavior of entire structure.

Step (v). Solution for the unknown nodal displacements

The global equations are to be modified to account for the boundary conditions of the problem. After specifying the boundary conditions, the equilibrium equations can be expressed as

$$[K_1] \{r_1\} = \{R_1\}. \tag{2.3}$$

For linear problems, the vector $\{r_1\}$ can be solved very easily.

Step (vi). Computation of element strains and stresses

From the known nodal displacements $\{r_1\}$, the elements strains and stresses can be computed by using predefined equations for structure.

The terminology used in the previous 6 steps has to be modified if we want to extend the concept to other fields. For example, put the field variable in place of displacement, the characteristic matrix in place of stiffness matrix, and the element resultants in place of element strains.

2.4 METHODS OF PRESCRIBING BOUNDARY CONDITIONS

There are 3 methods of prescribing boundary conditions.

2.4.1 Elimination Method

This method is useful when performing hand calculations. It poses difficulties in implementing in software. This method has been used in this book for solving the problems by finite element method using hand calculations and results in reduced sizes of matrices thus making it suitable for hand calculations. The method is explained below in brief. Consider the following set of global equations,

$$\begin{bmatrix} k_{11} & k_{12} & k_{13} & k_{14} \\ k_{21} & k_{22} & k_{23} & k_{24} \\ k_{31} & k_{32} & k_{33} & k_{34} \\ k_{41} & k_{42} & k_{43} & k_{44} \end{bmatrix} \begin{Bmatrix} u_1 \\ u_2 \\ u_3 \\ u_4 \end{Bmatrix} = \begin{Bmatrix} P_1 \\ P_2 \\ P_3 \\ P_4 \end{Bmatrix} \tag{2.4}$$

Let u_3 be prescribed, i.e., $u_3 = s$.
This condition is imposed as follows:

1. Eliminate the row corresponding to u_3 (3rd row).
2. Transfer the column corresponding to u_3 (3rd column) to right-hand side after multiplying it by "s". These steps result in the following set of modified equations,

$$\begin{bmatrix} k_{11} & k_{12} & k_{14} \\ k_{21} & k_{22} & k_{24} \\ k_{41} & k_{42} & k_{44} \end{bmatrix} \begin{Bmatrix} u_1 \\ u_2 \\ u_4 \end{Bmatrix} = \begin{Bmatrix} P_1 \\ P_2 \\ P_4 \end{Bmatrix} - s \begin{Bmatrix} k_{13} \\ k_{23} \\ k_{43} \end{Bmatrix}. \qquad (2.5)$$

This set of equations now may be solved for non-trivial solution.

2.4.2 Penalty Method

This is the method used in most of the commercial software because this method facilitates prescribing boundary conditions without changing the sizes of the matrices involved. This makes implementation easier.

2.4.3 Multipoint Constrains Method

This method is commonly used in functional analysis between nodes. For example, there are many applications in trusses where the end supports are on an inclined plane and do not coincide with the coordinate system used to describe the truss. Another application of the method is the functional relationship between the temperature at one node and temperature at one or more other nodes.

2.5 PRACTICAL APPLICATIONS OF FINITE ELEMENT ANALYSIS

There are 3 practical applications of finite element analysis:

- Analysis of new design
- Optimization projects
- Failure analysis

2.6 FINITE ELEMENT ANALYSIS SOFTWARE PACKAGE

There are 3 main steps involved in solving an engineering problem using any commercial software:

Step (i). Preprocessing
In this step, a CAD model of the system (component) is prepared and is meshed (discretized). Boundary conditions (support conditions and loads) are applied to the meshed model.

Step (ii). Processing
In this step, the software internally calculates the elements stiffness matrices, element load vectors, global stiffness matrix, global load vector, and solves after applying boundary conditions for primary unknowns (e.g., *displacements/temperatures etc.*) and secondary unknowns (e.g., stress/strain/heat flux etc.).

Step (iii). Post-processing
Post-processing involves sorting and plotting the output to make the interpretation of results easier.

2.7 FINITE ELEMENT ANALYSIS FOR STRUCTURE

There are several common methods in finite element analysis used for evaluating displacements, stresses, and strains in any structure under different boundary conditions and loads. They are summarized below:

1. *Displacement Method:* This method is the most commonly used method. The structure is subjected to applied loads or/and specific displacements. The primary unknowns are displacements found by using an inversion of the stiffness matrix, and the derived unknowns are stresses and strains. Indeed, the stiffness matrix for any element can be calculated by the variational principle.
2. *Force Method:* The structure is subjected to applied loads or/and specific displacements. The primary unknowns are member forces, found by using an inversion of the flexibility matrix, and the derived unknowns are stresses and strains. Indeed, the calculation of the flexibility matrix is possible only for discrete structural elements (e.g., piping, beams, and trusses).
3. *Mixed Method:* The structure is subjected to applied loads or/and specific displacements. This method uses very large stiffness coefficients and very small flexibility coefficients in the same matrix.

4. *Hybrid Method:* The structure is subjected to applied loads and stress boundary conditions. This hybrid method has the merit of the FEA method, i.e., the flexibility and sparse matrix of FEM for complicated inhomogeneous scatterers.

2.8 TYPES OF ELEMENTS

In general, the region in space is considered nonregular geometric. However, the FEA method divides the nonregular geometric region to small regular geometric regions. There are 3 types of elements in finite element.

1. *One-Dimensional Elements:* The objects are subdivided into short-line segments. A one-dimensional finite element expresses the object as a function of one independent variable such as one coordinate x. Finite elements use one-dimensional elements to solve systems that are governed by ordinary differential equations in terms of an independent variable. The number of node points in an element can vary from 2 up to any value needed. Indeed, increasing the number of nodes for an element increases the accuracy of the solution, but it also increases the complexity of calculations. When the elements have a polynomial approximation higher than first order, we call that *higher order elements.* Figure 2.1 shows one-dimensional elements. For example, the one-dimensional element is sufficient in dealing with heat dissipation in cooling fins.

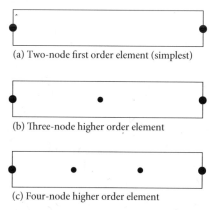

(a) Two-node first order element (simplest)

(b) Three-node higher order element

(c) Four-node higher order element

Figure 2.1. **One-dimensional elements.**

2. *Two-Dimensional Elements:* The objects can be divided into triangles, rectangles, quadrilaterals, or other suitable subregions. A two-dimensional finite element expresses the object as a function of 2 variables such as the 2 coordinates x and y. A finite element uses two-dimensional elements to solve systems that are governed by partial differential equations. The simplest two-dimensional element is the triangular element. Figure 2.2 shows two-dimensional elements. For example, a two-dimensional element is sufficient in plane stress or plane strain.

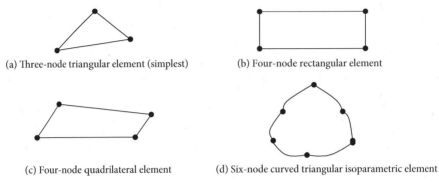

(a) Three-node triangular element (simplest) (b) Four-node rectangular element

(c) Four-node quadrilateral element (d) Six-node curved triangular isoparametric element

Figure 2.2. **Two-dimensional elements.**

3. *Three-Dimensional Elements:* The objects can be divided into tetrahedral elements, rectangular prismatic elements, pie-shaped elements, or other suitable shapes of elements. A three-dimensional finite element expresses the object as a function of 3 variables such as the 3 coordinates x, y, and z. A finite element uses three-dimensional elements to solve systems that are governed by differential equations. The simplest three-dimensional element is the tetrahedral element. Figure 2.3 shows three-dimensional elements.

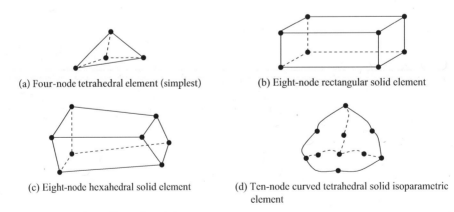

(a) Four-node tetrahedral element (simplest) (b) Eight-node rectangular solid element

(c) Eight-node hexahedral solid element (d) Ten-node curved tetrahedral solid isoparametric element

Figure 2.3. **Three-dimensional elements.**

These three types of elements are applied and discussed in the electromagnetics analysis chapter of this book.

2.9 DIRECT METHOD FOR LINEAR SPRING

Here, we will use the direct method in a one-dimension domain to derive the stiffness matrix for the linear spring element shown in Figure 2.4. Reference points 1 and 2 located at the ends of the linear spring element are the nodes. The symbols f_1 and f_2 are local nodal forces (or axial loads) associated with the local axis x. The symbols u_1 and u_2 are local nodal displacements (or degree of freedom at each node) for the spring element. u_i is the displacement of the spring due to a load f_i. The symbol k is the stiffness of the spring (or spring constant). k is load required to give the spring a unit displacement. The symbol L is the bar length. The local axis x acts in the same direction of the spring that can lead to direct measurement of forces and displacements along the spring.

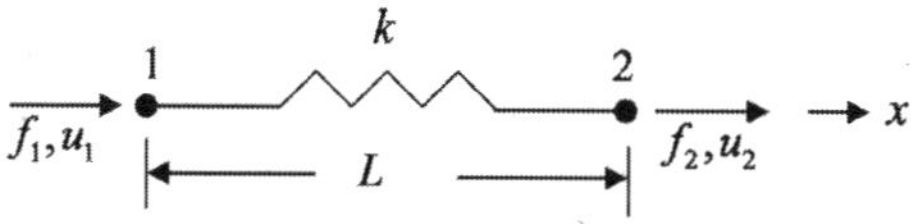

Figure 2.4. Linear spring element.

The displacements can be defined related to forces as

$$u = u_1 - u_2 \qquad (2.6)$$

$$f_1 = ku = k(u_1 - u_2). \qquad (2.7)$$

The equilibrium of forces gives

$$f_2 = -f_1. \qquad (2.8)$$

Based on equation (2.7), the above equation becomes

$$f_2 = k(u_2 - u_1). \qquad (2.9)$$

By combining equations (2.7) and (2.9) and writing the resulting equations in matrix form we get

$$\begin{Bmatrix} f_1 \\ f_2 \end{Bmatrix} = \begin{bmatrix} k & -k \\ -k & k \end{bmatrix} \begin{Bmatrix} u_1 \\ u_2 \end{Bmatrix}. \qquad (2.10)$$

or

$$\{f_i\} = [k]\{u_i\} \qquad (2.11)$$

where

$\{f_i\}$ = a vector of internal nodal forces = $\begin{Bmatrix} f_1 \\ f_2 \end{Bmatrix}$

$[k]$ = the elemental stiffness matrix = $\begin{bmatrix} k & -k \\ -k & k \end{bmatrix}$

$\{u_i\}$ = a vector of nodal displacements = $\begin{Bmatrix} u_1 \\ u_2 \end{Bmatrix}$.

For many interconnected spring elements, we can use the following:

$$\{Q_i\} = [K]\{u_i\} \qquad (2.12)$$

where
$\{Q_i\}$ = a vector of external nodal forces = $\Sigma\{f_i\}$
$[K]$ = the structural stiffness matrix = $\Sigma[k]$
$\{u\}$ = a vector of nodal displacements of the structure.

PROBLEMS

1. Define finite element analysis?
2. What are the advantages and features of finite element analysis?
3. What are the 3 common methods to solve any engineering problem?
4. What is procedure of finite element analysis (related to structural problems)?
5. What are the 2 methods for prescribing the boundary conditions?
6. Give the 3 practical applications of finite element analysis?
7. What are the 3 main steps involved in solving an engineering problem using any commercial software?
8. What are the 4 common methods in finite element analysis used for evaluating displacements, stresses, and strains in any structure under different boundary conditions and loads?
9. What is the primary variable in finite element method structural analysis?
10. Calculate the structural stiffness matrix of the system as shown in Figure 2.5.

Figure 2.5. Two springs in series structure.

REFERENCES

1. P. P. Silvester and R. L. Ferrari, "Finite Elements for Electrical Engineering," Cambridge Press, 1983.
2. B. Szabo and I. Babuska, "Finite Element Analysis," John Wiley & Sons, 1991.
3. R. D. Cook and D. S. Malkus, and M. E. Plesha, "Concepts and Applications of Finite Element Analysis," John Wiley and Sons, 1989.
4. O. C. Zienkiewicz, "The Finite Element Method, 3rd. ed.," McGraw-Hill, 1979.
5. W. B. Bickford, "A First Course in the Finite Element Method," Richard D. Irwin Publisher, 1989.
6. G. L. Narasaiah, "Finite Element Analysis," CRC Press, 2009.
7. P. E. Allaire, "Basics of the Finite Element Method: Solid Mechanics, Heat Transfer, and Fluid Mechanics," Wm. C. Brown Publishers, 1985.
8. CTF Ross, "Finite Element Methods in Structural Mechanics," Ellis Horwood Limited Publishers, 1985.

Chapter 3

Finite Element Analysis of Axially Loaded Members

3.1 INTRODUCTION

In this chapter, we will use the bar element in the analysis of rod-like axially loaded members. We start with the two popular bar elements using a two-node element and a three-node element as well as bars of constant cross-section area, bars of varying cross-section area, and the stepped bar.

Stress is an internal force that has been distributed over the area of the rod's cross section and it is defined as

$$\sigma = \frac{F}{A}, \qquad (3.1)$$

where σ is the stress, F is the force, and A is the cross-sectional area.

Thus, stress is a measure of force per unit area. When the stress tends to lengthen the rod, the stress is called *tension*, and $\sigma > 0$. When the stress tends to shortened the rod, the stress is called *compression*, and $\sigma < 0$. The orientations of forces in tension and compression are shown in Figure 3.1.

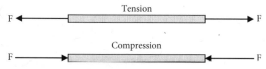

Figure 3.1. Directions of tensile and compressive forces.

The derived unit of stress is the pascal (Pa), where pascal is equal to newtons per square meter (N/m²), 1 Pa = 1 N/m², pascal is used in the SI units. The derived unit of stress is the dimension pound-per-square-inch (psi), where 1 psi = 1 lb/in². Psi is used in the USCS (U.S. Customary) units. In stresses, calculations are generally very large, therefore, they often use the prefixes kilo- (k), mega- (M), and giga- (G) for factors of 10^3, 10^6, and 10^9, respectively. Thus,

$$1 \text{ kPa} = 10^3 \text{ Pa}, \quad 1 \text{ MPa} = 10^6 \text{ Pa}, \quad 1 \text{ GPa} = 10^9 \text{ Pa}.$$

The numerical values for stresses unit conversion between the USCS and SI can be presented as

$$1 \text{ psi} = 6.895 \times 10^{-3} \text{ MPa}.$$

Strain (ε) is the amount of elongation that occurs per unit of the rod's original length and is calculated as

$$\varepsilon = \frac{\Delta L}{L}, \tag{3.2}$$

where, ΔL is the change in length of the rod (elongation).

Strain is a dimensionless quantity and is generally very small.

For each individual rod, the applied force and elongation are proportional to each other based on the following expression

$$F = k\Delta L, \tag{3.3}$$

where k is the stiffness.

Figure 3.2 shows force and elongation behaviors of rods at various cross-sectional areas and lengths.

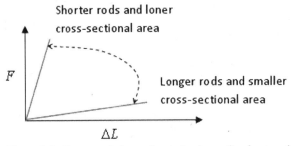

Figure 3.2. Force and elongation behaviors of rods at various cross-sectional areas and lengths.

Stress and strain are useful in mechanical engineering because they are scaled with respect to the rod's size.

Stress and strain are proportional to each other and presented as

$$\sigma = E\varepsilon, \quad (3.4)$$

where E is the elastic modulus (or Young's modulus).

The elastic modulus has the dimensions of force per unit area. The elastic modulus is a physical material property, and is the slope of the stress-strain curve for low strain.

By combining equations (3.1) and (3.2), we get

$$\Delta L = \frac{FL}{EA}. \quad (3.5)$$

With the stiffness in equation (3.3), it can be written as

$$k = \frac{EA}{L}. \quad (3.6)$$

Each rod formed of same material has similar stress-strain behavior as presented in Figure 3.3.

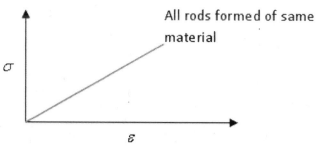

Figure 3.3. All rods formed of same material have similar stress-strain behavior.

When a system is motionless or has constant velocity, then the system has zero acceleration, and the system is to be in *equilibrium*. The *static equilibrium* is used for a system at rest. For equilibrium, the resultant of all forces and all moments acting on the system is balanced to zero resultant. That is, the sum of all force vectors (**F**) acting upon a system is zero and the sum of all moment vectors (**M**) acting upon a system is zero, and they can be written as

$$\sum \mathbf{F} = 0 \tag{3.7}$$

$$\sum \mathbf{M} = 0. \tag{3.8}$$

The total extension (or contraction) of a uniform bar in pure tension or compression is defined as

$$\delta = \frac{FL}{AE}. \tag{3.9}$$

The equation (3.9) does not apply to a long bar loaded in compression if there is a possibility of bucking.

3.1.1 Two-Node Bar Element

Figure 3.4. Two-node bar for rod-like axially loaded members.

This element has 2 end nodes and each node has 1 degree of freedom, namely translation along its length. Its formulation is based on linear interpolation. It gives accurate results only if loads are applied at nodes and the area is constant over the element. However, required accuracy for practical purposes in other cases can be obtained by taking a larger number of smaller elements. The interpolation equation, element stiffness matrix, strain-displacement matrix, element strain, and element stress for 2-node (linear) bar element are given by

$$\{u\} = [N_1 \quad N_2] \begin{Bmatrix} u_1 \\ u_2 \end{Bmatrix} \tag{3.10}$$

$$\{u\} = \begin{bmatrix} \dfrac{(x_2 - x)}{L} & \dfrac{(x - x_1)}{L} \end{bmatrix} \tag{3.11}$$

$$[k] = \frac{AE}{L} \begin{bmatrix} 1 & -1 \\ -1 & 1 \end{bmatrix} \tag{3.12}$$

$$[B] = \frac{1}{L}[-1 \quad 1] \qquad (3.13)$$

$$\{\varepsilon\} = [B]\{q\} \qquad (3.14)$$

$$\{\sigma\} = E[B]\{q\} \qquad (3.15)$$

where
u_1 and u_2 = nodal (unknown) displacements (degree of freedom) at node 1 and 2, respectively
$\{u\}$ = displacement matrix at the nodes
A = cross section of the area of the bar
$L = x_2 - x_1$ = length of the bar
E = Young's modulus (modulus of elasticity)
$\dfrac{AE}{L}$ = bar constant

$[k]$ = stiffness matrix of the element
$[B]$ = strain-displacement matrix
$\{\varepsilon\}$ = strain matrix
$\{\sigma\}$ = stress matrix.

Uniformly distributed load per unit length w acting on the element can be converted into equivalent loads using,

$$\{W\} = wL \begin{Bmatrix} \dfrac{1}{2} \\ \dfrac{1}{2} \end{Bmatrix}, \qquad (3.16)$$

$\{W\}$ = the potential energy of load system.

Thermal loads due to a change in temperature ΔT can be converted into equivalent nodal loads using

$$\{Q\} = EA\alpha(\Delta T) \begin{Bmatrix} -1 \\ 1 \end{Bmatrix}, \qquad (3.17)$$

where
a is the coefficient of thermal expansion.

3.1.2 Three-Node Bar Element

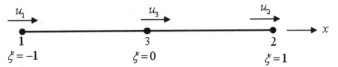

Figure 3.5. Three-node bar for rod-like axially loaded members.

This element has a midside node, in addition to 2 end nodes. Each node has 1 degree of freedom, namely translation along its length. Its formulation is based on quadratic interpolation and this element gives accurate results even with distributed loads and a linearly varying cross-sectional area. Coarse mesh with fewer of these elements can give the desired accuracy as compared to a fine mesh of 2-node bar element. The interpolation equation, element stiffness matrix, strain-displacement matrix, element strain, and element stress for the quadratic bar element are given by,

$$[u] = [N_1 \quad N_2 \quad N_3] \begin{Bmatrix} u_1 \\ u_2 \\ u_3 \end{Bmatrix} = \left[-\frac{\xi}{2} + \frac{\xi^2}{2} \quad \frac{\xi}{2} + \frac{\xi^2}{2} \quad 1 - \xi^2 \right] \quad (3.18)$$

$$[k] = \frac{AE}{3L} \begin{bmatrix} 7 & 1 & -8 \\ 1 & 7 & -8 \\ -8 & -8 & 16 \end{bmatrix} \quad (3.19)$$

$$[B] = \frac{2}{L} \left[-\frac{1-2\xi}{2}, \quad \frac{1+2\xi}{2}, \quad -2\xi \right] \quad (3.20)$$

$$\{\varepsilon\} = [B]\{q\} \quad (3.21)$$

$$\{\sigma\} = E[B]\{q\}. \quad (3.22)$$

Uniformly distributed load per unit length w, acting on the element, can be converted into equivalent loads using,

$$\{W\} = w\frac{L}{6} \begin{Bmatrix} 1 \\ 1 \\ 4 \end{Bmatrix}. \quad (3.23)$$

3.2 BARS OF CONSTANT CROSS-SECTION AREA

This section will demonstrate examples on bars of constant cross-sectional area using FEA.

Example 3.1

Consider a 2 m long steel bar of 50 mm² cross-sectional areas as shown in Figure 3.6. Use a two element mesh to model this problem. Find nodal displacements, element stresses, and reaction.

Take Young's modulus, $E = 2 \times \dfrac{10^5 \, N}{mm^2}$, P = 100 N.

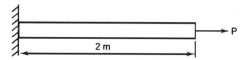

Figure 3.6. Bar with tip load for Example 3.1.

Solution

(I) Analytical method [Refer to Figure 3.6(a)]

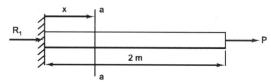

Figure 3.6(a). Analytical method for bar with tip load for Example 3.1.

Displacement calculation

Displacement at section a–a,

$$\delta = \frac{Px}{AE} = \frac{100x}{50*2*10^5} = 1*10^{-5} x.$$

Displacement at node 2,

$$\delta_{x=1000} = 1 \times 10^{-5} \times 1000 = 0.01 \text{ mm}.$$

Displacement at node 3,

$$\delta_{x=2000} = 1\times 10^{-5} \times 2000 = 0.02 \text{ mm}.$$

Stress calculation

Maximum stress in the bar $= \dfrac{P}{A} = \dfrac{100}{50} = 2$ N/mm^2 (Constant).

Reaction calculation

For reaction calculation, $\sum F_x = 0$

$R_1 + 100 = 0$

$R_1 = -100$ N (Direction is leftwards).

(II) FEM by hand calculations

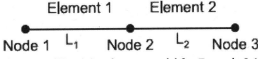

Figure 3.6(b). Finite element model for Example 3.1.

$$L_1 = L_2 = 1000 \text{ mm}$$
$$A = A_1 = A_2 = 50 \text{ mm}^2$$
$$E = E_1 = E_2 = 2\times 10^5 \text{ N/mm}^2$$

Stiffness matrix for element 1 is,

$$[k_1] = \frac{A_1 E_1}{L_1}\begin{bmatrix} 1 & -1 \\ -1 & 1 \end{bmatrix} = \frac{50\times 2\times 10^5}{1000}\begin{bmatrix} 1 & -1 \\ -1 & 1 \end{bmatrix} = 0.1\times 10^5 \begin{bmatrix} 1 & -1 \\ -1 & 1 \end{bmatrix} \begin{matrix} 1 \\ 2 \end{matrix}$$

$$[k_2] = \frac{A_2 E_2}{L_2}\begin{bmatrix} 1 & -1 \\ -1 & 1 \end{bmatrix} = \frac{50\times 2\times 10^5}{1000}\begin{bmatrix} 1 & -1 \\ -1 & 1 \end{bmatrix} = 0.1\times 10^5 \begin{bmatrix} 1 & -1 \\ -1 & 1 \end{bmatrix} \begin{matrix} 2 \\ 3 \end{matrix}$$

Global equation is,

$$[K]\{r\} = \{R\} \tag{3.24}$$

$$0.1 \times 10^5 \begin{bmatrix} 1 & -1 & 0 \\ -1 & 1+1 & -1 \\ 0 & -1 & 1 \end{bmatrix} \begin{Bmatrix} u_1 \\ u_2 \\ u_3 \end{Bmatrix} = \begin{Bmatrix} R_1 \\ 0 \\ 100 \end{Bmatrix}. \tag{3.25}$$

Boundary conditions are, at node 1, $u_1 = 0$.
By using elimination method, the above matrix reduces to,

$$0.1 \times 10^5 \begin{bmatrix} 2 & -1 \\ -1 & 1 \end{bmatrix} \begin{Bmatrix} u_2 \\ u_3 \end{Bmatrix} = \begin{Bmatrix} 0 \\ 100 \end{Bmatrix}.$$

By matrix multiplication, we get

$$0.1 \times 10^5 (2 \times u_2 - u_3) = 0 \tag{3.26}$$

$$0.1 \times 10^5 (-u_2 + u_3) = 100. \tag{3.27}$$

By solving equations (3.26) and (3.27), we get

$$u_2 = 0.01 \text{ mm}$$

$$u_3 = 0.02 \text{ mm}.$$

Stress (σ) calculation

Stress for element 1 is,

$$\{\sigma_1\} = \frac{E}{L_1}[-1 \ 1]\begin{Bmatrix} u_1 \\ u_2 \end{Bmatrix} = \frac{2 \times 10^5}{1000}[-1 \ 1]\begin{Bmatrix} 0 \\ 0.01 \end{Bmatrix} = 2 \text{ N/mm}^2.$$

Stress for element 2 is,

$$\{\sigma_2\} = \frac{E}{L_2}[-1 \ 1]\begin{Bmatrix} u_2 \\ u_3 \end{Bmatrix} = \frac{2 \times 10^5}{1000}[-1 \ 1]\begin{Bmatrix} 0.01 \\ 0.02 \end{Bmatrix} = 2 \text{ N/mm}^2.$$

Reaction calculation

From equation (3.25)

$$0.1 \times 10^5 (u_1 - u_2) = R_1$$
$$0.1 \times 10^5 (0 - 0.01) = R_1$$
$$R_1 = -100 \text{ N}.$$

(III) Software results

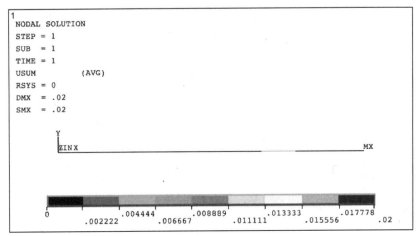

Figure 3.6(c). Deflection pattern for a bar (refer to Appendix C for color figures).

Deflection values as node *(Computer generated output)*

The following degree of freedom results are in global coordinates:

NODE	UX	UY	UZ	USUM
1	0.0000	0.0000	0.0000	0.0000
2	0.10000E-01	0.0000	0.0000	0.10000E-01
3	0.20000E-01	0.0000	0.0000	0.20000E-01

Maximum absolute values

NODE	3	0	0	3
VALUE	0.20000E-01	0.0000	0.0000	0.20000E-01

Finite Element Analysis of Axially Loaded Members

```
ELEMENT SOLUTION
STEP = 1
SUB  = 1
TIME = 1
LS1      (NOAVG)
DMX = .02
SMN = 2
SMX = 2
```

Figure 3.6(d). Stress pattern for a bar (refer to Appendix C for color figures).

Stress values at elements (Computer generated output)

STAT ELEM	CURRENT LS1
1	2.0000
2	2.0000

Reaction value (Computer generated output)

The following X, Y, Z solutions are in global coordinates

NODE	FX	FY
1	−100.00	0.0000

Answers for Example 3.1

Parameter	Analytical method	FEM-hand calculations	Software results
Displacement at node 2	0.01 mm	0.01 mm	0.01 mm
Displacement at node 3 (Maximum displacement)	0.02 mm	0.02 mm	0.02 mm
Maximum stress in element 1	2 N/mm^2	2 N/mm^2	2 N/mm^2
Maximum stress in element 2	2 N/mm^2	2 N/mm^2	2 N/mm^2
Reaction at fixed end	−100 N	−100 N	−100 N

Example 3.2

Bar under distributed and concentrated forces. Consider the bar shown in Figure 3.7 subjected to loading as shown below. Use 4 element mesh models and find nodal displacements, element stresses, and reaction at the fixed end. Take $E = 2 \times 10^5$ N/mm^2, $A = 50$ mm^2, $P = 100$ N.

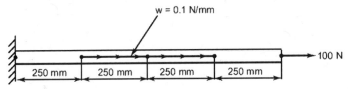

Figure 3.7. Bar under distributed and concentrated forces for Example 3.2.

Solution

(I) Analytical method [Refer Figure 3.7(a)]

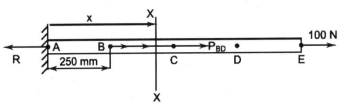

Figure 3.7(a). Analytical method for the Bar under distributed and concentrated forces for Example 3.2.

Reaction calculation

$$-R + w[L_2 + L_3] + P = 0$$

$$-R + (0.1) \times [250 + 250] + 100 = 0$$

$$R = 150 \text{ N}$$

Stress calculation

$$\sigma_{AB} = \frac{P}{A} = \frac{150}{50} = 3 \text{ N/mm}^2$$

$$\sigma_{DE} = \frac{P}{A} = \frac{100}{50} = 2 \text{ N/mm}^2$$

To find σ_{BD}, consider section XX

$$\sigma_{XX} = \frac{P_{BD}}{A_{BD}} = \frac{150-(x-250)\times 0.1}{50} = \frac{150-0.1x+25}{50} = \frac{175-0.1x}{50}$$

$$\sigma_{XX}\big|_{\text{at B}} = \sigma_{XX}\big|_{x=250} = \frac{175-0.1\times 250}{50} = 3 \text{ N/mm}^2$$

$$\sigma_{XX}\big|_{\text{at C}} = \sigma_{XX}\big|_{x=500} = \frac{175-0.1\times 500}{50} = 2.5 \text{ N/mm}^2$$

$$\sigma_{XX}\big|_{\text{at D}} = \sigma_{XX}\big|_{x=750} = \frac{175-0.1\times 750}{50} = 2 \text{ N/mm}^2$$

Displacement calculation
Displacement at E,

$$\delta_E = \Delta_{AB} + \Delta_{BD} + \Delta_{DE}$$

$$\delta_E = \frac{\sigma_{AB}L_{AB}}{E} + \int_{x=250}^{750} \frac{(175-0.1x)\,dx}{AE} + \left(\frac{\sigma_{DE}L_{DE}}{E}\right)$$

$$\delta_E = \frac{3\times 250}{2\times 10^5} + \frac{1}{50\times 2\times 10^5}\left(175x - 0.1\frac{x^2}{2}\right)_{250}^{750} + \frac{2\times 250}{2\times 10^5}$$

$$\delta_E = 0.00375 + 0.00625 + 0.0025 = 0.0125 \text{ mm}$$

Displacement at B,

$$\delta_B = \Delta_{AB} = \frac{3\times 250}{2\times 10^5} = 0.00375 \text{ mm}$$

Displacement at D,

$$\delta_D = \Delta_{AB} + \Delta_{BD} = \frac{3\times 250}{2\times 10^5} = 0.00375 + 0.00625 = 0.01 \text{ mm}$$

Displacement at C,

$$\delta_C = \Delta_{AB} + \int_{x=250}^{500} \frac{(175-0.1x)\,dx}{AE} = 0.00375 + \frac{1}{50\times2\times10^5}\left(175 - \frac{0.1x^2}{2}\right)_{250}^{500}$$

$$= 0.0072 \text{ mm}$$

(II) FEM by hand calculations

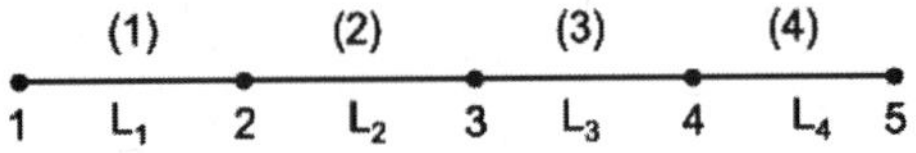

Figure 3.7(b). **Finite element model for Example 3.2.**

$$L_1 = L_2 = L_3 = L_4 = 250 \text{ mm}$$

$$A = A_1 = A_2 = A_3 = A_4 = 50 \text{ mm}^2$$

$$E = E_1 = E_2 = E_3 = E_4 = 2\times10^5 \text{ N/mm}^2$$

Stiffness matrix for elements is,

$$[k_1] = \frac{A_1E_1}{L_1}\begin{bmatrix} 1 & -1 \\ -1 & 1 \end{bmatrix} = \frac{50\times2\times10^5}{250}\begin{bmatrix} 1 & -1 \\ -1 & 1 \end{bmatrix} = 0.4\times10^5 \begin{array}{c} \begin{matrix} 1 & \;\;2 \end{matrix} \\ \begin{bmatrix} 1 & -1 \\ -1 & 1 \end{bmatrix} \end{array} \begin{matrix} 1 \\ 2 \end{matrix}$$

$$[k_2] = \frac{A_2E_2}{L_2}\begin{bmatrix} 1 & -1 \\ -1 & 1 \end{bmatrix} = \frac{50\times2\times10^5}{250}\begin{bmatrix} 1 & -1 \\ -1 & 1 \end{bmatrix} = 0.4\times10^5 \begin{array}{c} \begin{matrix} 2 & \;\;3 \end{matrix} \\ \begin{bmatrix} 1 & -1 \\ -1 & 1 \end{bmatrix} \end{array} \begin{matrix} 2 \\ 3 \end{matrix}$$

$$[k_3] = \frac{A_3E_3}{L_3}\begin{bmatrix} 1 & -1 \\ -1 & 1 \end{bmatrix} = \frac{50\times2\times10^5}{250}\begin{bmatrix} 1 & -1 \\ -1 & 1 \end{bmatrix} = 0.4\times10^5 \begin{array}{c} \begin{matrix} 3 & \;\;4 \end{matrix} \\ \begin{bmatrix} 1 & -1 \\ -1 & 1 \end{bmatrix} \end{array} \begin{matrix} 3 \\ 4 \end{matrix}$$

$$[k_4] = \frac{A_4E_4}{L_4}\begin{bmatrix} 1 & -1 \\ -1 & 1 \end{bmatrix} = \frac{50\times2\times10^5}{250}\begin{bmatrix} 1 & -1 \\ -1 & 1 \end{bmatrix} = 0.4\times10^5 \begin{array}{c} \begin{matrix} 4 & \;\;5 \end{matrix} \\ \begin{bmatrix} 1 & -1 \\ -1 & 1 \end{bmatrix} \end{array} \begin{matrix} 4 \\ 5 \end{matrix}$$

Nodal load calculation for elements 2 and 3,

$$W_1 = \begin{Bmatrix} \dfrac{wL_2}{2} \\ \dfrac{wL_2}{2} \end{Bmatrix} = \begin{Bmatrix} \dfrac{0.1 \times 250}{2} \\ \dfrac{0.1 \times 250}{2} \end{Bmatrix} = \begin{Bmatrix} 12.5 \\ 12.5 \end{Bmatrix} \begin{matrix} 2 \\ 3 \end{matrix}$$

$$W_2 = \begin{Bmatrix} \dfrac{wL_3}{2} \\ \dfrac{wL_3}{2} \end{Bmatrix} = \begin{Bmatrix} \dfrac{0.1 \times 250}{2} \\ \dfrac{0.1 \times 250}{2} \end{Bmatrix} = \begin{Bmatrix} 12.5 \\ 12.5 \end{Bmatrix} \begin{matrix} 3 \\ 4 \end{matrix}$$

Global equation is,

$$[K]\{r\} = \{R\} \tag{3.28}$$

$$0.4 \times 10^5 \begin{bmatrix} 1 & 1 & 0 & 0 & 0 \\ -1 & 1+1 & -1 & 0 & 0 \\ 0 & -1 & 1+1 & -1 & 0 \\ 0 & 0 & -1 & 1+1 & -1 \\ 0 & 0 & 0 & -1 & 1 \end{bmatrix} \begin{Bmatrix} u_1 \\ u_2 \\ u_3 \\ u_4 \\ u_5 \end{Bmatrix} = \begin{Bmatrix} R_1 \\ 12.5 \\ 12.5+12.5 \\ 12.5 \\ 100 \end{Bmatrix} \tag{3.29}$$

Boundary conditions are at node 1, $u_1 = 0$
By using the elimination method the above matrix reduces to,

$$0.4 \times 10^5 \begin{bmatrix} 2 & -1 & 0 & 0 \\ -1 & 2 & -1 & 0 \\ 0 & -1 & 2 & -1 \\ 0 & 0 & -1 & 1 \end{bmatrix} \begin{Bmatrix} u_2 \\ u_3 \\ u_4 \\ u_5 \end{Bmatrix} = \begin{Bmatrix} 12.5 \\ 25 \\ 12.5 \\ 100 \end{Bmatrix}.$$

By solving the above matrix and equations, we get

$$u_2 = 0.0038 \text{ mm}$$
$$u_3 = 0.0072 \text{ mm}$$
$$u_4 = 0.01 \text{ mm}$$
$$u_5 = 0.0125 \text{ mm}.$$

Stress (σ) calculation

$$\{\sigma_1\} = \frac{E}{L_1}[-1 \quad 1]\begin{Bmatrix} u_1 \\ u_2 \end{Bmatrix} = \frac{2 \times 10^5}{250}[-1 \quad 1]\begin{Bmatrix} 0 \\ 0.0038 \end{Bmatrix} = 3.04 \text{ N/mm}^2$$

$$\{\sigma_2\} = \frac{E}{L_2}[-1 \quad 1]\begin{Bmatrix} u_2 \\ u_3 \end{Bmatrix} = \frac{2 \times 10^5}{250}[-1 \quad 1]\begin{Bmatrix} 0.0038 \\ 0.0072 \end{Bmatrix} = 2.72 \text{ N/mm}^2$$

$$\{\sigma_3\} = \frac{E}{L_3}[-1 \quad 1]\begin{Bmatrix} u_3 \\ u_4 \end{Bmatrix} = \frac{2 \times 10^5}{250}[-1 \quad 1]\begin{Bmatrix} 0.0072 \\ 0.01 \end{Bmatrix} = 2.24 \text{ N/mm}^2$$

$$\{\sigma_4\} = \frac{E}{L_4}[-1 \quad 1]\begin{Bmatrix} u_4 \\ u_5 \end{Bmatrix} = \frac{2 \times 10^5}{250}[-1 \quad 1]\begin{Bmatrix} 0.01 \\ 0.0125 \end{Bmatrix} = 2 \text{ N/mm}^2$$

Reaction calculation: from equation (3.29)

$$0.4 \times 10^5 (u_1 - u_2) = R_1$$

$$0.4 \times 10^5 (0 - 0.0038) = R_1$$

$$R_1 = -152 \text{ N}$$

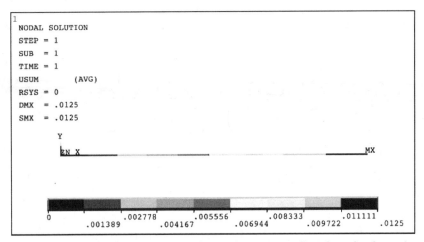

Figure 3.7(c). Deflection pattern for a bar (refer to Appendix C for color figures).

Finite Element Analysis of Axially Loaded Members

Deflection values at nodes (Computer generated output)

The following degree of freedom results are in global coordinates

NODE	UX	UY	UZ	USUM
1	0.0000	0.0000	0.0000	0.0000
2	0.37500E-02	0.0000	0.0000	0.37500E-02
3	0.71875E-02	0.0000	0.0000	0.71875E-02
4	0.10000E-01	0.0000	0.0000	0.10000E-01
5	0.12500E-01	0.0000	0.0000	0.12500E-01

Maximum absolute value

NODE	5	0	0	5
VALUE	0.12500E-01	0.0000	0.0000	0.12500E-01

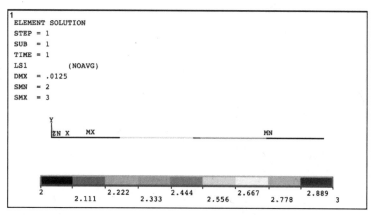

Figure 3.7(d). Stress pattern for a bar (refer to Appendix C for color figures).

Stress values at elements (Computer generated output)

STAT	CURRENT
ELEM	LS1
1	3.0000
2	2.7500
3	2.2500
4	2.0000

Reaction value *(Computer generated output)*
The following X, Y, Z SOLUTIONS are in global coordinates

NODE	FX	FY
1	−150.00	0.0000

Answers to Example 3.2

Parameter	Analytical method	FEM-hand calculations	Software results
Displacement at node 2	0.00375 mm	0.0038 mm	0.00375 mm
Displacement at node 3	0.0072 mm	0.0072 mm	0.00719 mm
Displacement at node 4	0.01 mm	0.01 mm	0.01 mm
Displacement at node 5	0.0125 mm	0.0125 mm	0.0125 mm
Stress in element 1	3 N/mm²	3.04 N/mm²	3 N/mm²
Stress in element 2	3 N/mm² to 2.5 N/mm²	2.72 N/mm²	2.75 N/mm²
Stress in element 3	2.5 N/mm² to 2 N/mm²	2.24 N/mm²	2.25 N/mm²
Stress in element 4	2 N/mm²	2 N/mm²	2 N/mm²
Reaction at fixed end	−1.50 N	−152 N	−150 N

Example 3.3

A and P = 80 kN is applied as shown in Figure 3.8. Determine the nodal displacements, element stresses, and support reactions in the bar. Take $E = 20 \times 10^3$ N/mm².

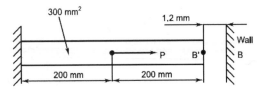

Figure 3.8. Example 3.3.

Solution

(I) Analytical method [Refer to Figure 3.8(a)]

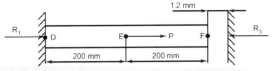

Figure 3.8(a). Analytical method for Example 3.3.

Let R_3 be the reaction developed at the wall after contact.

$$\frac{P_{DE}L}{AE} + \frac{P_{EF}L}{AE} = 1.2.$$

$$\frac{R_1 \times 200}{300 \times 20 \times 10^3} + \frac{(-R_3) \times 200}{300 \times 20 \times 10^3} = 1.2. \qquad (3.30)$$

$$\sum F_x = 0 \;\Rightarrow\; R_1 + R_3 = P = 80 \times 10^3. \qquad (3.31)$$

Solving equations (3.30) and (3.31)

$$R_1 = 58018 \text{ N}$$

$$R_3 = 21982 \text{ N}$$

Stresses are, $\quad \sigma_{DE} = \dfrac{R_1}{A} = \dfrac{58018}{300} = 193.39 \text{ N/mm}^2$

$$\sigma_{EF} = -\frac{R_3}{A} = -\frac{21982}{300} = -73.27 \text{ N/mm}^2.$$

Deflections are, $\quad \delta_2 = \delta_E = \Delta_{DE} = \dfrac{\sigma_{DE} L}{E} = \dfrac{193.39 \times 200}{20 \times 10^3} = 1.934 \text{ mm}$

$$\delta_3 = 1.2 \text{ mm}.$$

(II) FEM by hand calculations

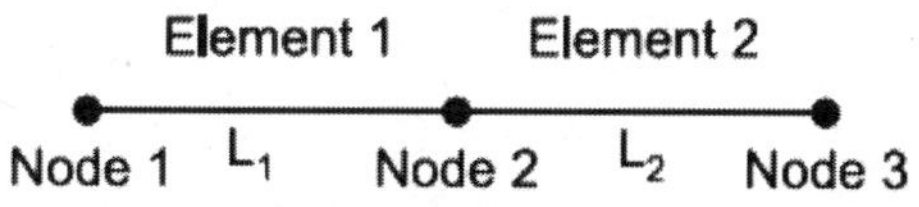

Figure 3.8(b). Finite element model for Example 3.3.

$$L_1 = 200 \text{ mm}, \; L_2 = 200 \text{ mm}$$

First, we should check whether contact occurs between the bar and the wall. For this, assume that the wall does not exist. The solution to the problem is as below. (Consider the two element model.)

Stiffness matrices are,

$$[k_1] = \frac{A_1 E_1}{L_1}\begin{bmatrix} 1 & -1 \\ -1 & 1 \end{bmatrix} = \frac{300 \times 20 \times 10^3}{200}\begin{bmatrix} 1 & -1 \\ -1 & 1 \end{bmatrix} = 30 \times 10^3 \begin{bmatrix} 1 & -1 \\ -1 & 1 \end{bmatrix}\begin{matrix} 1 \\ 2 \end{matrix}$$

$$[k_2] = \frac{A_2 E_2}{L_2}\begin{bmatrix} 1 & -1 \\ -1 & 1 \end{bmatrix} = \frac{300 \times 20 \times 10^3}{200}\begin{bmatrix} 1 & -1 \\ -1 & 1 \end{bmatrix} = 30 \times 10^3 \begin{bmatrix} 1 & -1 \\ -1 & 1 \end{bmatrix}\begin{matrix} 2 \\ 3 \end{matrix}$$

Global equation is,

$$[K]\{r\} = \{R\} \tag{3.32}$$

$$30 \times 10^3 \begin{bmatrix} 1 & -1 & 0 \\ -1 & 1+1 & -1 \\ 0 & -1 & 1 \end{bmatrix}\begin{Bmatrix} u_1 \\ u_2 \\ u_3 \end{Bmatrix} = \begin{Bmatrix} R_1 \\ 80 \times 10^3 \\ 0 \end{Bmatrix}.$$

Boundary conditions are at node 1, $u_1 = 0$.

By using the elimination method, the above matrix reduces to,

$$30 \times 10^3 \begin{bmatrix} 2 & -1 \\ -1 & 1 \end{bmatrix}\begin{Bmatrix} u_2 \\ u_3 \end{Bmatrix} = \begin{Bmatrix} 80 \times 10^3 \\ 0 \end{Bmatrix}.$$

By matrix multiplication, we get

$$30 \times 10^3 (2 \times u_2 - 1 \times u_3) = 80 \times 10^3 \tag{3.33}$$

$$30 \times 10^3 (-1 \times u_2 + 1 \times u_3) = 0 \tag{3.34}$$

By solving equations (3.33) and (3.34), we get, $u_2 = 2.67$ mm and $u_2 = 2.67$ mm.

Since displacement at node 3 is 2.67 mm (greater than 1.2 mm), we can say that contact does occur. The problem has to be resolved since the boundary conditions are now different. The displacement at B' is specified to be 1.2 mm as shown in Figure 3.8.

Global element equation is,

$$[K]\{r\} = \{R\} \tag{3.35}$$

Finite Element Analysis of Axially Loaded Members

$$30 \times 10^3 \begin{bmatrix} 1 & -1 & 0 \\ -1 & 1+1 & -1 \\ 0 & -1 & 1 \end{bmatrix} \begin{matrix} 1 \\ 2 \\ 3 \end{matrix} \begin{Bmatrix} u_1 \\ u_2 \\ u_3 \end{Bmatrix} = \begin{Bmatrix} R_1 \\ 80 \times 10^3 \\ 0 \end{Bmatrix} \quad (3.36)$$

Boundary conditions are at node 1, $u_1 = 0$ and at node 3, $u_2 = 1.2$. By using the elimination method, the above matrix reduces to,

$$30 \times 10^3 [2]\{u_2\} = \{80 \times 10^3\} - 1.2[30 \times 10^3 \times -1]$$

$$30 \times 10^3 \times 2 \times u_2 = 80 \times 10^3 + 36 \times 10^3$$

$$u_2 = 1.933 \text{ mm.}$$

Stress (σ) calculation: stress for element 1 is,

$$\{\sigma_1\} = \frac{E}{L_1}[-1 \ 1]\begin{Bmatrix} u_1 \\ u_2 \end{Bmatrix} = \frac{20 \times 10^3}{200}[-1 \ 1]\begin{Bmatrix} 0 \\ 1.933 \end{Bmatrix} = 193.3 \text{ N/mm}^2.$$

Stress for element 2 is,

$$\{\sigma_2\} = \frac{E}{L_2}[-1 \ 1]\begin{Bmatrix} u_2 \\ u_3 \end{Bmatrix} = \frac{20 \times 10^3}{200}[-1 \ 1]\begin{Bmatrix} 1.933 \\ 1.2 \end{Bmatrix} = -73.3 \text{ N/mm}^2.$$

Reaction calculation: from equation (3.36)

$$30 \times 10^3 (u_1 - u_2) = R_1$$

$$30 \times 10^3 (0 - 1.933) = R_1$$

$$R_1 = -57990 \text{ N (Direction is leftwards).}$$

We know that,

$$R_1 + P + R_3 = 0$$

$$-57990 + 80 \times 10^3 + R_3 = 0$$

$$R_3 = -22010 \text{ N (Direction is leftwards).}$$

(III) Software results

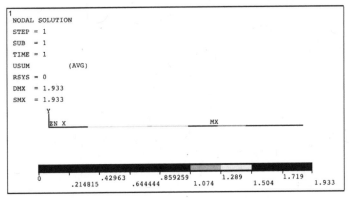

Figure 3.8(c). Deflection pattern for a bar (refer to Appendix C for color figures).

Deflection values at nodes

The following degree of freedom results are in global coordinates

NODE	UX	UY	UX	USUM
1	0.0000	0.0000	0.0000	0.0000
2	1.9333	0.0000	0.0000	1.9333
3	1.2000	0.0000	0.0000	1.2000

Maximum absolute values

NODE	2	0	0	2
VALUE	1.9333	0.0000	0.0000	1.9333

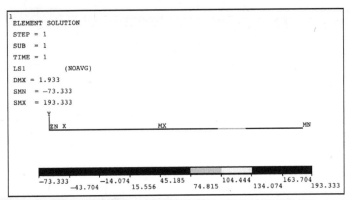

Figure 3.8(d). Stress pattern for a bar (refer to Appendix C for color figures).

Stress values at elements

STAT ELEM	CURRENT LS1
1	193.33
2	−73.333

Reaction value

The following X, Y, Z solutions are in global coordinates

NODE	FX	FY
1	−58000	0.0000
3	−22000	

Answers to Example 3.3

Parameter	Analytical method	FEM-hand calculations	Software results
Displacement at node 2	1.934 mm	1.933 mm	1.933 mm
Displacement at node 3	1.2 mm	1.2 mm	1.2 mm
Stress in element 1	193.39 N/mm²	193.3 N/mm²	193.33 N/mm²
Stress in element 2	−73.27 N/mm²	−73.3 N/mm²	−73.333 N/mm²
Reaction at fixed end	58.02 kN	57.94 kN	−58 kN
Reaction at wall	−21.98 kN	−22.01 kN	−22 kN

Example 3.4

A bar is subjected to self weight. Determine the nodal displacement for the bar hanging under its own weight as shown in Figure 3.9. Use two equal length elements. Let $E = 2 \times 10^{11}$ N/mm², mass density $\rho = 7800$ kg/m³, Area $A = 1000$ mm². Consider length of rod L= 2 m.

Figure 3.9. Bar under self weight for Example 3.4.

Solution

(I) Analytical method [Refer Figure 3.9(a)]

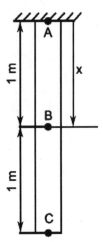

Figure 3.9(a). Analytical method for Example 3.4.

$$\delta_3 = \delta_C = \frac{\rho g L^2}{2E} = \frac{7800 \times 9.81 \times (2)^2}{2 \times 2 \times 10^{11}} = 7.6518 \times 10^{-7} \text{ m}$$

$$\delta_2 = \delta_B = \int_1^2 \frac{xA\rho g}{AE} dx = \frac{\rho g}{E} \left(\frac{x^2}{2}\right)_1^2 = \frac{7800 \times 9.81}{2 \times 10^{11}} \left(\frac{(2)^2}{2} - \frac{(1)^2}{2}\right) = 5.7389 \times 10^{-7} \text{ m}$$

(II) FEM by hand calculations

Figure 3.9(b). Finite element model for Example 3.4.

Finite Element Analysis of Axially Loaded Members

$$L_1 = L_2 = \frac{L}{2} = \frac{2}{2} = 1 \text{ m}$$

The element stiffness matrices are,
For element 1,

$$[k_1] = \frac{AE}{L_1}\begin{bmatrix} 1 & -1 \\ -1 & 1 \end{bmatrix} = \frac{1\times 10^{-3} \times 2\times 10^{11}}{1}\begin{bmatrix} 1 & -1 \\ -1 & 1 \end{bmatrix} = 2\times 10^8 \begin{matrix} & 1 & 2 \\ & \begin{bmatrix} 1 & -1 \\ -1 & 1 \end{bmatrix} & \begin{matrix} 1 \\ 2 \end{matrix} \end{matrix}.$$

For element 2,

$$[k_2] = \frac{AE}{L_2}\begin{bmatrix} 1 & -1 \\ -1 & 1 \end{bmatrix} = \frac{1\times 10^{-3} \times 2\times 10^{11}}{1}\begin{bmatrix} 1 & -1 \\ -1 & 1 \end{bmatrix} = 2\times 10^8 \begin{matrix} & 2 & 3 \\ & \begin{bmatrix} 1 & -1 \\ -1 & 1 \end{bmatrix} & \begin{matrix} 2 \\ 3 \end{matrix} \end{matrix}.$$

Nodal load vector due to weight is,

$$F_1 = \begin{Bmatrix} \frac{\rho A g L_1}{2} \\ \frac{\rho A g L_1}{2} \end{Bmatrix} = \begin{Bmatrix} \frac{7800 \times 1 \times 10^{-3} \times 9.81 \times 1}{2} \\ \frac{7800 \times 1 \times 10^{-3} \times 9.81 \times 1}{2} \end{Bmatrix} = \begin{Bmatrix} 38.26 \\ 38.26 \end{Bmatrix} \begin{matrix} 1 \\ 2 \end{matrix}$$

$$F_2 = \begin{Bmatrix} \frac{\rho A g L_2}{2} \\ \frac{\rho A g L_2}{2} \end{Bmatrix} = \begin{Bmatrix} \frac{7800 \times 1 \times 10^{-3} \times 9.81 \times 1}{2} \\ \frac{7800 \times 1 \times 10^{-3} \times 9.81 \times 1}{2} \end{Bmatrix} = \begin{Bmatrix} 38.26 \\ 38.26 \end{Bmatrix} \begin{matrix} 2 \\ 3 \end{matrix}.$$

Global equation is,

$$[K]\{r\} = \{R\} \tag{3.37}$$

$$2\times 10^8 \begin{matrix} & 1 & 2 & 3 & \\ \begin{matrix}1\\2\\3\end{matrix} & \begin{bmatrix} 1 & -1 & 0 \\ -1 & 1+1 & -1 \\ 0 & -1 & 1 \end{bmatrix} \end{matrix} \begin{Bmatrix} u_1 \\ u_2 \\ u_3 \end{Bmatrix} = \begin{Bmatrix} 38.26 \\ 76.52 \\ 38.26 \end{Bmatrix} \tag{3.38}$$

Boundary conditions are at node 1, $u_1 = 0$.
By using the elimination method, the above matrix reduces to,

$$2\times 10^8 \begin{bmatrix} 2 & -1 \\ -1 & 1 \end{bmatrix} \begin{Bmatrix} u_2 \\ u_3 \end{Bmatrix} = \begin{Bmatrix} 76.52 \\ 38.26 \end{Bmatrix}.$$

By matrix multiplication, we get

$$2\times 10^8 \left(2\times u_2 - 1\times u_3\right) = 76.52 \tag{3.39}$$

$$2\times 10^8 \left(-1\times u_2 + 1\times u_3\right) = 38.26. \tag{3.40}$$

By solving equations (3.39) and (3.40), we get

$$u_2 = 5.739\times 10^{-7} \text{ m}$$

$$u_3 = 7.652\times 10^{-7} \text{ m}.$$

(III) Software results

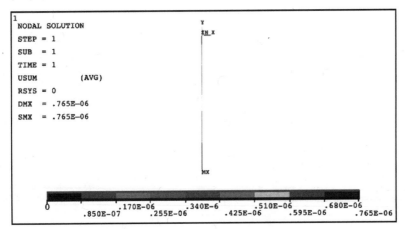

Figure 3.9(c). **Deflection pattern for a bar (refer to Appendix C for color figures).**

Deflection values at nodes

The following degree of freedom results are in global coordinates

NODE	UX	UY	UZ	USUM
1	0.0000	0.0000	0.0000	0.0000
2	0.0000	0.57389E-06	0.0000	0.57389E-06
3	0.0000	0.76518E-06	0.0000	0.76518E-06

NODE	0	3	0	3
VALUE	0.0000	0.76518E-06	0.0000	0.76518E-06

Maximum absolute values

Answers of Example 3.4

Parameter	Analytical method	FEM-hand calculations	Software results
Displacement at node 2	5.7389×10^{-7} m	5.7389×10^{-7} m	5.7389×10^{-7} m
Displacement at node 3	7.6518×10^{-7} m	7.6518×10^{-7} m	7.6518×10^{-7} m

Example 3.5

A rod rotating at a constant angular velocity $\omega = 45$ rad/sec is shown in Figure 3.10. Determine the nodal displacements and stresses in the rod. Consider only the centrifugal force. Ignore the bending of the rod. Use two quadratic elements. Take $A = 350$ mm^2, $E = 70$ GPa, Mass density $\rho = 7850$ kg/m^3, Length of the rod $L = 1$ m.

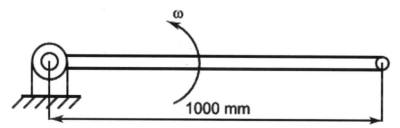

Figure 3.10. Rod rotation at a constant angular velocity for Example 3.5.

Solution

(I) Analytical method [Refer to Figure 3.10(a)]

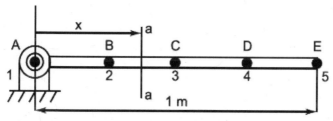

Figure 3.10(a). Analytical method for rod rotation at a constant angular velocity for Example 3.5.

$$L = 1 \text{ m}$$

Stress calculation: Stress at section a-a,

$$\sigma = \frac{mr\omega^2}{\text{Area}} = \frac{\rho \times A \times (L-x)\left(x + \frac{(L-x)}{2}\right)\omega^2}{A} = \frac{\rho\omega^2}{2}\left(L^2 - x^2\right) \quad (3.41)$$

$$\sigma_2 = \sigma_B = \sigma\big|_{x=\frac{L}{4}} = \frac{\rho\omega^2}{2}\left(L^2 - \frac{L^2}{16}\right) = \frac{7850\times(45)^2}{2}\left((1)^2 - \frac{(1)^2}{16}\right) = 7.45 \text{ MPa}$$

$$\sigma_3 = \sigma_C = \sigma\big|_{x=\frac{L}{2}} = 5.96 \text{ MPa}$$

$$\sigma_4 = \sigma_D = \sigma\big|_{x=\frac{3L}{4}} = 3.48 \text{ MPa}$$

$$\sigma_1 = \sigma_A = \sigma_{x=0} = 7.95 \text{ MPa}.$$

Displacement at section a–a = change in length of x,

$$\Delta x = \int_0^x \frac{P_x}{AE}dx = \int_0^x \frac{A\rho\omega^2(L^2-x^2)}{2AE}dx = \frac{\rho\omega^2}{2E}\left(L^2 x - \frac{x^3}{3}\right)_0^x$$

$$\delta_1 = \delta_A = \Delta x_{x=0} = 0$$

$$\delta_2 = \delta_B = \Delta x\big|_{x=\frac{L}{4}} = \frac{\rho\omega^2}{2E}\left(L^2 x - \frac{x^3}{3}\right)_0^{L/4} = \frac{7850\times(45)^2}{2\times70\times10^9}\left(L^2 \times \frac{L}{4} - \frac{L^3}{64\times 3}\right)$$

$$\delta_2 = \frac{7850\times(45)^2}{2\times70\times10^9}\left(\frac{(1)^3}{4} - \frac{(1)^3}{192}\right) = 2.78\times10^{-5} \text{ m} = 0.0278 \text{ mm}$$

$$\delta_3 = \delta_C = \Delta x\big|_{x=\frac{L}{2}} = 0.052 \text{ mm}$$

$$\delta_4 = \delta_D = \Delta x\big|_{x=\frac{3L}{4}} = 0.069 \text{ mm}$$

$$\delta_5 = \delta_E = \Delta x\big|_{x=L} = 0.076 \text{ mm}.$$

Reaction calculation

$$\sum F_x = 0$$

$$R_1 + AL\rho\frac{L}{2}\omega^2 = 0$$

$$\therefore \quad R_1 = -\frac{AL^2 \rho \omega^2}{2} = -\frac{3.5 \times 10^{-4} \times (1)^2 \times 7850 \times (45)^2}{2} - 2781.84 \text{ N}.$$

(II) FEM by hand calculations

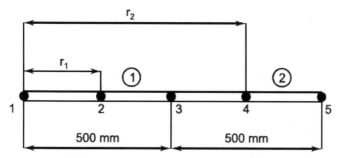

Figure 3.10(b). Finite element model for Example 3.5 (with two quadratic elements).

A finite element model of the rod, with two quadratic elements, is shown in Figure 3.10(b). The element stiffness matrices are,

$$L = L_1 = L_2 = 0.5 \text{ m}$$

$$[k_1] = \frac{A_1 E_1}{3L_1} \begin{bmatrix} 7 & 1 & -8 \\ 1 & 7 & -8 \\ -8 & -8 & 16 \end{bmatrix} = \frac{35 \times 10^{-4} \times 70 \times 10^9}{3 \times 0.5} \begin{bmatrix} 7 & 1 & -8 \\ 1 & 7 & -8 \\ -8 & -8 & 16 \end{bmatrix}$$

$$[k_1] = 163.33 \times 10^5 \begin{bmatrix} 1 & 3 & 2 \\ 7 & 1 & -8 \\ 1 & 7 & -8 \\ -8 & -8 & 16 \end{bmatrix} \begin{matrix} 1 \\ 3 \\ 2 \end{matrix}$$

$$[k_2] = 163.33 \times 10^5 \begin{bmatrix} 3 & 5 & 4 \\ 7 & 1 & -8 \\ 1 & 7 & -8 \\ -8 & -8 & 16 \end{bmatrix} \begin{matrix} 3 \\ 5 \\ 4 \end{matrix}$$

Thus, the global stiffness matrix is,

$$[K] = 163.33 \times 10^5 \begin{bmatrix} 7 & -8 & 1 & 0 & 0 \\ -8 & 16 & -8 & 0 & 0 \\ 1 & -8 & 14 & -8 & 1 \\ 0 & 0 & -8 & 16 & -8 \\ 0 & 0 & 1 & -8 & 7 \end{bmatrix} \begin{matrix} 1 \\ 2 \\ 3 \\ 4 \\ 5 \end{matrix}$$

The centrifugal force or body force F_c (kg/m^3) is given by,

$$F_c = \frac{\rho r \omega^2}{g}. \qquad (3.42)$$

Note that F is a function of the distance r from the pin. Taking the average values of F over each element, we have,

$$F_1 = \frac{\rho r_1 \omega^2}{g} = \frac{7850 \times 0.25 \times (45)^2}{9.81} = 405103.2 \text{ kg/m}^3$$

$$F_2 = \frac{\rho r_2 \omega^2}{g} = \frac{7850 \times 0.75 \times (45)^2}{9.81} = 1215309.6 \text{ kg/m}^3.$$

Thus, the element body force vectors are,

$$f_1 = A \times L_1 \times F_1 \begin{Bmatrix} \frac{1}{6} \\ \frac{1}{6} \\ \frac{2}{3} \end{Bmatrix} = 3.5 \times 10^{-4} \times 0.5 \times 405103.2 \begin{Bmatrix} \frac{1}{6} \\ \frac{1}{6} \\ \frac{2}{3} \end{Bmatrix}$$

$$= 70.89 \begin{Bmatrix} \frac{1}{6} \\ \frac{1}{6} \\ \frac{2}{3} \end{Bmatrix} = \begin{Bmatrix} 11.815 \\ 11.815 \\ 47.26 \end{Bmatrix} \begin{matrix} 1 \\ 2 \\ 3 \end{matrix} \quad \longleftarrow \text{Global dof}$$

$$f_2 = A \times L_2 \times F_2 \begin{Bmatrix} \dfrac{1}{6} \\ \dfrac{1}{6} \\ \dfrac{2}{3} \end{Bmatrix} = 3.5 \times 10^{-4} \times 0.5 \times 1215309.6 \begin{Bmatrix} \dfrac{1}{6} \\ \dfrac{1}{6} \\ \dfrac{2}{3} \end{Bmatrix}$$

$$= 212.68 \begin{Bmatrix} \dfrac{1}{6} \\ \dfrac{1}{6} \\ \dfrac{2}{3} \end{Bmatrix} = \begin{Bmatrix} 35.45 \\ 35.45 \\ 141.79 \end{Bmatrix} \begin{matrix} 3 \\ 5 \\ 4 \end{matrix} \quad \longleftarrow \text{Global dof}$$

Assembling f_1 and f_2, we obtain,

$$F = \begin{Bmatrix} 11.815 \\ 47.26 \\ 47.26 \\ 141.79 \\ 35.45 \end{Bmatrix} \times 9.81 = \begin{Bmatrix} 115.91 \\ 463.62 \\ 463.62 \\ 1390.96 \\ 347.76 \end{Bmatrix} \text{ N.}$$

The global equation is,

$$[K]\{r\} = \{R\} \tag{3.43}$$

$$163.33 \times 10^5 \begin{bmatrix} 7 & -8 & 1 & 0 & 0 \\ -8 & 16 & -8 & 0 & 0 \\ 1 & -8 & 14 & -8 & 1 \\ 0 & 0 & -8 & 16 & -8 \\ 0 & 0 & 1 & -8 & 7 \end{bmatrix} \begin{matrix} 1 \\ 2 \\ 3 \\ 4 \\ 5 \end{matrix} \begin{Bmatrix} u_1 \\ u_2 \\ u_3 \\ u_4 \\ u_4 \end{Bmatrix} = \begin{Bmatrix} 115.91 \\ 463.62 \\ 463.62 \\ 1390.96 \\ 347.76 \end{Bmatrix} \tag{3.44}$$

Boundary conditions are at node 1, $u_1 = 0$.

By using the elimination method, the above matrix reduces to,

$$163.33 \times 10^5 \begin{bmatrix} 16 & -8 & 0 & 0 \\ -8 & 14 & -8 & 1 \\ 0 & -8 & 16 & -8 \\ 0 & 1 & -8 & 7 \end{bmatrix} \begin{Bmatrix} u_2 \\ u_3 \\ u_4 \\ u_5 \end{Bmatrix} = \begin{Bmatrix} 463.62 \\ 463.62 \\ 1390.96 \\ 347.76 \end{Bmatrix}.$$

By solving the above matrix and equations, we get

$$u_2 = 2.661 \times 10^{-5} \text{ mm} = 0.0266 \text{ mm}$$
$$u_3 = 0.0497 \text{ mm}$$
$$u_4 = 0.0657 \text{ mm}$$
$$u_5 = 0.0709 \text{ mm}.$$

The stress at node 1 in element 1 is given by,

$$\sigma_{11} = \frac{2E}{L_1}[-1.5 \;\; -0.5 \;\; 2] \begin{Bmatrix} u_1 \\ u_3 \\ u_2 \end{Bmatrix} = \frac{2 \times 70 \times 10^3}{500}[-1.5 \;\; -0.5 \;\; 2] \begin{Bmatrix} 0 \\ 0.0497 \\ 0.0266 \end{Bmatrix} = 7.924 \text{ MPa}.$$

The stress at node 2 in element 1 is given by,

$$\sigma_{12} = \frac{2E}{L_1}[-0.5 \;\; 0.5 \;\; 0] \begin{Bmatrix} u_1 \\ u_3 \\ u_2 \end{Bmatrix} = \frac{2 \times 70 \times 10^3}{500}[-0.5 \;\; 0.5 \;\; 0] \begin{Bmatrix} 0 \\ 0.0497 \\ 0.0266 \end{Bmatrix} = 6.972 \text{ MPa}.$$

The stress at node 3 in element 1 is given by,

$$\sigma_{13} = \frac{2E}{L_1}[0.5 \;\; 1.5 \;\; -2] \begin{Bmatrix} u_1 \\ u_3 \\ u_2 \end{Bmatrix} = \frac{2 \times 70 \times 10^3}{500}[0.5 \;\; 1.5 \;\; -2] \begin{Bmatrix} 0 \\ 0.0497 \\ 0.0266 \end{Bmatrix} = 5.992 \text{ MPa}.$$

The stress at node 1 in element 2 is given by,

$$\sigma_{21} = \frac{2E}{L_2}[-1.5 \quad -0.5 \quad 2]\begin{Bmatrix} u_3 \\ u_5 \\ u_4 \end{Bmatrix} = \frac{2\times70\times10^3}{500}[-1.5 \quad -0.5 \quad 2]\begin{Bmatrix} 0.0497 \\ 0.0709 \\ 0.0657 \end{Bmatrix} = 5.992 \text{ MPa}.$$

The stress at node 2 in element 2 is given by,

$$\sigma_{22} = \frac{2E}{L_2}[-0.5 \quad 0.5 \quad 0]\begin{Bmatrix} u_3 \\ u_5 \\ u_4 \end{Bmatrix} = \frac{2\times70\times10^3}{500}[-0.5 \quad 0.5 \quad 0]\begin{Bmatrix} 0.0497 \\ 0.0709 \\ 0.0657 \end{Bmatrix} = 2.968 \text{ MPa}.$$

The stress at node 3 in element 2 is given by,

$$\sigma_{23} = \frac{2E}{L_2}[0.5 \quad 1.5 \quad -2]\begin{Bmatrix} u_3 \\ u_5 \\ u_4 \end{Bmatrix} = \frac{2\times70\times10^3}{500}[0.5 \quad 1.5 \quad -2]\begin{Bmatrix} 0.0497 \\ 0.0709 \\ 0.0657 \end{Bmatrix} = -0.056 \text{ MPa}.$$

(III) Software results

While solving the problem using software, 4 linear bar elements are taken instead of 2 quadratic elements.

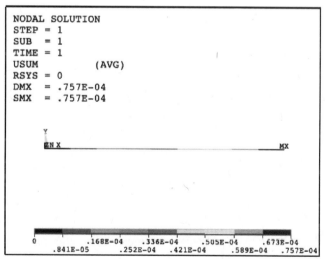

Figure 3.10(c). Deflection pattern for a rod (refer to Appendix C for color figures).

Deflection values at nodes (in m)

The following degree of freedom results are in global coordinates

NODE	UX	UY	UZ	USUM
1	0.0000	0.0000	0.0000	0.0000
2	0.27795E-04	0.0000	0.0000	0.27795E-04
3	0.52041E-04	0.0000	0.0000	0.52041E-04
4	0.69191E-04	0.0000	0.0000	0.69191E-04
5	0.75696E-04	0.0000	0.0000	0.75696E-04

Maximum absolute values

NODE	5	0	0	5
VALUE	0.75696E-04	0.0000	0.0000	0.75696E-04

Reaction value

The following X, Y, Z solutions are in global coordinates

NODE	FX	FY
1	−2781.8	−3.3691

Answers of Example 3.5

Parameter	Analytical method	FEM-hand calculations	Software results
Displacement at node 2	0.0278 mm	0.0266 mm	0.0278 mm
Displacement at node 3	0.052 mm	0.0497 mm	0.052 mm
Displacement at node 4	0.069 mm	0.0657 mm	0.069 mm
Displacement at node 5	0.076 mm	0.0709 mm	0.076 mm
Stress in node 1 of element 1	7.95 MPa	7.924 MPa	---
Stress in node 2 of element 1	7.45 MPa	6.972 MPa	---
Stress in node 3 of element 1	5.96 MPa	5.992 MPa	---
Stress in node 1 of element 2	5.96 MPa	5.992 MPa	---
Stress in node 2 of element 2	3.48 MPa	2.968 MPa	---
Stress in node 3 of element 2	0 MPa	−0.056 MPa	---
Reaction at fixed end	−2781.84 N	---	−2781.8 N

Each problem given in this book uses a different procedure for solving using software. For familiarizing, procedure for one problem is given from each chapter using software. Other problems are left to the user to explore the software for solving the problems.

Procedure for solving the problem using ANSYS® 11.0 academic teaching software For Example 3.3

PREPROCESSING

1. **Main Menu > Preprocessor > Element Type > Add/Edit/Delete > Add > Structural Link > 2D spar 1 > OK > Close**

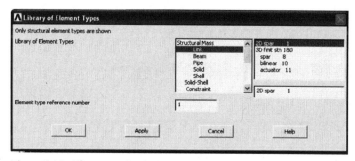

Figure 3.11. Element selection.

2. **Main Menu > Preprocessor > Real Constants > Add/Edit/Delete > Add > OK**

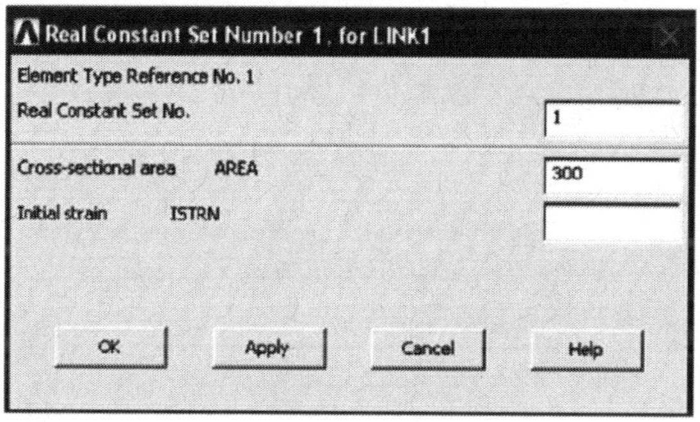

Figure 3.12. Enter the cross-sectional area.

Cross-sectional area AREA > **Enter 300 > OK > Close**
Enter the material properties

3. **Main Menu > Preprocessor > Material Props > Material Models**
   Material Model Number 1, click **Structural> Linear > Elastic > Isotropic**
   Enter **EX= 2E4 and PRXY=0.3 > OK**
   (Close the Define Material Model Behavior window.)
   Create the nodes and elements. Create 3 nodes 2 elements.
4. **Main Menu > Preprocessor > Modeling > Create > Nodes > In Active CS**
   Enter the coordinates of node 1 > **Apply**
   Enter the coordinate of node 2 > **Apply**
   Enter the coordinates of node 3 > **OK**

Node locations		
Node number	X coordinate	Y coordinate
1	0	0
2	200	0
3	400	0

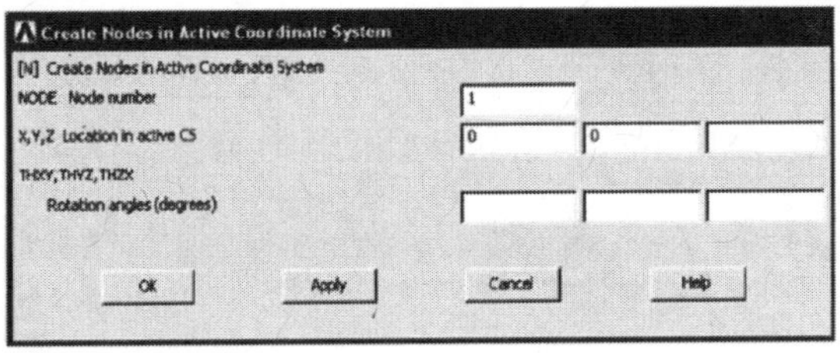

Figure 3.13. Enter the node coordinates.

5. **Main Menu > Preprocessor > Modeling > Create > Elements > Auto Numbered > Thru node** Pick the 1st and 2nd node > **Apply** Pick 2nd and 3rd node > **OK**

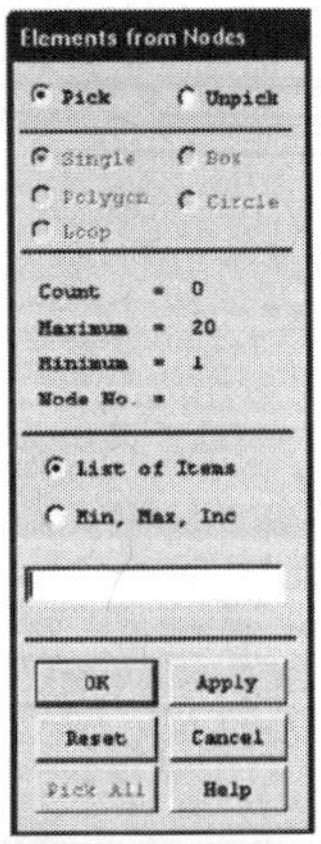

Figure 3.14. Pick the nodes to create elements.

Finite Element Analysis of Axially Loaded Members

Apply the displacement boundary conditions and loads.

6. **Main Menu > Preprocessor > Loads > Apply > Structural > Displacement > On Nodes** Pick the 1st node > **Apply > All DOF=0 > OK**

7. **Main Menu > Preprocessor > Loads > Define Loads > Apply > Structural > Displacement > On Nodes** Pick the 3rd node > **Apply > Select UX and enter displacement value = 1.2 > OK**

8. **Main Menu > Preprocessor > Loads > Define Loads > Apply > Structural > Force/Moment > On Nodes** Pick the 2nd > **OK > Force. Moment value = 80e3 > OK**

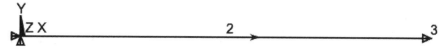

Figure 3.15. Model with loading and displacement boundary conditions.

The model-building step is now complete, and we can proceed to the solution. First to be safe, save the model.

Solution

The interactive solution proceeds.

9. **Main Menu > Solution > Solve > Current LS > OK**
   The **STATUS Command** window displays the problem parameters and the **Solve Current Load Step** window is shown. Check the solution options in the **/STATUS** window and if all is OK, select **File > Close**
   In the **Solve Current Load Step** WINDOW, Select **OK**, and the solution is complete, **close** the **'Solution is Done!'** window.

POST-PROCESSING

We can now plot the results of this analysis and also list the computed values.

10. **Main Menu > General Postproc > Plot Results > Contour Plot > Nodal Solu > DOF Solution > Displacement vector sum > OK**
    The result is shown in Figure 3.8(c).
    To find the axial stress, the following procedure is followed.

11. **Main Menu > General Postproc > Element Table > Define Table > Add**

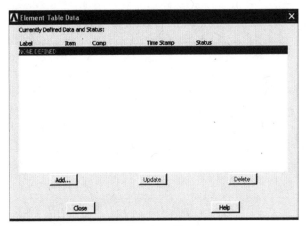

Figure 3.16. Define the element table.

Select **By sequence num and LS** and type **1 after LS** as shown in Figure 3.17.

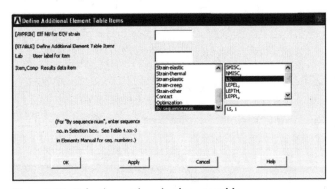

Figure 3.17. Selecting options in element table.

OK > Close

12. **Main Menu > General Postproc > Plot Results > Contour Plot > Elem Table > Select > LS1 > OK**

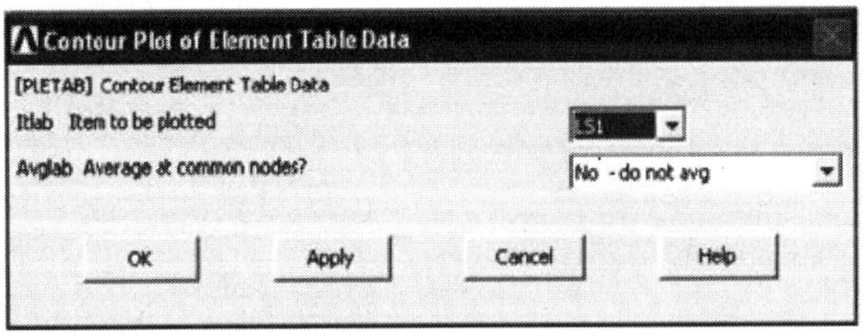

Figure 3.18. Selecting options for finding out axial stress.

The result is shown in Figure 3.8(d).

3.3 BARS OF VARYING CROSS-SECTION AREA

This section will demonstrate thorough examples explaining FEA on bars of varying cross-section area.

Example 3.6

Solve for displacement and stress given in Figure 3.19 using 2 finite elements model. Take Young's modulus E = 200 GPa.

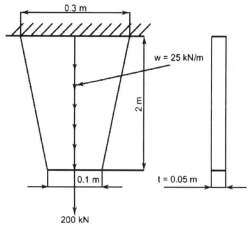

Figure 3.19. Example 3.6.

Solution

(I) Analytical method [Refer Figure 3.19(a)]

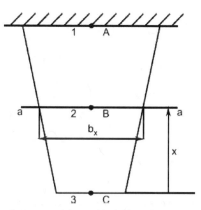

Figure 3.19(a). Analytical method for Example 3.6.

$$L = 2 \text{ m.}$$

Stress calculation

$$b_x = 0.1 + (0.3 - 0.1)\frac{x}{2} = 0.1 + 0.1x$$

$$A_x = b_x \times t = (0.1 + 0.1x)0.05$$

$$\sigma_x = \frac{P_x}{A_x} = \frac{200 \times 10^3 + 25 \times 10^3 \times x}{(0.1 + 0.1x)0.05}$$

$$\sigma_1 = \sigma_A = \sigma_x\big|_{x=2} = \frac{200 \times 10^3 + 25 \times 10^3 \times 2}{(0.1 + 0.1 \times 2)0.05} = 16.67 \text{ MPa}$$

$$\sigma_2 = \sigma_B = \sigma_x\big|_{x=1} = \frac{200 \times 10^3 + 25 \times 10^3 \times 1}{(0.1 + 0.1 \times 1)0.05} = 22.5 \text{ MPa}$$

$$\sigma_3 = \sigma_C = \sigma_x\big|_{x=0} = \frac{200 \times 10^3 + 25 \times 10^3 \times 0}{(0.1 + 0.1 \times 0)0.05} = 40 \text{ MPa}.$$

Displacement Calculation

Displacement at section a-a = change in length of $(L-x)$

$$\delta = \int \frac{\sigma_x}{E} = \int_{L-x}^{L} \frac{P_x}{A_x E} dx = \int_{L-x}^{L} \left(\frac{200 \times 10^3 + 25 \times 10^3 \times x}{(0.1 + 0.1x)0.05 \times 200 \times 10^9} \right) dx$$

$$\delta_3 = \delta_C = \int_{L-2}^{2} \left(\frac{200 \times 10^3 + 25 \times 10^3 \times x}{(0.1 + 0.1x)0.05 \times 200 \times 10^9} \right) dx = \int_{0}^{2} \left(\frac{200 \times 10^3 + 25 \times 10^3 \times x}{(0.1 + 0.1x)0.05 \times 200 \times 10^9} \right) dx$$

$$= 0.2423 \text{ mm}$$

$$\delta_2 = \delta_B = \int_{L-1}^{2} \left(\frac{200 \times 10^3 + 25 \times 10^3 \times x}{(0.1 + 0.1x)0.05 \times 200 \times 10^9} \right) dx = \int_{1}^{2} \left(\frac{200 \times 10^3 + 25 \times 10^3 \times x}{(0.1 + 0.1x)0.05 \times 200 \times 10^9} \right) dx$$

$$= 0.096 \text{ mm}.$$

(II) FEM by hand calculations

Using 2 elements each of 1 m length, we obtain the finite element model as shown in Figure 3.19(c). We can write the equivalent model as shown in Figure 3.19(b). At the middle of the bar width is,

$$\frac{(0.3 + 0.1)}{2} = 0.2 \text{ m}.$$

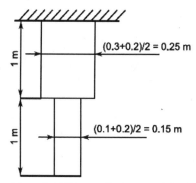

Figure 3.19(b). Equivalent model of Finite element model for Example 3.6.

Figure 3.19(c). Finite element model for Example 3.6.

$$A_1 = 0.25 \times 0.05 = 0.0125 \text{ m}^2$$
$$A_2 = 0.15 \times 0.05 = 0.0075 \text{ m}^2$$
$$E_1 = E_2 = 200 \times 10^9 \text{ N/m}^2$$
$$L_1 = L_2 = 1 \text{ m}.$$

Stiffness matrix for element 1 is,

$$[k_1] = \frac{A_1 E_1}{L_1}\begin{bmatrix} 1 & -1 \\ -1 & 1 \end{bmatrix} = \frac{0.0125 \times 200 \times 10^9}{1}\begin{bmatrix} 1 & -1 \\ -1 & 1 \end{bmatrix} = 2.5 \times 10^9 \begin{bmatrix} 1 & -1 \\ -1 & 1 \end{bmatrix}\begin{matrix} 1 \\ 2 \end{matrix}.$$

Stiffness matrix for element 2 is,

$$[k_2] = \frac{A_2 E_2}{L_2}\begin{bmatrix} 1 & -1 \\ -1 & 1 \end{bmatrix} = \frac{0.0075 \times 200 \times 10^9}{1}\begin{bmatrix} 1 & -1 \\ -1 & 1 \end{bmatrix} = 1.5 \times 10^9 \begin{bmatrix} 1 & -1 \\ -1 & 1 \end{bmatrix}\begin{matrix} 2 \\ 3 \end{matrix}.$$

Distributes load calculation for elements 1 and 2,

$$W_1 = \begin{Bmatrix} \dfrac{w \times L_1}{2} \\ \dfrac{w \times L_1}{2} \end{Bmatrix} = \begin{Bmatrix} \dfrac{25 \times 1}{2} \\ \dfrac{25 \times 1}{2} \end{Bmatrix} \times 10^3 = \begin{Bmatrix} 12.5 \\ 12.5 \end{Bmatrix} \begin{matrix} 1 \\ 2 \end{matrix} \times 10^3$$

$$W_2 = \begin{Bmatrix} \dfrac{w \times L_2}{2} \\ \dfrac{w \times L_2}{2} \end{Bmatrix} = \begin{Bmatrix} \dfrac{25 \times 1}{2} \\ \dfrac{25 \times 1}{2} \end{Bmatrix} \times 10^3 = \begin{Bmatrix} 12.5 \\ 12.5 \end{Bmatrix} \begin{matrix} 2 \\ 3 \end{matrix} \times 10^3.$$

Global equation is,

$$[K]\{r\} = \{R\}. \tag{3.45}$$

$$10^9 \begin{bmatrix} 2.5 & -2.5 & 0 \\ -2.5 & 2.5+1.5 & -1.5 \\ 0 & -1.5 & 1.5 \end{bmatrix} \begin{matrix} 1 \\ 2 \\ 3 \end{matrix} \begin{Bmatrix} u_1 \\ u_2 \\ u_3 \end{Bmatrix} = \begin{Bmatrix} 12.5 + R_1 \\ 25 \\ 12.5 + 200 \end{Bmatrix} \times 10^3. \tag{3.46}$$

Boundary conditions are at node 1, $u_1 = 0$.
By using the elimination method, the above matrix reduces to,

$$10^9 \begin{bmatrix} 4 & -1.5 \\ -1.5 & 1.5 \end{bmatrix} \begin{Bmatrix} 25 \\ 212.5 \end{Bmatrix} \times 10^3.$$

By matrix multiplication, we get

$$10^9 (4 \times u_2 - 1.5 \times u_3) = 25 \times 10^3 \tag{3.47}$$

$$10^9 (-1.5 \times u_2 + 1.5 \times u_3) = 212.5 \times 10^3. \tag{3.48}$$

By solving equations (3.47) and (3.48), we get

$$u_2 = 9.5 \times 10^{-5} \text{ m} = 0.095 \text{ mm}$$

$$u_3 = 2.37 \times 10^{-4} \text{ m} = 0.237 \text{ mm}.$$

Stress (σ) calculation

Stress in element 1,

$$\{\sigma_1\} = \frac{E_1}{L_1}\begin{bmatrix}-1 & 1\end{bmatrix}\begin{Bmatrix}u_1\\u_2\end{Bmatrix} = \frac{2\times 10^5}{1000}\begin{bmatrix}-1 & 1\end{bmatrix}\begin{Bmatrix}0\\0.095\end{Bmatrix} = 19 \text{ MPa}.$$

Stress in element 2,

$$\{\sigma_2\} = \frac{E_2}{L_2}\begin{bmatrix}-1 & 1\end{bmatrix}\begin{Bmatrix}u_2\\u_3\end{Bmatrix} = \frac{2\times 10^5}{1000}\begin{bmatrix}-1 & 1\end{bmatrix}\begin{Bmatrix}0.095\\0.237\end{Bmatrix} = 28.4 \text{ MPa}.$$

(III) Software results

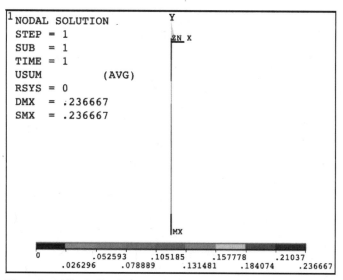

Figure 3.19(d). Deflection pattern for a tapered bar (refer to Appendix C for color figures).

Deflection values at node

The following degree of freedom results are in global coordinates

NODE	UX	UY	UX	USUM
1	0.0000	0.0000	0.0000	0.0000
2	0.0000	−0.95000E-01	0.0000	0.95000E-01
3	0.0000	−0.23667	0.0000	0.23667

Maximum absolute values

NODE	0	3	0	3
VALUE	0.0000	−0.23667	0.0000	0.23667

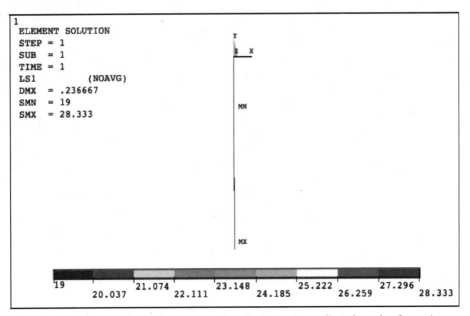

Figure 3.19(e). Stress pattern for a tapered bar (refer to Appendix C for color figures).

Stress values at elements

STAT ELEM	CURRENT LS1
3	19.000
4	28.333

Reaction value

The following X, Y, Z solutions are in global coordinates

NODE	FX	FY
1	0.0000	0.25000E+06

Answers of Example 3.6

Parameter	Analytical method	FEM-hand calculations	Software results
Displacement at node 2	0.096 mm	0.095 mm	0.095 mm
Displacement at node 3	0.2423 mm	0.237 mm	0.23667 mm
Stress in element 1	16.67 MPa to 22.5 MPa	19 MPa	19 MPa
Stress in element 2	22.5 MPa to 40 MPa	28.4 MPa	28.33 MPa

In the above example, 2 elements are used for solving the problem by hand calculation and by software. To get the convergence of the solution with the analytical method a higher number of elements are to be used.

Example 3.7

Find the displacement and stress distribution in the tapered bar shown in Figure 3.20 using 2 finite elements under an axial load of P = 100 N.
  Cross-sectional area at fixed end = 22 mm²
  Cross-sectional are at free end =100 mm²
  Young's modulus E = 200 GPa

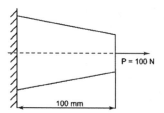

Figure 3.20. Example 3.7

Solution

(I) Analytical method [refer to Figure 3.20(a)]

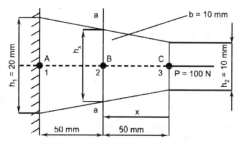

Figure 3.20(a). Analytical method for Example 3.7.

Assume, b = thickness = 10 mm
Area at section a-a = $b \times h_x$

$$h_x = \left(h_2 + \left(\frac{h_1 - h_2}{L}\right)x\right)$$

$$A_x = b \times h_x = 10\left(10 + \left(\frac{20-10}{100}\right)x\right) = 10(10+0.1x).$$

Stress calculation

$$\sigma_x = \frac{P}{A_x} = \frac{100}{10(10+0.1x)}$$

$$\sigma_1 = \sigma_{x=100} = \frac{100}{10(10+0.1 \times 100)} = 0.5 \text{ MPa}$$

$$\sigma_2 = \sigma_{x=50} = \frac{100}{10(10+0.1 \times 50)} = 0.667 \text{ MPa}$$

$$\sigma_3 = \sigma_{x=0} = \frac{100}{10(10+0.1 \times 0)} = 1 \text{ MPa}.$$

Displacement calculation

$$\delta_3 = \delta_C = \frac{PL}{Eb(h_1-h_2)}\ln\frac{h_1}{h_2} = \frac{100 \times 100}{2 \times 10^5 \times 10(20-10)}\ln\frac{20}{10} = 3.47 \times 10^{-4} \text{ mm}$$

$$\delta_2 = \delta_B = \frac{100 \times 50}{2 \times 10^5 \times 10(20-15)}\ln\frac{20}{15} = 1.44 \times 10^{-4} \text{ mm}.$$

(II) FEM by hand calculations

Using 2 elements each of 50 mm length, we obtain the finite element model as shown in Figure 3.20(c). We can write the equivalent model as shown in Figure 3.20(b) at the middle, area of cross-section of bar is $\frac{(200+100)}{2} = 150$ mm^2.

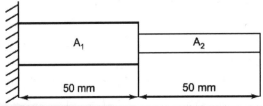

Figure 3.20(b). Equivalent model of finite element model for Example 3.7.

Finite Element Analysis of Axially Loaded Members

```
      Element 1      Element 2
 •————————•————————•
Node 1  L₁   Node 2  L₂  Node 3
```
Figure 3.20(c). Finite element model for Example 3.7.

$$A_1 = \frac{(200+150)}{2} = 175 \text{ mm}^2$$

$$A_2 = \frac{(150+100)}{2} = 125 \text{ mm}^2$$

$$L_1 = L_2 = 50 \text{ mm}$$

$$E_1 = E_2 = 2 \times 10^5 \text{ N/mm}^2.$$

Stiffness matrix for element 1 is,

$$[k_1] = \frac{A_1 E_1}{L_1} \begin{bmatrix} 1 & -1 \\ -1 & 1 \end{bmatrix} = \frac{0.175 \times 2 \times 10^5}{50} \begin{bmatrix} 1 & -1 \\ -1 & 1 \end{bmatrix} = 7 \times 10^5 \begin{bmatrix} 1 & -1 \\ -1 & 1 \end{bmatrix} \begin{matrix} 1 \\ 2 \end{matrix}.$$

Stiffness matrix for element 2 is,

$$[k_2] = \frac{A_2 E_2}{L_2} \begin{bmatrix} 1 & -1 \\ -1 & 1 \end{bmatrix} = \frac{0.125 \times 2 \times 10^5}{50} \begin{bmatrix} 1 & -1 \\ -1 & 1 \end{bmatrix} = 5 \times 10^5 \begin{bmatrix} 1 & -1 \\ -1 & 1 \end{bmatrix} \begin{matrix} 2 \\ 3 \end{matrix}.$$

Global equation is,

$$[K]\{r\} = \{R\} \qquad (3.49)$$

$$10^5 \begin{bmatrix} 7 & -7 & 0 \\ -7 & 7+5 & -5 \\ 0 & -5 & 5 \end{bmatrix} \begin{Bmatrix} u_1 \\ u_2 \\ u_3 \end{Bmatrix} = \begin{Bmatrix} R_1 \\ 0 \\ 100 \end{Bmatrix}. \qquad (3.50)$$

Boundary conditions are at node 1, $u_1 = 0$.
By using the elimination method, the above matrix reduces to,

$$10^5 \begin{bmatrix} 12 & -5 \\ -5 & 5 \end{bmatrix} \begin{Bmatrix} u_1 \\ u_2 \end{Bmatrix} = \begin{Bmatrix} 0 \\ 100 \end{Bmatrix}.$$

By matrix multiplication, we get

$$10^5 (12 \times u_2 - 5 \times u_3) = 0 \qquad (3.51)$$

$$10^5 (-5 \times u_2 + 5 \times u_3) = 100. \qquad (3.52)$$

By solving equations (3.51) and (3.52), we get

$$u_2 = 1.429 \times 10^{-4} \text{ mm}$$

$$u_3 = 3.429 \times 10^{-4} \text{ mm}.$$

Stress calculation

Stress in element 1,

$$\{\sigma_1\} = \frac{E_1}{L_1}[-1 \quad 1]\begin{Bmatrix} u_1 \\ u_2 \end{Bmatrix} = \frac{2 \times 10^5}{50}[-1 \quad 1]\begin{Bmatrix} 0 \\ 1.429 \times 10^{-4} \end{Bmatrix} = 0.5716 \text{ MPa}.$$

Stress in element 2,

$$\{\sigma_2\} = \frac{E_2}{L_2}[-1 \quad 1]\begin{Bmatrix} u_2 \\ u_3 \end{Bmatrix} = \frac{2 \times 10^5}{50}[-1 \quad 1]\begin{Bmatrix} 1.429 \times 10^{-4} \\ 3.429 \times 10^{-4} \end{Bmatrix} = 0.8 \text{ MPa}.$$

(III) Software results

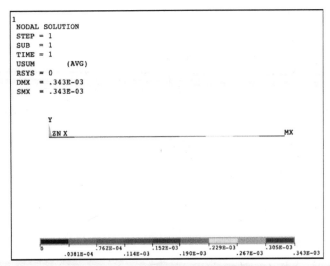

Figure 3.20(d). Deflection pattern for a tapered bar (refer to Appendix C for color figures).

Deflection values at nodes

The following degree of freedom results are in global coordinates

NODE	UX	UY	UZ	USUM
1	0.0000	0.0000	0.0000	0.0000
2	0.14286E-03	0.0000	0.0000	0.14286E-03
3	0.34286E-03	0.0000	0.0000	0.34286E-03

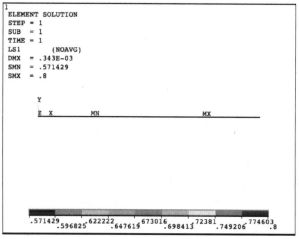

Figure 3.20(e). Stress pattern for a tapered bar (refer to Appendix C for color figures).

Stress values at elements

STAT ELEM	CURRENT LS1
1	0.57143
2	0.80000

Answer for Example 3.7

Parameter	Analytical method	FEM-hand calculations	Software results
Displacement at node 2	1.44×10^{-4} mm	1.429×10^{-4} mm	1.4286×10^{-4} mm
Displacement at node 3	3.47×10^{-4} mm	3.429×10^{-4} mm	3.4286×10^{-4} mm
Stress in element 1	0.5 MPa to 0.667 MPa	0.5716 MPa	0.57143 MPa
Stress in element 2	0.667 MPa to 1 MPa	0.8 MPa	0.8 MPa

Example 3.8

Find the nodal displacements, element stresses, and reaction in the tapered bar subjected to a load of 6000 N as shown in Figure 3.21. Further the member experiences a temperature increase of 30°C. Use 3 equal length elements for finite element model. Take E = 200 GPa, $\nu = 0.3$, and $\alpha = 7 \times 10^{-6}/°C$.

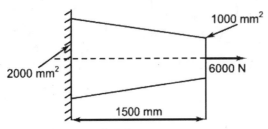

Figure 3.21. Example 3.8

Solution

(I) FEM by hand calculations

We obtain the finite element model as shown in Figure 3.21(b). We can write the equivalent model as shown in Figure 3.21(a).

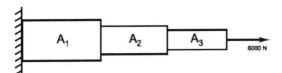

Figure 3.21(a). Equivalent model of the finite element model for Example 3.8.

```
    L₁        L₂        L₃
●─────────●─────────●─────────●
Node 1   Node 2    Node 3   Node 4
```

Figure 3.21(b). Finite element model for Example 3.8.

$$L_1 = L_2 = L_3 = 500 \text{ mm}$$

$$\Delta T = 30°C$$

$$A_1 = 2000 \text{ mm}^2$$

$$A_2 = \frac{2000 + 1000}{2} = 1500 \text{ mm}^2$$

$$A_3 = 1000 \text{ mm}^2.$$

Finite Element Analysis of Axially Loaded Members

Element stiffness matrices are,

$$[k_1] = \frac{A_1 E}{L_1}\begin{bmatrix} 1 & -1 \\ -1 & 1 \end{bmatrix} = \frac{2000 \times 2 \times 10^5}{500}\begin{bmatrix} 1 & -1 \\ -1 & 1 \end{bmatrix} = 8 \times 10^5 \begin{bmatrix} 1 & 2 \\ 1 & -1 \\ -1 & 1 \end{bmatrix}\begin{matrix}1\\2\end{matrix}$$

$$[k_2] = \frac{A_2 E}{L_2}\begin{bmatrix} 1 & -1 \\ -1 & 1 \end{bmatrix} = \frac{1500 \times 2 \times 10^5}{500}\begin{bmatrix} 1 & -1 \\ -1 & 1 \end{bmatrix} = 6 \times 10^5 \begin{bmatrix} 2 & 3 \\ 1 & -1 \\ -1 & 1 \end{bmatrix}\begin{matrix}2\\3\end{matrix}$$

$$[k_3] = \frac{A_3 E}{L_3}\begin{bmatrix} 1 & -1 \\ -1 & 1 \end{bmatrix} = \frac{1000 \times 2 \times 10^5}{500}\begin{bmatrix} 1 & -1 \\ -1 & 1 \end{bmatrix} = 4 \times 10^5 \begin{bmatrix} 3 & 4 \\ 1 & -1 \\ -1 & 1 \end{bmatrix}\begin{matrix}3\\4\end{matrix}.$$

Nodal loads due to thermal effect are,

$$\{Q_{1(th)}\} = EA_1 a(\Delta T)\begin{Bmatrix}-1\\1\end{Bmatrix} = 2\times 10^5 \times 2000 \times 7 \times 10^{-6} \times 30 \begin{Bmatrix}-1\\1\end{Bmatrix} = 84 \times 10^3 \begin{Bmatrix}-1\\1\end{Bmatrix}\begin{matrix}1\\2\end{matrix}$$

$$\{Q_{2(th)}\} = EA_2 a(\Delta T)\begin{Bmatrix}-1\\1\end{Bmatrix} = 2\times 10^5 \times 1500 \times 7 \times 10^{-6} \times 30 \begin{Bmatrix}-1\\1\end{Bmatrix} = 63 \times 10^3 \begin{Bmatrix}-1\\1\end{Bmatrix}\begin{matrix}2\\3\end{matrix}$$

$$\{Q_{3(th)}\} = EA_3 a(\Delta T)\begin{Bmatrix}-1\\1\end{Bmatrix} = 2\times 10^5 \times 1000 \times 7 \times 10^{-6} \times 30 \begin{Bmatrix}-1\\1\end{Bmatrix} = 42 \times 10^3 \begin{Bmatrix}-1\\1\end{Bmatrix}\begin{matrix}3\\4\end{matrix}.$$

Global forced vector

$$\{R\} = \begin{Bmatrix} -84 \times 10^3 \\ 84 \times 10^3 - 63 \times 10^3 \\ 63 \times 10^3 - 42 \times 10^3 \\ 42 \times 10^3 \end{Bmatrix}\begin{matrix}1\\2\\3\\4\end{matrix} = \begin{Bmatrix} -84 \times 10^3 \\ 21 \times 10^3 \\ 21 \times 10^3 \\ 42 \times 10^3 \end{Bmatrix}\begin{matrix}1\\2\\3\\4\end{matrix}.$$

Global equation is,

$$10^5 \begin{matrix} 1 & 2 & 3 & 4 \\ \end{matrix}$$
$$10^5 \begin{bmatrix} 8 & -8 & 0 & 0 \\ -8 & 8+6 & -6 & 0 \\ 0 & -6 & 6+4 & -4 \\ 0 & 0 & -4 & 4 \end{bmatrix} \begin{matrix} 1 \\ 2 \\ 3 \\ 4 \end{matrix} \begin{Bmatrix} u_1 \\ u_2 \\ u_3 \\ u_4 \end{Bmatrix} = \begin{Bmatrix} -84 \times 10^3 + R_1 \\ 21 \times 10^3 \\ 21 \times 10^3 \\ 42 \times 10^3 + 6000 \end{Bmatrix} \begin{matrix} 1 \\ 2 \\ 3 \\ 4 \end{matrix}. \quad (3.53)$$

Using the elimination method of applying boundary conditions, i.e., $u_1 = 0$. The equation (3.53) reduces to,

$$10^5 \begin{bmatrix} 14 & -6 & 0 \\ -6 & 10 & -4 \\ 0 & -4 & 4 \end{bmatrix} \begin{Bmatrix} u_2 \\ u_3 \\ u_4 \end{Bmatrix} = \begin{Bmatrix} 21 \times 10^3 \\ 21 \times 10^3 \\ 48 \times 10^3 \end{Bmatrix}.$$

Solving the above matrix and equations, we get

$$u_2 = 0.1125 \text{ mm}$$
$$u_3 = 0.2275 \text{ mm}$$
$$u_4 = 0.3475 \text{ mm}.$$

Stress calculation

$$\sigma_1 = \frac{E}{L_1}[-1 \quad 1]\begin{Bmatrix} u_1 \\ u_2 \end{Bmatrix} - Ea(\Delta T) = \frac{2 \times 10^5}{500}[-1 \quad 1]\begin{Bmatrix} 0 \\ 0.1125 \end{Bmatrix}$$
$$- 2 \times 10^5 \times 7 \times 10^{-6} \times 30 = 3 \text{ MPa}$$

$$\sigma_2 = \frac{E}{L_2}[-1 \quad 1]\begin{Bmatrix} u_2 \\ u_3 \end{Bmatrix} - Ea(\Delta T) = \frac{2 \times 10^5}{500}[-1 \quad 1]\begin{Bmatrix} 0.1125 \\ 0.2275 \end{Bmatrix}$$
$$- 2 \times 10^5 \times 7 \times 10^{-6} \times 30 = 4 \text{ MPa}$$

$$\sigma_3 = \frac{E}{L_3}[-1 \quad 1]\begin{Bmatrix} u_3 \\ u_4 \end{Bmatrix} - Ea(\Delta T) = \frac{2 \times 10^5}{500}[-1 \quad 1]\begin{Bmatrix} 0.2275 \\ 0.3475 \end{Bmatrix}$$
$$- 2 \times 10^5 \times 7 \times 10^{-6} \times 30 = 6 \text{ MPa}.$$

Reaction calculation: from equation (3.53),

$$8 \times 10^5 u_1 - 8 \times 10^5 u_2 = -84 \times 10^3 + R_1$$
$$R_1 = 6000 \text{ N.}$$

(III) Software results

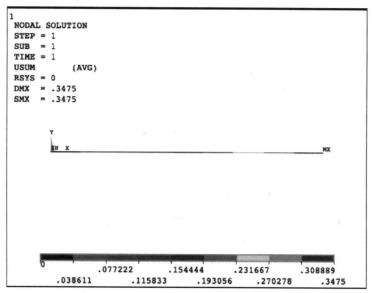

Figure 3.21(c). Deflection pattern for a tapered bar (refer to Appendix C for color figures).

Deflection values at nodes

The following degree of freedom results are in global coordinates

NODE	UX	UY	UX	USUM
1	0.0000	0.0000	0.0000	0.0000
2	0.11250	0.0000	0.0000	0.11250
3	0.22750	0.0000	0.0000	0.22750
4	0.34750	0.0000	0.0000	0.34750

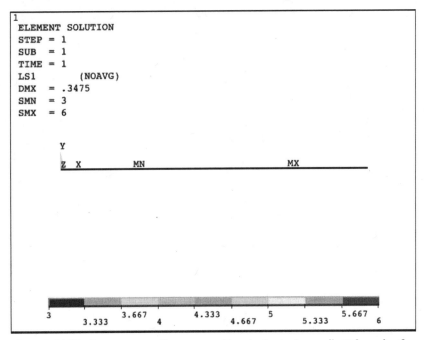

Figure 3.21(d). Stress pattern for a tapered bar (refer to Appendix C for color figures).

Stress values at elements

STAT ELEM	CURRENT LS1
1	3.0000
2	4.0000
3	6.0000

Reaction value

The following X, Y, Z solutions are in global coordinates

NODE	FX	FY
1	−6000.0	0.0000

Answers for Example 3.8

Parameter	FEM-hand calculations	Software results
Displacement at node 2	0.1125 mm	0.1125 mm
Displacement at node 3	0.2275 mm	0.2275 mm
Displacement at node 4	0.3475 mm	0.3475 mm
Stress in node 1 of element 1	3 MPa	3 MPa
Stress in node 2 of element 1	4 MPa	4 MPa
Stress in node 3 of element 1	6 MPa	6 MPa
Reaction at fixed end	−6000 N	−6000 N

Procedure for solving the problem using ANSYS® 11.0 academic teaching software
For Example 3.6

PROCESSING

1. Main Menu > Preprocessor > Element Type > Add/Edit > Delete > Add > Structural Link > 2D spar 1 > OK > Close

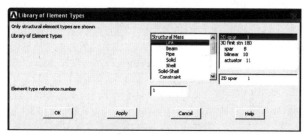

Figure 3.22. Element selection.

2. Main Menu > Preprocessor > Real Constants > Add/Edit/Delete > Add > OK

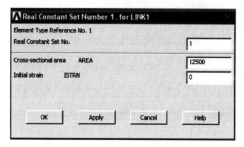

Figure 3.23. Enter the cross-sectional area of 1st element.

Cross-sectional area > Enter 12500 > OK > Add > OK

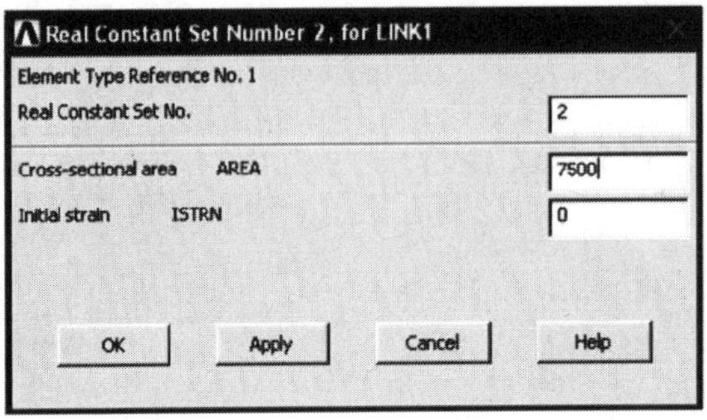

Figure 3.24. Enter the cross-sectional area of 2nd element.

Cross-sectional area AREA > Enter 7500 > OK > Close
Enter the material properties.
3. **Main Menu > Preprocessor > Material Props > Material Models**
   Material Model Number 1, click **Structural > Linear > Elastic > Isotropic**
   Enter **EX=200E3 and PRXY=0.3 > OK**
   (**Close** the Define Material Model Behavior window.)
   Create the nodes and element. As stated in the example, use 2 element model. Hence create 3 nodes and 2 elements.
4. **Main Menu > Preprocessor > Modeling > Create > Nodes > In Active CS**
   Enter the coordinates of node 1 > **Apply** Enter the coordinates of node 2 > **Apply** Enter the coordinates of node 3 > **OK**.

Node locations		
Node number	X coordinate	Y coordinate
1	0	0
2	0	−1000
3	0	−2000

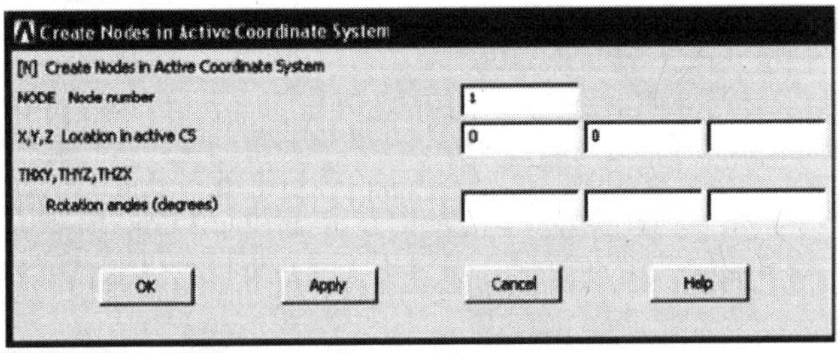

Figure 3.25. Enter the node coordinates.

Finite Element Analysis of Axially Loaded Members

5. **Main Menu > Preprocessor > Modeling > Create > Elements > Elem Attributes > OK > Auto Numbered > Thru nodes** Pick the 1st and 2nd node > **OK**

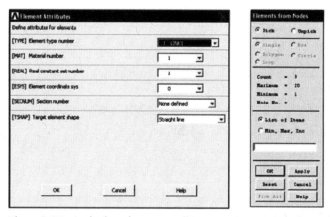

Figure 3.26. Assigning element attributes to element 1 and creating element 1.

Elem Attributes > change the Real constant set number to 2 > OK > Auto Numbered > Thru nodes Pick the 2nd and 3rd node > **OK**

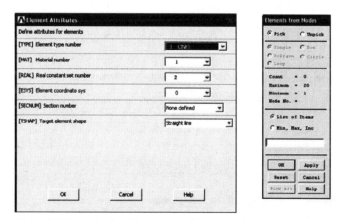

Figure 3.27. Assigning element attributes to element 2 and creating element 2.

Apply the displacement boundary conditions and loads.

6. **Main Menu > Preprocessor > Loads > Define Loads > Apply > Structural > Displacement > On Nodes** Pick the 1st node > **Apply > All DOF=0 > OK**
7. **Main Menu > Preprocessor > Loads > Define Loads > Apply > Structural > Force/Moment > On Nodes** Pick the 2nd node > **OK > Force/Moment value=−25e3 in FY direction > OK > Force/Moment > On Nodes** Pick the 1st node > **OK > Force/Moment value=−12.5e3 in FY direction > OK**

Figure 3.28. Model with loading and displacement boundary conditions.

The mode-building step is now complete, and we can proceed to the solution. First to be safe, save the model.

Solution. The interactive solution proceeds.

8. **Main Menu > Solution > Solve > Current LS > OK**

   The **/STATUS Command** window displays the problem parameters and the **Solve Current Load Step** window. Select **OK**, and when the solution is complete, close the **'Solution is DONE!'** window.

POST-PROCESSING

We can now plot the results of this analysis and also list the computed values.

9. **Main Menu > General Postproc > Plot Results > Contour Plot > Nodal Solu > DOF Solution > Displacement vector sum > OK**

   This result is shown in Figure 3.16(d).

   To find the axial stress, the following procedure is followed.

10. **Main Menu > General Postproc > Element Table > Define Table > Add**

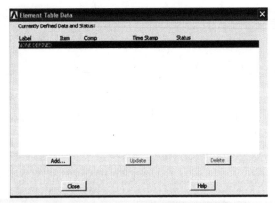

Figure 3.29. Defining the element table.

Finite Element Analysis of Axially Loaded Members

Select **By sequence num LS** and type **1 after LS** as shown in Figure 3.27.

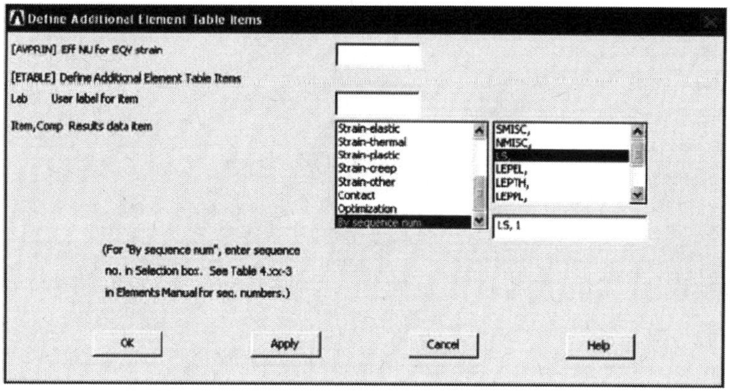

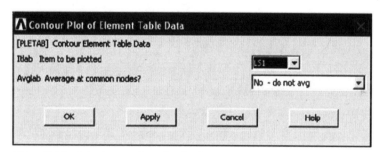

Figure 3.30. Selecting options in the element table.

OK

11. Main Menu > General PostProc > Plot Results > Contour Plot > Elem Table > Select **LS1** > **OK**

Figure 3.31. Selecting options for finding out axial stress.

This result is shown in Figure 3.19(e).

3.4 STEPPED BAR

This section will demonstrate examples on stepped bar using FEA.

Example 3.9

Find the nodal displacements, stresses in each element, and reaction at the fixed end for the Figure 3.32 shown below. Take $A_1 = 200$ mm^2, $A_2 = 200$ mm^2, and $E_1 = E_2 = 200$ GPa.

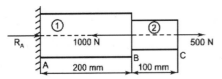

Figure 3.32. Example 3.9

Solution

(I) Analytical method [refer to Figure 3.32]

Displacement calculation

$$d_C = D_{AC} = D_{AB} + D_{BC} = \frac{-P_{AB}L_{AB}}{A_{AB}E} + \frac{P_{BC}L_{BC}}{A_{BC}E} = \frac{-500 \times 200}{200 \times 2 \times 10^5} + \frac{500 \times 100}{200 \times 2 \times 10^5}$$

$$d_C = -2.5 \times 10^{-3} + 2.5 \times 10^{-3} = 0$$

$$d_B = D_{AB} = \frac{-P_{AB}L_{AB}}{A_{AB}E} = \frac{-500 \times 200}{200 \times 2 \times 10^5} = -2.5 \times 10^{-3} \text{ mm.}$$

Stress calculation

$$\sigma_{AB} = \frac{P_{AB}}{A_{AB}} = \frac{-500}{200} = -2.5 \text{ MPa} \quad \text{(Compressive)}$$

$$\sigma_{BC} = \frac{P_{BC}}{A_{BC}} = \frac{500}{100} = 5 \text{ MPa} \quad \text{(Tensile).}$$

Reaction calculation

$$\sum F_x = 0$$

$$R_A - 1000 + 500 = 0$$

$$R_A = 500 \text{ N.}$$

(II) FEM by hand calculations

Figure 3.32(a). Finite element model for Example 3.9.

$$L_1 = 200 \text{ mm}$$
$$L_2 = 100 \text{ mm.}$$

Displacement calculation

Stiffness matrices for elements 1 and 2 are,

$$[k_1] = \frac{A_1 E_1}{L_1} \begin{bmatrix} 1 & -1 \\ -1 & 1 \end{bmatrix} = \frac{200 \times 2 \times 10^5}{200} \begin{bmatrix} 1 & -1 \\ -1 & 1 \end{bmatrix} = 2 \times 10^5 \begin{bmatrix} 1 & -1 \\ -1 & 1 \end{bmatrix} \begin{matrix} 1 \\ 2 \end{matrix}$$

$$[k_2] = \frac{A_2 E_2}{L_2} \begin{bmatrix} 1 & -1 \\ -1 & 1 \end{bmatrix} = \frac{100 \times 2 \times 10^5}{100} \begin{bmatrix} 1 & -1 \\ -1 & 1 \end{bmatrix} = 2 \times 10^5 \begin{bmatrix} 1 & -1 \\ -1 & 1 \end{bmatrix} \begin{matrix} 2 \\ 3 \end{matrix}.$$

Global equation is,

$$2 \times 10^5 \begin{bmatrix} 1 & -1 & 0 \\ -1 & 1+1 & -1 \\ 0 & -1 & 1 \end{bmatrix} \begin{Bmatrix} u_1 \\ u_2 \\ u_3 \end{Bmatrix} = \begin{Bmatrix} R_1 \\ -1000 \\ 500 \end{Bmatrix}. \tag{3.54}$$

Using the elimination method and applying boundary conditions at node 1, $u_1 = 0$. The equation (3.54) reduces to

$$2 \times 10^5 \begin{bmatrix} 2 & -1 \\ -1 & 1 \end{bmatrix} \begin{Bmatrix} u_2 \\ u_3 \end{Bmatrix} = \begin{Bmatrix} -1000 \\ 500 \end{Bmatrix}.$$

By solving the above matrix and equations, we get,

$$u_2 = -2.5 \times 10^{-3} \text{ mm}$$
$$u_3 = 0.$$

Stress calculations

$$\sigma_1 = \frac{E_1}{L_1}[-1 \quad 1]\begin{Bmatrix} u_1 \\ u_2 \end{Bmatrix} = \frac{2 \times 10^5}{200}[-1 \quad 1]\begin{Bmatrix} 0 \\ -2.5 \times 10^{-3} \end{Bmatrix} = -2.5 \text{ MPa} \quad \text{(Compressive)}$$

$$\sigma_2 = \frac{E_2}{L_2}[-1 \quad 1]\begin{Bmatrix} u_2 \\ u_3 \end{Bmatrix} = \frac{2 \times 10^5}{100}[-1 \quad 1]\begin{Bmatrix} -2.5 \times 10^{-3} \\ 0 \end{Bmatrix} = 5 \text{ MPa} \quad \text{(Tensile)}.$$

Reaction calculation

From equation (i)

$$2\times10^5\left(u_1 - u_2\right) = R_1$$
$$2\times10^5\left(0-\left(-2.5\times10^{-3}\right)\right) = R_1$$
$$R_1 = 500 \text{ N.}$$

(III) Software results

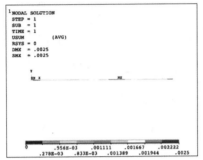

Figure 3.32(b). Deflection pattern for a stepped bar (refer to Appendix C for color figures).

Deflection values at nodes

The following degree of freedom results are in global coordinates

NODE	UX	UY	UZ	USUM
1	0.0000	0.0000	0.0000	0.0000
2	−0.25000E-02	0.0000	0.0000	0.25000E-02
3	0.0000	0.0000	0.0000	0.0000

Figure 3.32(c). Stress for a stepped bar (refer to Appendix C for color figures).

Stress value at elements

STAT ELEM	CURRENT LS1
1	-2.5000
3	5.0000

Reaction value

The following X, Y, Z solutions are in global coordinates

NODE	FX	FY
1	500.00	0.0000

Answers for Example 3.9

Parameter	Analytical method	FEM-hand calculations	Software results
Displacement at node 2	-2.5×10^{-3} mm	-2.5×10^{-3} mm	-2.5×10^{-3} mm
Displacement at node 3	0	0	0
Stress in element 1	−2.5 MPa	−2.5 MPa	−2.5 MPa
Stress in element 2	5 MPa	−5 MPa	5 MPa
Reaction at fixed end	500 N	500 N	500 N

Example 3.10

Find the nodal displacements, stress in each element, and reaction of the fixed end for Figure 3.33 shown below. Take $E_1 = 2 \times 10^5$ N/mm² and $E_2 = 1 \times 10^5$ N/mm².

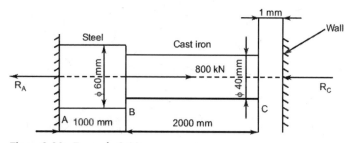

Figure 3.33. Example 3.10

Solution

(1) Analytical method [refer to Figure 3.33]

$$A_{AB} = \frac{\pi}{4} d_1^2 = \frac{\pi}{4}(60)^2 = 2827.43 \text{ mm}^2$$

$$A_{BC} = \frac{\pi}{4} d_1^2 = \frac{\pi}{4}(40)^2 = 1256.64 \text{ mm}^2.$$

In the absence of the right wall,

$$\Delta_L = \Delta_{AB} = \frac{P_{AB} \times L_{AB}}{A_{AB} \times E_{AB}} = \frac{800 \times 10^3 \times 1000}{2827.43 \times 2 \times 10^5} = 1.415 \text{ mm}.$$

Hence, the contact does occur with the right will since $u_3 = 1.415$ mm.
Let R_A and R_C be the reactions developed due to constraint.

$$R_A + R_C = 800 \times 10^3 \tag{3.55}$$

$$\frac{R_A \times L_{AB}}{A_{AB} \times E_{AB}} + \frac{(-R_C) \times L_{BC}}{A_{BC} \times E_{BC}} = 1$$

$$\frac{R_A \times (1000)}{2827.43 \times 2 \times 10^5} + \frac{(-R_C) \times (2000)}{1256.64 \times 1 \times 10^5} = 1$$

$$1.7684 \times 10^{-6} \times R_A - 1.5915 \times 10^{-5} \times R_C = 1. \tag{3.56}$$

By solving equations (3.55) and (3.56),
we get

$$R_A = 776547.49 \text{ N}$$
$$R_C = 23452.51 \text{ N}.$$

Displacement calculation

$$\delta_B = \Delta_{AB} = \frac{R_A \times L_{AB}}{A_{AB} \times E_{AB}} = \frac{776547.49 \times 1000}{2827.43 \times 2 \times 10^5} = 1.373 \text{ mm}$$

$$\delta_C = 1 \text{ mm}.$$

Stress calculation

$$\sigma_{AB} = \frac{R_A}{A_{AB}} = \frac{776547.49}{2827.43} = 274.65 \text{ MPa}$$

$$\sigma_{BC} = \frac{(-R_C)}{A_{BC}} = \frac{-23452.51}{1256.64} = -18.66 \text{ MPa.}$$

(II) FEM by hand calculations

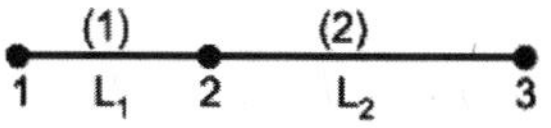

Figure 3.33(a). Finite element model for Example 3.10.

$$L_1 = 1000 \text{ mm}$$
$$L_2 = 2000 \text{ mm.}$$

In this example, first determine whether contact occurs between the bar and the wall. To do this, assume that the wall does not exist. Then the solution to the problem is (consider the 2 element model),
Stiffness matrix for element 1 is,

$$[k_1] = \frac{A_1 E_1}{L_1}\begin{bmatrix} 1 & -1 \\ -1 & 1 \end{bmatrix} = \frac{2827.43 \times 2 \times 10^5}{1000}\begin{bmatrix} 1 & -1 \\ -1 & 1 \end{bmatrix} = 5.655 \times 10^5 \begin{matrix} & 1 & 2 \\ & \begin{bmatrix} 1 & -1 \\ -1 & 1 \end{bmatrix} & \begin{matrix} 1 \\ 2 \end{matrix} \end{matrix}.$$

Stiffness matrix for element 2 is,

$$[k_2] = \frac{A_2 E_2}{L_2}\begin{bmatrix} 1 & -1 \\ -1 & 1 \end{bmatrix} = \frac{1256.64 \times 1 \times 10^5}{2000}\begin{bmatrix} 1 & -1 \\ -1 & 1 \end{bmatrix} = 0.628 \times 10^5 \begin{matrix} & 2 & 3 \\ & \begin{bmatrix} 1 & -1 \\ -1 & 1 \end{bmatrix} & \begin{matrix} 2 \\ 3 \end{matrix} \end{matrix}.$$

Global equation is,

$$[K]\{r\} = \{R\} \quad (3.57)$$

$$10^5 \begin{bmatrix} 5.655 & -5.655 & 0 \\ -5.655 & 5.655+0.628 & -0.628 \\ 0 & -0.628 & 0.628 \end{bmatrix} \begin{matrix} 1 \\ 2 \\ 3 \end{matrix} \begin{Bmatrix} u_1 \\ u_2 \\ u_3 \end{Bmatrix} = \begin{Bmatrix} R_1 \\ 800 \times 10^3 \\ 0 \end{Bmatrix}. \quad (3.58)$$

Boundary conditions are at node 1, $u_1 = 0$.

By using the elimination method, the above matrix reduces to,

$$10^5 \begin{bmatrix} 6.283 & -0.628 \\ -0.628 & 0.628 \end{bmatrix} \begin{Bmatrix} u_2 \\ u_3 \end{Bmatrix} = \begin{Bmatrix} 800 \times 10^3 \\ 0 \end{Bmatrix}.$$

By matrix multiplication, we get

$$10^5 (6.283 \times u_2 - 0.628 \times u_3) = 800 \times 10^3 \qquad (3.59)$$

$$10^5 (0.628 \times u_2 + 0.628 \times u_3) = 0 \qquad (3.60)$$

By solving equations (3.59) and (3.60)
we get

$$u_2 = 1.415 \text{ mm and } u_3 = 1.415 \text{ mm}.$$

Since the displacement of node 3 is 1.415 mm, we can say that contact does occur. The problem has to be resolved since the boundary conditions are now different. The displacement at node 3 is given as 1 mm.

Global equation is,

$$10^5 \begin{bmatrix} 5.655 & -5.655 & 0 \\ -5.655 & 5.655+0.628 & -0.628 \\ 0 & -0.628 & 0.628 \end{bmatrix} \begin{matrix} 1 \\ 2 \\ 3 \end{matrix} \begin{Bmatrix} u_1 \\ u_2 \\ u_3 \end{Bmatrix} = \begin{Bmatrix} R_1 \\ 800 \times 10^3 \\ 0 \end{Bmatrix}. \qquad (3.61)$$

Boundary conditions at node 1, $u_1 = 0$ and at node 3, $u_2 = 1$ mm.
By using the elimination method, the above matrix reduces to,

$$10^5 [6.283]\{u_2\} = [800 \times 10^3] - 1[10^5 \times (-0.628)]$$

$$10^5 [6.283]\{u_2\} = 800 \times 10^3 + 0.628 \times 10^5$$

$$u_2 = 1.373 \text{ mm}.$$

Stress calculation

$$\sigma_1 = \frac{E_1}{L_1}[-1 \quad 1]\begin{Bmatrix} u_1 \\ u_2 \end{Bmatrix} = \frac{2 \times 10^5}{1000}[-1 \quad 1]\begin{Bmatrix} 0 \\ 1.373 \end{Bmatrix} = 274.6 \text{ MPa}$$

$$\sigma_2 = \frac{E_2}{L_2}[-1 \quad 1]\begin{Bmatrix} u_2 \\ u_3 \end{Bmatrix} = \frac{1 \times 10^5}{2000}[-1 \quad 1]\begin{Bmatrix} 1.373 \\ 1 \end{Bmatrix} = 18.65 \text{ MPa}.$$

Reaction calculation

From equation (3.60)

$$5.655 \times 10^5 \times u_1 - 5.655 \times 10^5 \times u_2 = R_1$$

$$0 - 5.655 \times 10^5 \times 1.373 = R_1$$

$$R_1 = -776431.5 \text{ N} \quad \text{(Direction is leftwards).}$$

We know that,

$$R_1 + P + R_3 = 0$$

$$-776431.5 + 800 \times 10^3 + R_3 = 0$$

$$R_3 = -23568.5 \text{ N} \quad \text{(Direction is leftwards).}$$

(III) Software results

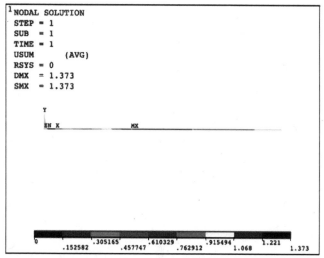

Figure 3.33(b). Deflection pattern for a stepped bar (refer to Appendix C for color figures).

Deflection values at nodes

The following degree of freedom results are in global coordinates

NODE	UX	UY	UZ	USUM
1	0.0000	0.0000	0.0000	0.0000
2	1.3732	0.0000	0.0000	1.3732
3	1.0000	0.0000	0.0000	1.0000

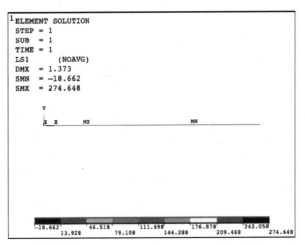

Figure 3.33(c). Stress pattern for a stepped bar (refer to Appendix C for color figures).

Stress values at elements

STAT ELEM	CURRENT LS1
1	247.65
2	−18.662

Reaction value

The following *X*, *Y*, *Z* solutions are in global coordinates

NODE	FX	FY
1	−0.77655E +06	0.0000
3	−23451.	

Answers for Example 3.10

Parameter	Analytical method	FEM-hand calculations	Software results
Displacement at node 2	1.373 mm	1.373 mm	1.3732 mm
Displacement at node 3	1 mm	1 mm	1 mm
Stress in element 1	274.65 MPa	274.65 MPa	−274.65 MPa
Stress in element 2	−18.66 MPa	−5 MPa	5 MPa
Reaction at fixed end	−776.5 kN	−776.4 kN	−776.55 kN
Reaction at wall	−23.45 kN	−23.57 kN	−23.451 kN

Example 3.11

Find the nodal displacement, stress in each element, and reaction at fixed ends for Figure 3.34 as shown below. If the structure is subjected to an increase in temperature, $\Delta T = 75°C$, $P_1 = 50$ kN, $P_2 = 75$ kN.

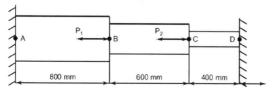

Figure 3.34 Example 3.11

	Bronze	Aluminum	Steel
	$A = 2400$ mm²	1200 mm²	600 mm²
	$E = 83$ GPa	70 GPa	200 GPa
	$\alpha = 18.9 \times 10^{-6}/°C$	$23 \times 10^{-6}/°C$	$11.7 \times 10^{-6}/°C$

Solution

(I) Analytical method [refer to Figure 3.34]

Problem can be solved by method of superposition by considering load and temperature separately.

Step 1: Consider only the loads P_1, P_2, and neglect rise in temperature.

$$R_1' + R_2' = 125 \times 10^3 \quad (R_1' \text{ and } R_2' \text{ are reactions due to } P_1 \text{ and } P_2) \quad (3.62)$$

$$\frac{(-P_{AB}) \times 800}{2400 \times 83 \times 10^3} + \frac{(-P_{BC}) \times 600}{1200 \times 70 \times 10^3} + \frac{(P_{CD}) \times 400}{600 \times 200 \times 10^3} = 0.$$

But $P_{AB} = R_1'$, $P_{CD} = R_2'$ and $P_{BC} = R_1' - 50 \times 10^3$

$$\frac{(-R_1') \times 800}{2400 \times 83 \times 10^3} - \frac{(R_1' - 50 \times 10^3) \times 600}{1200 \times 70 \times 10^3} + \frac{(P_{CD}) \times 400}{600 \times 200 \times 10^3} = 0. \quad (3.63)$$

Solving equations (3.61) and (3.62)

$$R_1' = 53.39 \text{ kN}$$

$$R_2' = 71.61 \text{ kN}$$

$$\therefore \sigma_{AB}' = \frac{(-53.39 \times 10^3)}{2400} = -22.25 \text{ MPa}$$

$$\therefore \sigma'_{BC} = \frac{-(53.39 \times 10^3 - 50 \times 10^3)}{1200} = -2.825 \text{ MPa}$$

$$\therefore \sigma'_{CD} = \frac{(71.61 \times 10^3)}{600} = 119.35 \text{ MPa}$$

$$\Delta'_{AB} = \frac{-22.25 \times 800}{83 \times 10^3} = -0.2144 \text{ mm}$$

$$\Delta'_{BC} = \frac{-2.825 \times 600}{70 \times 10^3} = -0.0242 \text{ mm}$$

$$\Delta'_{CD} = \frac{119.35 \times 400}{200 \times 10^3} = 0.2395 \text{ mm}.$$

Step 2: Consider only the rise in temperature and neglect P_1 and P_2. Free expansions due to $\Delta T = 75°C$ are

$$(\Delta L_{AB})_T = a \times L \times (\Delta T) = 18.9 \times 10^{-6} \times 800 \times 75 = 1.134 \text{ mm}$$

$$(\Delta L_{BC})_T = a \times L \times (\Delta T) = 23 \times 10^{-6} \times 600 \times 75 = 1.035 \text{ mm}$$

$$(\Delta L_{CD})_T = a \times L \times (\Delta T) = 11.7 \times 10^{-6} \times 400 \times 75 = 0.351 \text{ mm}$$

$$\text{Total } (\Delta L)_T = 1.134 + 1.035 + 0.351 = 2.52 \text{ mm}.$$

For equilibrium

$$\frac{(-R''_1) \times 800}{2400 \times 83 \times 10^3} + \frac{(-R''_1) \times 600}{1200 \times 70 \times 10^3} + \frac{(-R''_1) \times 400}{600 \times 200 \times 10^3} = - \text{ Total } (\Delta L)_T$$

$$\frac{(-R''_1) \times 800}{2400 \times 83 \times 10^3} + \frac{(-R''_1) \times 600}{1200 \times 70 \times 10^3} + \frac{(-R''_1) \times 400}{600 \times 200 \times 10^3} = -2.52.$$

Solving,

$$R''_1 = 173.89 \text{ kN}$$

$$\sigma''_{AB} = \frac{-173890}{2400} = -72.45 \text{ MPa}$$

$$\sigma''_{BC} = \frac{-173890}{1200} = -144.91 \text{ MPa}$$

$$\sigma''_{CD} = \frac{-173890}{600} = -289.82 \text{ MPa}$$

$$(\Delta L_{AB})_{Load} = \frac{-72.45 \times 800}{83 \times 10^3} = -0.698 \text{ mm}$$

$$(\Delta L_{BC})_{Load} = \frac{-144.91 \times 600}{70 \times 10^3} = -1.242 \text{ mm}$$

$$(\Delta L_{CD})_{Load} = \frac{-289.82 \times 400}{200 \times 10^3} = -0.5796 \text{ mm}$$

$$\Delta''_{AB} = 1.134 - 0.698 = 0.436 \text{ mm}$$

$$\Delta''_{BC} = 1.035 - 1.242 = -0.207 \text{ mm}$$

$$\Delta''_{CD} = 0.351 - 0.5796 = -0.2286 \text{ mm}.$$

Step 3: Use method of superposition and combine steps (1) and (2).

Stresses are, $\sigma_{AB} = \sigma'_{AB} + \sigma''_{AB} = -22.45 - 72.45 = -94.7$ MPa.

Similarity, $\sigma_{BC} = \sigma'_{BC} + \sigma''_{BC} = -147.74$ MPa

$$\sigma_{CD} = \sigma'_{CD} + \sigma''_{CD} = -170.47 \text{ MPa}.$$

Change in lengths are,

$$\Delta_{AB} = \Delta'_{AB} + \Delta''_{AB} = -0.2144 + 0.436 = 0.2216 \text{ mm}$$

$$\Delta_{BC} = \Delta'_{BC} + \Delta''_{BC} = -0.0242 - 0.207 = -0.2312 \text{ mm}$$

$$\Delta_{CD} = \Delta'_{CD} + \Delta''_{CD} = 0.2395 - 0.2286 = 0.0109 \text{ mm}$$

$$u_2 = \Delta_{AB} = 0.2216 \text{ mm}$$

$$u_3 = \Delta_{CD} = 0.0109 \text{ mm}.$$

Reactions are,

$$R_1 = R'_1 + R''_1 = 53.39 + 173.89 = 227.28 \text{ kN}$$

$$R_2 = R'_2 + R''_2 = 71.61 - 173.89 = -102.28 \text{ kN}.$$

(II) FEMby hand calculations

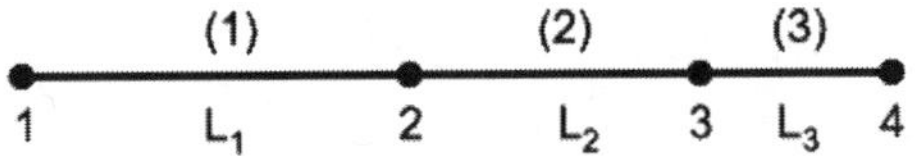
(1)
(2)
(3)
1
L1
2
L2
3
L3
4

Figure 3.34(a). **Finite element model for Example 3.11.**

$$L_1 = 800 \text{ mm}, \quad L_2 = 600 \text{ mm}, \quad L_3 = 400 \text{ mm}$$

$$A_1 = 2400 \text{ mm}^2, \quad A_2 = 1200 \text{ mm}^2, \quad A_3 = 600 \text{ mm}^2$$

$$E_1 = 83\times10^3 \text{ N/mm}^2, \quad E_2 = 70\times10^3 \text{ N/mm}^2, \quad E_3 = 200\times10^3 \text{ N/mm}^2$$

$$\alpha_1 = 18.9\times10^{-6}/^\circ\text{C}, \quad \alpha_2 = 23\times10^{-6}/^\circ\text{C}, \quad \alpha_3 = 11.7\times10^{-6}/^\circ\text{C}.$$

Element stiffness matrices are,

$$[k_1] = \frac{A_1E_1}{L_1}\begin{bmatrix} 1 & -1 \\ -1 & 1 \end{bmatrix} = \frac{2400\times83\times10^3}{800}\begin{bmatrix} 1 & -1 \\ -1 & 1 \end{bmatrix} = 249\times10^3 \begin{array}{c} \begin{matrix} 1 & 2 \end{matrix} \\ \begin{bmatrix} 1 & -1 \\ -1 & 1 \end{bmatrix} \end{array}\begin{matrix} 1 \\ 2 \end{matrix}$$

$$[k_2] = \frac{A_2E_2}{L_2}\begin{bmatrix} 1 & -1 \\ -1 & 1 \end{bmatrix} = \frac{1200\times70\times10^3}{600}\begin{bmatrix} 1 & -1 \\ -1 & 1 \end{bmatrix} = 140\times10^3 \begin{array}{c} \begin{matrix} 2 & 3 \end{matrix} \\ \begin{bmatrix} 1 & -1 \\ -1 & 1 \end{bmatrix} \end{array}\begin{matrix} 2 \\ 3 \end{matrix}$$

$$[k_3] = \frac{A_3E_3}{L_3}\begin{bmatrix} 1 & -1 \\ -1 & 1 \end{bmatrix} = \frac{600\times200\times10^3}{400}\begin{bmatrix} 1 & -1 \\ -1 & 1 \end{bmatrix} = 300\times10^3 \begin{array}{c} \begin{matrix} 3 & 4 \end{matrix} \\ \begin{bmatrix} 1 & -1 \\ -1 & 1 \end{bmatrix} \end{array}\begin{matrix} 3 \\ 4 \end{matrix}.$$

Effect of temperature and thermal loads are,

$$\{Q_{1(th)}\} = E_1A_1\alpha_1(\Delta T)\begin{Bmatrix} -1 \\ 1 \end{Bmatrix} = 83\times10^3\times2400\times18.9\times10^{-6}\times75\begin{Bmatrix} -1 \\ 1 \end{Bmatrix}$$

$$= 282.37\times10^3\begin{Bmatrix} -1 \\ 1 \end{Bmatrix}\begin{matrix} 1 \\ 2 \end{matrix}$$

$$\{Q_{2(th)}\} = E_2 A_2 \alpha_2 (\Delta T) \begin{Bmatrix} -1 \\ 1 \end{Bmatrix} = 70 \times 10^3 \times 1200 \times 23 \times 10^{-6} \times 75 \begin{Bmatrix} -1 \\ 1 \end{Bmatrix}$$

$$= 144.9 \times 10^3 \begin{Bmatrix} -1 \\ 1 \end{Bmatrix} \begin{matrix} 2 \\ 3 \end{matrix}$$

$$\{Q_{3(th)}\} = E_3 A_3 \alpha_3 (\Delta T) \begin{Bmatrix} -1 \\ 1 \end{Bmatrix} = 200 \times 10^3 \times 600 \times 11.7 \times 10^{-6} \times 75 \begin{Bmatrix} -1 \\ 1 \end{Bmatrix}$$

$$= 105.3 \times 10^3 \begin{Bmatrix} -1 \\ 1 \end{Bmatrix} \begin{matrix} 3 \\ 4 \end{matrix}.$$

Global Force vector

$$\{R\} = \begin{Bmatrix} -282.37 \times 10^3 \\ 282.37 \times 10^3 - 144.9 \times 10^3 \\ 144.9 \times 10^3 - 105.3 \times 10^3 \\ 105.3 \times 10^3 \end{Bmatrix} \begin{matrix} 1 \\ 2 \\ 3 \\ 4 \end{matrix} = \begin{Bmatrix} -282.37 \times 10^3 \\ 137.47 \times 10^3 \\ 39.6 \times 10^3 \\ 105.3 \times 10^3 \end{Bmatrix} \begin{matrix} 1 \\ 2 \\ 3 \\ 4 \end{matrix}.$$

$$10^3 \begin{bmatrix} 249 & -249 & 0 & 0 \\ -249 & 249+140 & -140 & 0 \\ 0 & -140 & 140+300 & -300 \\ 0 & 0 & -300 & 300 \end{bmatrix} \begin{matrix} 1 \\ 2 \\ 3 \\ 4 \end{matrix} \begin{Bmatrix} u_1 \\ u_2 \\ u_3 \\ u_4 \end{Bmatrix} = 10^3 \begin{Bmatrix} -282.37 + R_1 \\ 137.47 - 50 \\ 39.6 - 75 \\ 105.3 + R_4 \end{Bmatrix}. \quad (3.64)$$

Using the elimination method and applying boundary conditions,

i.e., $\qquad u_1 = u_4 = 0.$

The equation (3.63) reduces to,

$$10^3 \begin{bmatrix} 389 & -140 \\ -140 & 440 \end{bmatrix} \begin{Bmatrix} u_2 \\ u_3 \end{Bmatrix} = 10^3 \begin{Bmatrix} 87.47 \\ -35.4 \end{Bmatrix}.$$

By solving the above matrix and equation,

we get $u_2 = 0.2212$ mm and $u_3 = -0.0101$ mm.

Stress calculation

$$\sigma_1 = \frac{E_1}{L_1}[-1 \ 1]\begin{Bmatrix} u_1 \\ u_2 \end{Bmatrix} - E_1 a_1 (\Delta T)$$

$$= \frac{83 \times 10^3}{800}[-1 \ 1]\begin{Bmatrix} 0 \\ 0.2212 \end{Bmatrix} - 83 \times 10^3 \times 18.9 \times 10^{-6} \times 75 = -94.7 \text{ MPa}$$

$$\sigma_2 = \frac{E_2}{L_2}[-1 \ 1]\begin{Bmatrix} u_2 \\ u_3 \end{Bmatrix} - E_2 a_2 (\Delta T)$$

$$= \frac{70 \times 10^3}{600}[-1 \ 1]\begin{Bmatrix} 0.2212 \\ -0.0101 \end{Bmatrix} - 70 \times 10^3 \times 23 \times 10^{-6} \times 75 = -145.38 \text{ MPa}$$

$$\sigma_3 = \frac{E_3}{L_3}[-1 \ 1]\begin{Bmatrix} u_3 \\ u_4 \end{Bmatrix} - E_3 a_3 (\Delta T)$$

$$= \frac{200 \times 10^3}{400}[-1 \ 1]\begin{Bmatrix} -0.0101 \\ 0 \end{Bmatrix} - 200 \times 10^3 \times 11.7 \times 10^{-6} \times 75 = -170.45 \text{ MPa}.$$

Reaction calculation

$$249 \times 10^3 \times u_1 - 249 \times 10^3 \times u_2 = -282.37 \times 10^3 + R_1$$
$$0 - 249 \times 10^3 \times 0.2212 = -282.37 \times 10^3 + R_1$$
$$R_1 = 227.29 \text{ kN}$$
$$-300 \times 10^3 \times u_3 + 300 \times 10^3 \times u_4 = 105.3 \times 10^3 + R_4$$
$$-300 \times 10^3 \times (-0.0101) + 0 = 105.3 \times 10^3 + R_4$$
$$R_4 = -102.27 \text{ kN}.$$

(III) Software results

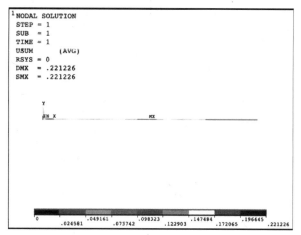

Figure 3.34(b). Deflection pattern for a stepped bar (refer to Appendix C for color figures).

Deflection values at nodes

The following degree of freedom results are in global coordinates

NODE	UX	UY	UZ	USUM
1	0.0000	0.0000	0.0000	0.0000
2	0.22123	0.0000	0.0000	0.22123
3	−0.10064E-01	0.0000	0.0000	0.10064E-01
4	0.0000	0.0000	0.0000	0.0000

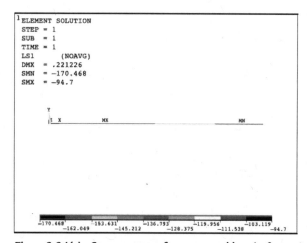

Figure 3.34(c). Stress pattern for a stepped bar (refer to Appendix C for color figures).

Stress values at elements

STAT ELEM	CURRENT LS1
1	−94.700
2	−147.73
3	−170.47

Reaction values

The following X, Y, Z solutions are in global coordinates

NODE	FX	FY
1	0.22728E +06	0.0000
4	−0.10228E+06	0.0000

Answers for Example 3.11

Parameter	Analytical method	FEM-hand calculations	Software results
Displacement at node 2	0.2216 mm	0.2212 mm	0.22123 mm
Displacement at node 3	−0.0109 mm	−0.0101 mm	−0.010064 mm
Stress in element 1	−94.7 MPa	−94.7 MPa	−94.7 MPa
Stress in element 2	−147.74 MPa	−145.38 MPa	−147.73 MPa
Stress in element 3	−170.47 MPa	−170.45 MPa	−170.47 MPa
Reaction at fixed end	227.28 kN	227.2912 kN	227.28 kN
Reaction at wall	−102.28 kN	−102.27 kN	−102.28 kN

Procedure for solving the example using ANSYS® 11.0 academic teaching software
For Example 3.11

PREPROCESSING

1. Main Menu > Preprocessor > Element Type > Add/Edit/Delete > Add > Structural Link > 2D spar 1 > Ok > Close

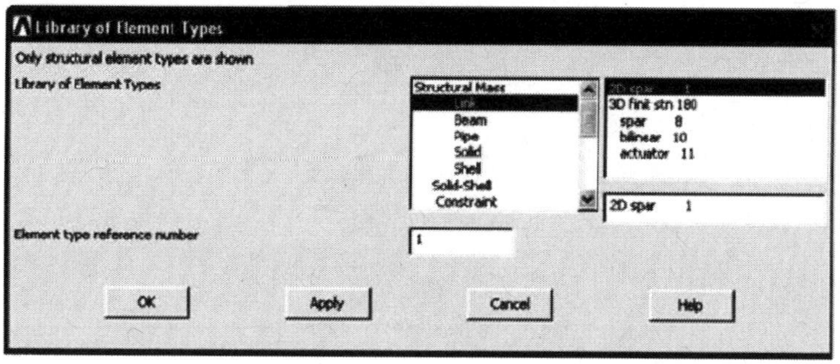

Finite Element Analysis of Axially Loaded Members

Figure 3.35. Element selection.

2. **Main Menu > Preprocessor > Real Constants > Add/Edit/Delete > Add > OK**

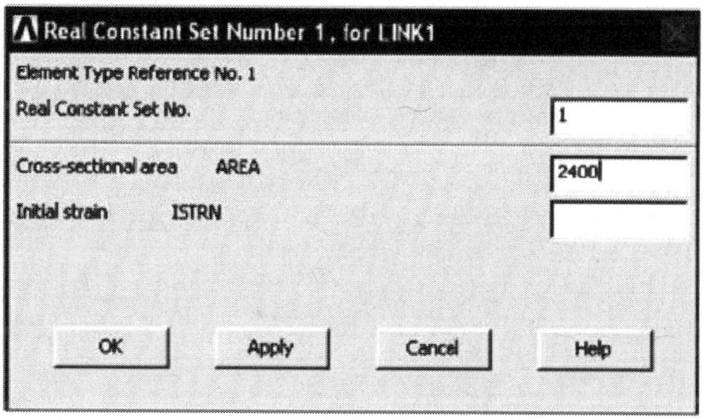

Figure 3.36. Enter the cross-sectional area of 1st element.

Cross-sectional area AREA > **Enter 2400 > OK > Add > OK**

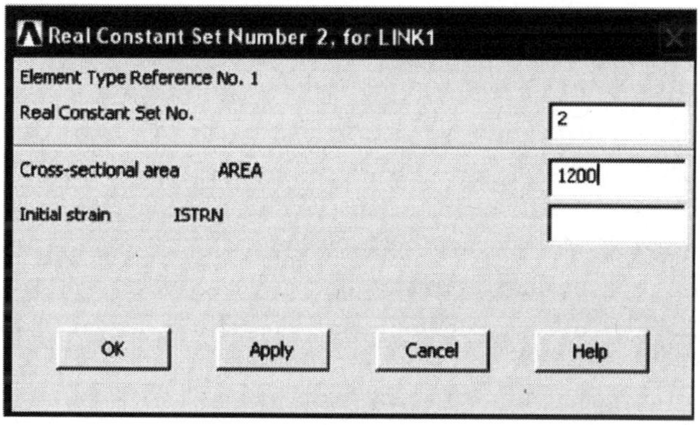

Figure 3.37. Enter the cross-sectional area of 2nd element.

Cross-sectional area AREA > **Enter 1200 > OK > Add > OK**

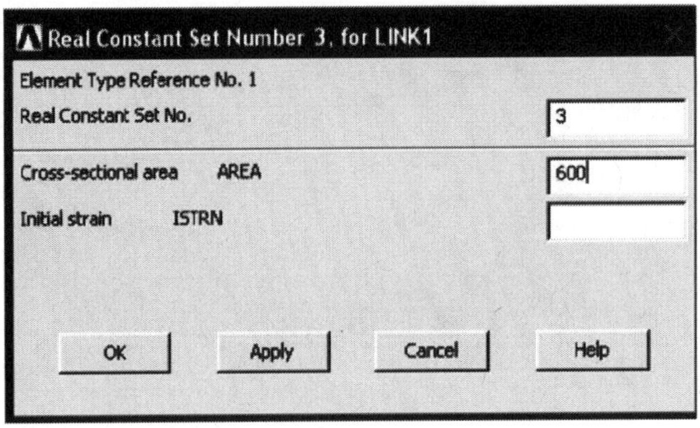

Figure 3.38. Enter the cross-sectional area of 3rd element.

Cross-sectional area AREA > **Enter 600** > **OK** > **Add** > **OK** > **Close**
Enter the material properties.
3. **Main Menu** > **Preprocessor** > **Material Props** > **Material Models**
   Material Model Number 1,
   Click **Structural** > **Linear** > **Elastic** > **Isotropic**
   Enter **EX = 0.83E5 and PRXY = 0.34** > **OK**
   Enter the coefficient of thermal expansion α
   Click **Structural** > **Thermal Expansion** > **Secant coefficient** > **Isotropic**
   Enter **ALPX – 18.9E-6** > **OK**
   Then in the material model window click on **Material menu** > **New Model** > **OK**
   Material Model Number 2,
   Click **Structural** > **Linear** > **Elastic** > **Isotropic**
   Enter **EX = 0.7E5 AND PRXY -0.35** > **OK**
   Enter the coefficient of thermal expansion α
   Click **Structural** > **Thermal Expansion** > **Secant coefficient** > **Isotropic**
   Enter **ALPX = 23E-6** > **OK**
   Then in the material model window click on **Material menu** > **New Model** > **OK**
   Material Model Number 3,
   Click **Structural** > **Linear** > **Elastic** > **Isotropic**
   Enter **EX =2E5 and PRXY = 0.3** > **OK**
   Enter the coefficient of thermal expansion α
   Click **Structural** > **Thermal Expansion** > **Secant coefficient** > **Isotropic**
   Enter **ALPX = 11.7E-6** > **OK**
   (**Close** the Define Material Model Behavior window.)
   Create the nodes and elements. Use 3 element models. Hence create 4 nodes and 3 elements.

Finite Element Analysis of Axially Loaded Members

4. **Main Menu > Preprocessor > Modeling > Create > Nodes > In Active CS** Enter the coordinates of node 1 > **Apply** Enter the coordinates of node > **Apply** Enter the coordinates of node 3 > **Apply** > Enter the coordinates of node 4 > **OK**.

Node locations		
Node number	X COORDINATE	Y COORDINATE
1	0	0
2	800	0
3	1400	0
4	1800	0

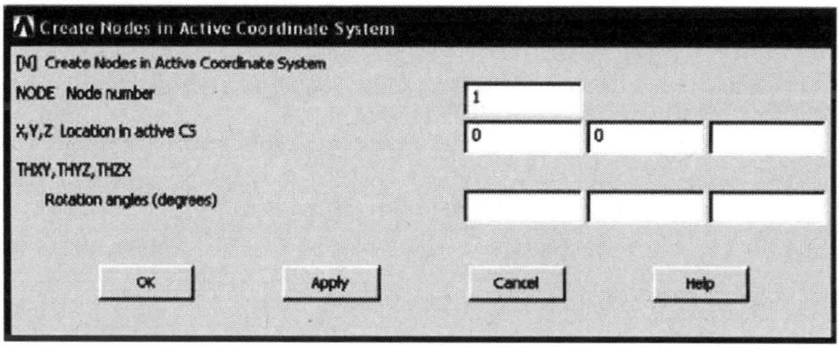

Figure 3.39. Enter the node coordinates.

5. **Main Menu > Preprocessor > Modeling > Create > Elements > Elem Attributes > OK > Auto Numbered > Thru nodes** Pick the 1st and 2nd node > **OK**

Figure 3.40. Assigning element attributes to element 1 and creating element 1.

Elem Attributes > change the material number to 2 > change the Real constant set number to 2 > **OK** > **Auto Numbered** > **Thru nodes** Pick the 2nd and 4th node > **OK**

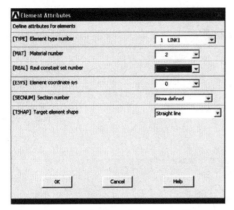

Figure 3.41. Assigning element attributes to element 2 and creating element 2.

Elem Attributes > change the material number to 3 > change the Real constant set number to 3 > **OK** > **Auto Numbered** > **Thru nodes** Pick the 3rd and 4th node > **OK**

Figure 3.42. Assigning element attributes to element 3 and creating element 3.

Apply the displacement boundary conditions, load, and temperature.

6. **Main Menu** > **Preprocessor** > **Loads** > **Define Loads** > **Apply** > **Structural** > **Displacement** > **On Nodes** Pick the 1st and 4th node > **Apply** > **All DOF = 0.** > **OK**

Finite Element Analysis of Axially Loaded Members 139

7. **Main Menu > preprocessor > Loads > Define Loads > Apply > Structural > Force/Moment > On Nodes** Pick the 2nd node **> OK > Force/Moment value = -50e3 in FX direction > OK > Force/Moment > On Nodes** Pick the 3rd node **> OK > Force/Moment value =-75e3 in FX direction > OK**
8. **Main Menu > Preprocessor > Loads > Define Loads > Apply > Structural > Temperature > On Elements** Pick the element, 2nd element and 3rd element **> OK** Enter **Temperature at location N = 75** as shown in Figure 3.38.

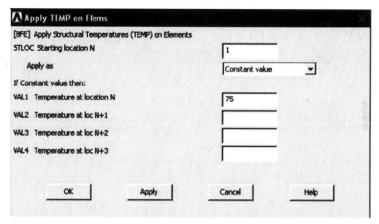

Figure 3.43. Enter the rise in temperature on elements.

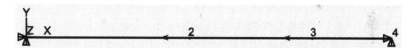

Figure 3.44. Model with loading and displacement boundary conditions.

The model-building step is now complete, and we can proceed to the solution. First to be safe, save the model.

Solution

The interactive solution proceeds.
9. **Main Meni > Solution > Solve > Current LS > OK**
The **/STATUS Command** window displays the problem parameters and **the Solve Current Load Step** window and if all is OK, select **FILE > CLOSE**
In the Solve **Current Load Step** window, Select **OK,** and when the solution is complete, **close** the **'Solution is Done!'** window.

POST-PROCESSING

We can now plot the results of this analysis and also list the computed values.

10. **Main Menu > General Postproc > Plot Results > Contour Plot > Nodal Solu > DOF Solution > Displacement vector sum > OK**

    This result is shown in Figure 3.34(b).

    To find the axial stress, the following procedure is followed.

11. **Main Menu > General Postproc > Element Table > Define Table > Add**

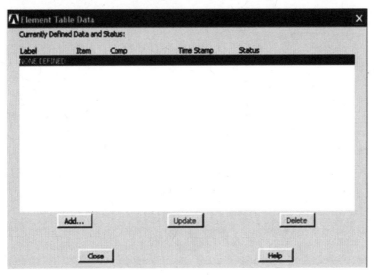

Figure 3.45. Defining the element table.

Select **By sequence num and LS** and type **1 after LS** as shown in Figure 3.43.

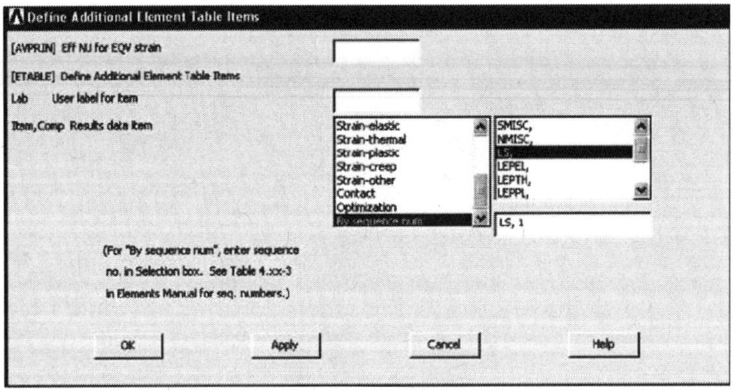

Figure 3.46. Selecting options in element table.

>**OK**

12. Main Menu > General Postproc > Plot Results > Contour Plot > Elem Table > Select **LS1** > **OK**

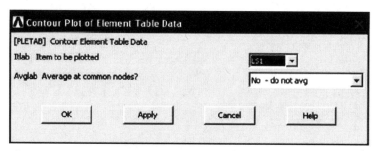

Figure 3.47. Selecting options for finding out axial stress.

This result is shown in Figure 3.34(c).

PROBLEMS

1. Determine the nodal displacement and element stress for the bar shown in Figure 3.48. Take 3 elements finite element model. Take E = 70 GPa.

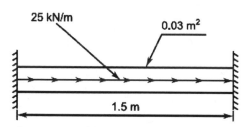

Figure 3.48. Problem 1

2. Determine the nodal displacements and stresses in the element for the axial distributed loading shown in Figure 3.49. Take one element model. Take E = 200 GPa, $A = 5 \times 10^{-4}$ m^2.

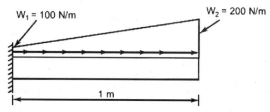

Figure 3.49. Problem 2

3. For the bar assembly shown in Figure 3.50, determine the nodal displacements, stresses in each element, and reactions. Take E = 210 GPa, $A = 5 \times 10^{-4}$ m².

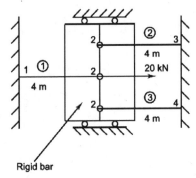

Figure 3.50. Problem 3

4. Find the deflection at the free end under its own weight for a tapered bar shown in Figure 3.51. Use 2 element models. Take E = 200 GPa, weight density $\rho = 7800$ kg/m³.

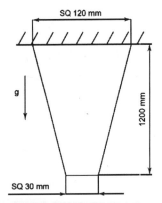

Figure 3.51. Problem 4

5. Determine the displacement, element stresses, and reactions for the tapered bar shown in Figure 3.52. Use 2 elements finite element models. Take E = 200 GPa, $A_1 = 2000$ mm², $A_2 = 4000$ mm².

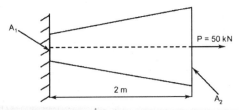

Figure 3.52. Problem 5

Finite Element Analysis of Axially Loaded Members

6. Consider the bar shown in Figure 3.53. An axial load P = 500 kN is applied as shown. Determine the
   (a) Nodal displacement (b) Stresses in each material (c) Reaction forces.

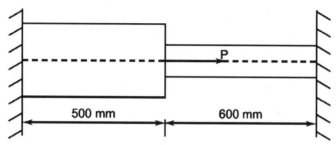

Figure 3.53. Problem 6

Aluminum
$A_1 = 300$ mm²
$E_1 = 70$ GPa

Steel
$A_2 = 1000$ mm²
$E_2 = 200$ GPa

7. In Figure 3.54, determine displacements at 2 and 3 stresses in the members and reactions if the temperature is increased by 60°.

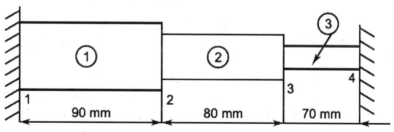

Figure 3.54. Problem 7

Member	Area A (mm²)	Youngs modulus E (GPa)	Thermal expansion coefficient α (/°C)
1	1000	70	23×10^{-6}
2	500	100	19×10^{-6}
3	300	200	12×10^{-6}

8. For the vertical bar shown in Figure 3.55, for the deflection at 2 and 3 and stress distribution. Take E = 25 GPa and density, ρ = 2100 kg/m³. Take self-weight of the bar into consideration and solve the problem using 2 elements.

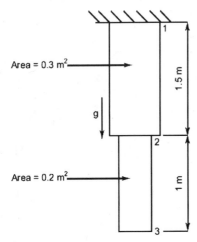

Figure 3.55. Problem 8

9. Find displacement and stresses shown in Figure 3.56. Take E = 200 GPa.

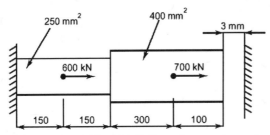

Figure 3.56. Problem 9 (all dimensions are in mm).

REFERENCES

1. J. Wickert, "An Introduction to Mechanical Engineering, Second Edition," Thomson Publisher, 2006.
2. J. M. Gere and S. P. Timoshenko, "Mechanics of Materials, Fourth Edition," PWS Publishing, 1997.
3. R. G. Budynas and J. K. Nisbett, "Shigley's Mechanical Engineering Design, Eighth Edition," McGraw-Hill Higher Education, 2008.
4. H. C. Martin and G. F. Carey, "Introduction to Finite Element Analysis: Theory and Applications," McGraw-Hill Book Company, 1973.
5. W. B. Bickford, "A First Course in the Finite Element Method, Second Edition," Richard D. Irwin, INC., 1990.

Chapter 4
Finite Element Analysis Trusses

4.1 INTRODUCTION

This chapter introduces the basic concepts in finite element formulation of trusses and provides the illustration of its ANSYS program.

4.2 TRUSS

Truss, by definition, is a load bearing structure formed by connecting members using pin joints. Truss element is used in the analysis of 2-D trusses.

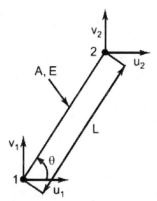

Figure 4.1. A 2-D Truss.

145

The element has two nodes, each having two degrees of freedom namely translations along the x- and y-axes.

The element stiffness matrix and element stress for a truss element are given by,

$$[k] = \frac{AE}{L} \begin{bmatrix} \cos^2\theta & \cos^2\theta \times \sin\theta & -\cos^2\theta & -\cos\theta \times \sin\theta \\ \cos\theta \times \sin\theta & \sin^2\theta & -\cos\theta \times \sin\theta & -\sin^2\theta \\ -\cos^2\theta & -\cos\theta \times \sin\theta & \cos^2\theta & \cos\theta \times \sin\theta \\ -\cos\theta \times \sin\theta & -\sin^2\theta & \cos\theta \times \sin\theta & \sin^2\theta \end{bmatrix} \quad (4.1)$$

$$= \frac{AE}{L} \begin{bmatrix} c^2 & cs & -c^2 & -cs \\ cs & s^2 & -cs & -s^2 \\ -c^2 & -cs & c^2 & cs \\ -cs & -s^2 & cs & s^2 \end{bmatrix}$$

$$\{\sigma\} = \frac{E}{L}[-\cos\theta \quad -\sin\theta \quad \cos\theta \quad \sin\theta]\{q\}, \text{ where } \{q\} = \begin{Bmatrix} u_1 \\ v_1 \\ u_2 \\ v_2 \end{Bmatrix}. \quad (4.2)$$

θ = angle of truss element at node 1 with positive x-axis (in degrees).

Example 4.1

Determine the nodal displacements, element stresses, and support reactions for the 3 member truss shown in Figure 4.2. Take $A = 800$ mm^2 and $E = 200$ GPa for all members.

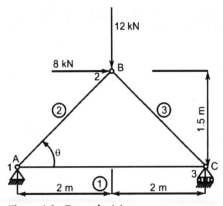

Figure 4.2. Example 4.1

Solution

(I) Analytical method [refer to Figure 4.2]

$$AB = AC = \sqrt{(2)^2 + (1.5)^2} = 2.5 \text{ m}$$

$$\sin\theta = \frac{3}{2}, \quad \cos\theta = \frac{4}{5}.$$

Consider equilibrium of joint B,

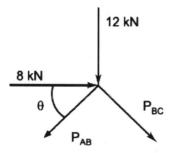

Figure 4.2(a). Analytical method for joint B in Example 4.1.

$$\sum F_x = 0 \text{ and } \sum F_y = 0$$

$$8 - P_{AB}\cos\theta + P_{BC}\cos\theta = 0 \qquad (4.3)$$

$$-12 - P_{AB}\sin\theta - P_{BC}\sin\theta = 0. \qquad (4.4)$$

Solving equations (4.3) and (4.4)

$$P_{AB} = -5 \text{ kN and } P_{BC} = -15 \text{ kN} \quad (P_{AB} \text{ and } P_{BC} \text{ are compressive}).$$

Consider equilibrium of joint A,

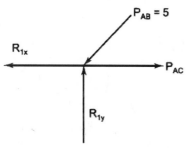

Figure 4.2(b). Analytical method for joint A in Example 4.1.

$$\sum F_x = 0 \text{ and } \sum F_y = 0$$
$$-R_{1x} + P_{AC} - P_{AB} \cos\theta = 0 \qquad (4.5)$$
$$R_{1y} - P_{AB} \sin\theta = 0.$$

Consider equilibrium of joint C,

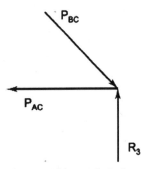

Figure 4.2(c). Analytical method for joint C in Example 4.1.

$$\sum F_x = 0 \text{ and } \sum F_y = 0$$
$$P_{AC} = P_{BC} \cos\theta = 15 \times \frac{4}{5} = 12 \text{ kN}$$
$$R_3 = P_{BC} \sin\theta = 15 \times \frac{3}{5} = 9 \text{ kN}.$$

For equation (4.5)

$$R_{1x} = P_{AC} - P_{AB} \cos\theta = 12 - 5 \times \frac{4}{5} = 8 \text{ kN}$$

$$\sigma_{AB} = \sigma_2 = \frac{P_{AB}}{A_{AB}} = \frac{-5 \times 10^3}{800} = -6.25 \text{ MPa (Compressive)}$$

$$\sigma_{BC} = \sigma_3 = \frac{P_{BC}}{A_{BC}} = \frac{-15 \times 10^3}{800} = -18.75 \text{ MPa (Compressive)}$$

$$\sigma_{AC} = \sigma_1 = \frac{P_{AC}}{A_{AC}} = \frac{12 \times 10^3}{800} = 15 \text{ MPa (Tensile)}$$

$$\Delta_{AB} = \frac{P_{AB} L_{AB}}{A_{AB} E_{AB}} = \frac{-5 \times 10^3 \times 2500}{800 \times 2 \times 10^5} = -0.078125 \text{ mm}$$

$$\Delta_{BC} = \frac{P_{BC}L_{BC}}{A_{BC}E_{BC}} = \frac{-15\times 10^3 \times 2500}{800\times 2\times 10^5} = -0.234375 \text{ mm}$$

$$\Delta_{AC} = \frac{P_{AC}L_{AC}}{A_{AC}E_{AC}} = \frac{12\times 10^3 \times 4000}{800\times 2\times 10^5} = 0.3 \text{ mm}.$$

Calculation of nodal displacements u_2, v_2, and u_3

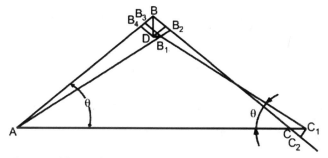

Figure 4.2(d). Analytical method for Calculation of nodal displacements u_2, v_2, and u_3 in Example 4.1.

$$DB_1 = u_2, \ BD = v_2 \text{ and } CC_1 = u_3$$
$$BB_3 = \Delta_{AB}$$
$$BB_2 = \Delta_{BC}$$
$$CC_1 = u_3 = \Delta_{AC} = 0.3 \text{ mm}$$
$$CC_2 = CC_1 \cos\theta = \Delta_{AC}\cos\theta = 0.3\times\frac{4}{3} = 0.24 \text{ mm}.$$

From geometry [refer to Figure 4.2(d)].

$$BB_3 = \Delta_{AB} = BD\sin\theta - DB_1\cos\theta = v_2\sin\theta - u_2\cos\theta \tag{4.6}$$
$$BB_2 = BC - B_2C = BC - (B_2C_2 - CC_2) = (BC - B_2C_2) + CC_2$$
$$BB_2 = \Delta_{BC} + CC_2 = BD\sin\theta + DB_1\cos\theta$$
$$\Delta_{BC} + CC_2 = v_2\sin\theta + u_2\cos\theta. \tag{4.7}$$

Substituting in equations (4.6) and (4.7)

$$0.078125 = v_2 \times \frac{3}{5} - u_2 \times \frac{4}{5} \tag{4.8}$$

$$0.234375 + 0.24 = v_2 \times \frac{3}{5} + u_2 \times \frac{4}{5}$$

$$0.474375 = v_2 \times \frac{3}{5} + u_2 \times \frac{4}{5}. \tag{4.9}$$

Solving equations (4.8) and (4.9), we get,

v_2 = 0.4604 mm (since point B moves downwards). Hence v_2 = −0.6404 mm.

(III) FEM by hand calculation [refer to Figure 4.2]

Elements	Node numbers		θ	cos θ	sin θ	L (mm)
	Local 1	Local 2				
1	1	3	0	1	0	4000
2	1	2	36.87	0.8	0.6	2500
3	2	3	−36.87	0.8	−0.6	2500

Angle calculation
For element 2

$$\sin\theta = \frac{1.5}{2.5} \Rightarrow \theta = 36.87°.$$

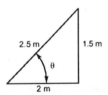

Figure 4.2(e). Angle calculation for element 2 in Example 4.1.

For element 3

$$\sin\theta = \frac{1.5}{2.5} \Rightarrow \theta = -36.87°.$$

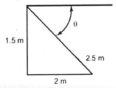

Figure 4.2(f). Angle calculation for element 3 in Example 4.1.

Finite Element Analysis Trusses

Element stiffness matrix for element 1 is,

$$[k_1] = \frac{AE}{L} \begin{bmatrix} \cos^2\theta & \cos^2\theta \times \sin\theta & -\cos^2\theta & -\cos\theta \times \sin\theta \\ \cos\theta \times \sin\theta & \sin^2\theta & -\cos\theta \times \sin\theta & -\sin^2\theta \\ -\cos^2\theta & -\cos\theta \times \sin\theta & \cos^2\theta & \cos\theta \times \sin\theta \\ -\cos\theta \times \sin\theta & -\sin^2\theta & \cos\theta \times \sin\theta & \sin^2\theta \end{bmatrix}$$

$$[k_1] = \frac{800 \times 200 \times 10^3}{4000} \begin{bmatrix} \cos^2 0 & \cos^2 0 \times \sin 0 & -\cos^2 0 & -\cos 0 \times \sin 0 \\ \cos 0 \times \sin 0 & \sin^2 0 & -\cos 0 \times \sin 0 & -\sin^2 0 \\ -\cos^2 0 & -\cos 0 \times \sin 0 & \cos^2 0 & \cos 0 \times \sin 0 \\ -\cos 0 \times \sin 0 & -\sin^2 0 & \cos 0 \times \sin 0 & \sin^2 0 \end{bmatrix}$$

$$[k_1] = 40 \times 10^3 \begin{matrix} & u_1 & v_1 & u_3 & v_3 & \\ & \begin{bmatrix} 1 & 0 & -1 & 0 \\ 0 & 0 & 0 & 0 \\ -1 & 0 & 1 & 0 \\ 0 & 0 & 0 & 0 \end{bmatrix} & \begin{matrix} u_1 \\ v_1 \\ u_3 \\ v_3 \end{matrix} \end{matrix}.$$

Element stiffness matrix for element 2 is,

$$[k_2] = \frac{AE}{L} \begin{bmatrix} c^2 & cs & -c^2 & -cs \\ cs & s^2 & -cs & -s^2 \\ -c^2 & -cs & c^2 & cs \\ -cs & -s^2 & cs & s^2 \end{bmatrix}$$

$$= \frac{800 \times 200 \times 10^3}{2500} \begin{bmatrix} (0.8)^2 & 0.8 \times 0.6 & -(0.8)^2 & -0.8 \times 0.6 \\ 0.8 \times 0.6 & (0.6)^2 & -0.8 \times 0.6 & -(0.6)^2 \\ -(0.8)^2 & -0.8 \times 0.6 & (0.8)^2 & 0.8 \times 0.6 \\ -0.8 \times 0.6 & -(0.6)^2 & 0.8 \times 0.6 & (0.6)^2 \end{bmatrix}$$

$$[k_2] = 64 \times 10^3 \begin{matrix} & u_1 & v_1 & u_2 & v_2 & \\ & \begin{bmatrix} 0.64 & 0.48 & -0.64 & -0.48 \\ 0.48 & 0.36 & -0.48 & -0.36 \\ -0.64 & 0.48 & 0.64 & 0.48 \\ -0.48 & -0.36 & 0.48 & 0.36 \end{bmatrix} & \begin{matrix} u_1 \\ v_1 \\ u_2 \\ v_2 \end{matrix} \end{matrix}.$$

Element stiffness matrix for element 3 is,

$$[k_3] = \frac{AE}{L}\begin{bmatrix} c^2 & cs & -c^2 & -cs \\ cs & s^2 & -cs & -s^2 \\ -c^2 & -cs & c^2 & cs \\ -cs & -s^2 & cs & s^2 \end{bmatrix} = 64 \times 10^3 \begin{bmatrix} 0.64 & -0.48 & -0.64 & 0.48 \\ -0.48 & 0.36 & 0.48 & -0.36 \\ -0.64 & 0.48 & 0.64 & -0.48 \\ 0.48 & -0.36 & -0.48 & 0.36 \end{bmatrix} \begin{matrix} u_2 \\ v_2 \\ u_3 \\ v_3 \end{matrix}$$

(columns: u_2, v_2, u_3, v_3)

Global stiffness matrix is,

$$[K] = 10^3 \begin{bmatrix} 40+40.96 & 30.72 & -40.96 & -30.72 & -40 & 0 \\ 30.72 & 23.04 & -30.72 & -23.04 & 0 & 0 \\ -40.96 & -30.72 & 40.96+40.96 & 30.72-30.72 & -40.96 & 30.72 \\ -30.72 & -23.04 & 30.72-30.72 & 23.04+23.04 & 30.72 & -23.04 \\ -40 & 0 & -40.96 & 30.72 & 40+40.96 & -30.72 \\ 0 & 0 & 30.72 & -23.04 & -30.72 & 23.04 \end{bmatrix} \begin{matrix} u_1 \\ v_1 \\ u_2 \\ v_2 \\ u_3 \\ v_3 \end{matrix}$$

(columns: u_1, v_1, u_2, v_2, u_3, v_3)

$$[K] = 10^3 \begin{bmatrix} 80.96 & 30.72 & -40.96 & -30.72 & -40 & 0 \\ 30.72 & 23.04 & -30.72 & -23.04 & 0 & 0 \\ -40.96 & -30.72 & 81.92 & 0 & -40.96 & 30.72 \\ -30.72 & -23.04 & 0 & 46.08 & 30.72 & -23.04 \\ -40 & 0 & -40.96 & 30.72 & 80.96 & -30.72 \\ 0 & 0 & 30.72 & -23.04 & -30.72 & 23.04 \end{bmatrix} \begin{matrix} u_1 \\ v_1 \\ u_2 \\ v_2 \\ u_3 \\ v_3 \end{matrix}$$

(columns: u_1, v_1, u_2, v_2, u_3, v_3)

Global equation is,

$$[K] = 10^3 \begin{bmatrix} 80.96 & 30.72 & -40.96 & -30.72 & -40 & 0 \\ 30.72 & 23.04 & -30.72 & -23.04 & 0 & 0 \\ -40.96 & -30.72 & 81.92 & 0 & -40.96 & 30.72 \\ -30.72 & -23.04 & 0 & 46.08 & 30.72 & -23.04 \\ -40 & 0 & -40.96 & 30.72 & 80.96 & -30.72 \\ 0 & 0 & 30.72 & -23.04 & -30.72 & 23.04 \end{bmatrix} \begin{Bmatrix} u_1 \\ v_1 \\ u_2 \\ v_2 \\ u_3 \\ v_3 \end{Bmatrix} = \begin{Bmatrix} R_{1x} \\ R_{1y} \\ 8 \\ -12 \\ 0 \\ R_{3y} \end{Bmatrix} \times 10^3.$$

Finite Element Analysis Trusses

Using the elimination method for applying boundary conditions,

$$u_1 = v_1 = v_3 = 0.$$

Then the above matrix reduces to,

$$10^3 \begin{bmatrix} 81.92 & 0 & -40.96 \\ 0 & 46.08 & 30.72 \\ -40.96 & 30.72 & 80.96 \end{bmatrix} \begin{bmatrix} u_2 \\ v_2 \\ u_3 \end{bmatrix} = \begin{bmatrix} 8 \\ -12 \\ 0 \end{bmatrix} \times 10^3.$$

Solving the above matrix and equations,

we get $\quad u_2 = 0.2477$ mm, $v_2 = -0.4604$ mm, and $v_3 = 0.3$ mm.

Stress calculation

Stress in element 1 is,

$$\sigma_1 = \frac{E}{L_1}[-\cos\theta \quad -\sin\theta \quad \cos\theta \quad \sin\theta]\begin{bmatrix} u_1 \\ v_1 \\ u_3 \\ v_3 \end{bmatrix} = \frac{200\times 10^3}{4000}[-c \quad -s \quad c \quad s]\begin{bmatrix} u_1 \\ v_1 \\ u_3 \\ v_3 \end{bmatrix}$$

$$\sigma_1 = \frac{200\times 10^3}{4000}[-1 \quad 0 \quad 1 \quad 0]\begin{bmatrix} 0 \\ 0 \\ 0.3 \\ 0 \end{bmatrix} = 15 \text{ MPa.}$$

Stress in element 2 is,

$$\sigma_2 = \frac{E}{L_2}[-\cos\theta \quad -\sin\theta \quad \cos\theta \quad \sin\theta]\begin{bmatrix} u_1 \\ v_1 \\ u_2 \\ v_2 \end{bmatrix} = \frac{200\times 10^3}{2500}[-c \quad -s \quad c \quad s]\begin{bmatrix} u_1 \\ v_1 \\ u_2 \\ v_2 \end{bmatrix}$$

$$\sigma_2 = \frac{200\times 10^3}{2500}[-0.8 \quad -0.6 \quad 0.8 \quad 0.6]\begin{bmatrix} 0 \\ 0 \\ 0.2477 \\ -0.4604 \end{bmatrix} = -6.249 \text{ MPa.}$$

Stress in element 3 is,

$$\sigma_2 = \frac{E}{L_3}[-\cos\theta \quad -\sin\theta \quad \cos\theta \quad \sin\theta]\begin{bmatrix}u_2\\v_2\\u_3\\v_3\end{bmatrix} = \frac{200\times 10^3}{2500}[-c \quad -s \quad c \quad s]\begin{bmatrix}u_2\\v_2\\u_3\\v_3\end{bmatrix}$$

$$\sigma_3 = \frac{200\times 10^3}{2500}[-0.8 \quad 0.6 \quad 0.8 \quad -0.6]\begin{bmatrix}0.2477\\-0.4604\\0.3\\0\end{bmatrix} = -18.752 \text{ MPa.}$$

Reaction calculation

From global equation,

$$-40.96\times u_2 - 30.72\times v_2 - 40\times u_3 = R_{1x}$$
$$-40.96\times 0.2477 - 30.72\times(-0.4604) - 40\times 0.3 = R_{1x}$$
$$R_{1x} = -8 \text{ kN}$$

$$-30.72\times u_2 - 23.04\times v_2 = R_{1y}$$
$$-30.72\times 0.2477 - 23.04\times(-0.4604) = R_{1y}$$
$$R_{1y} = 3 \text{ kN}$$

$$-30.72\times u_2 - 23.04\times v_2 - 30.72\times u_3 = R_{3y}$$
$$-30.72\times 0.2477 - 23.04\times(-0.4604) - 30.72\times(0.3) = R_{3y}$$
$$R_{3y} = 9 \text{ kN.}$$

(III) Software results

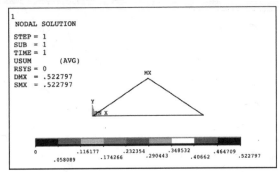

Figure 4.2(g). Deflection pattern for a truss for Example 4.1 (refer to Appendix C for color figures).

Deflection value at nodes

The following degree of freedom results are in global coordinates

NODE	UX	UY	UZ	USUM
1	0.0000	0.0000	0.0000	0.0000
2	0.24766	−0.46042	0.0000	0.52280
3	0.30000	0.0000	0.0000	0.30000

Maximum absolute values

NODE	3	2	0	2
VALUE	0.30000	−0.46042	0.0000	0.52280

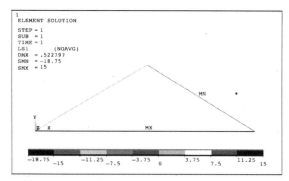

Figure 4.2(h). Stress pattern for a truss for Example 4.1 (refer to Appendix C for color figures).

Stress values of elements

STAT ELEM	CURRENT LS1
1	15.000
2	−6.2500
3	−18.750

Reaction values

The following X, Y, Z solutions are in global coordinates

NODE	FX	FY
1	−8000.0	3000.0
3		9000.0

Answers for Example 4.1

Parameter	Analytical method	FEM-Hand calculations	Software results
Displacement of node 2 in			
x-direction	0.2477 mm	0.2477 mm	0.24766 mm
y-direction	−0.4604 mm	−0.4604 mm	−0.46042 mm
Displacement of node 3 in			
x-direction	0.3 mm	0.3 mm	0.3 mm
Stress in			
Element 1	15 MPa	15 MPa	15 MPa
Element 2	−6.25 MPa	−6.248 MPa	−6.25 MPa
Element 3	−18.75 MPa	−18.752 MPa	−18.75 MPa
Reaction			
At 1 in x-direction	−8 kN	−8 kN	−8 kN
At 1 in y-direction	3 kN	3 kN	3 kN
At 3 in y-direction	9 kN	9 kN	9 kN

Example 4.2

For the truss shown in Figure 4.3, determine nodal displacements and stresses in each member. All elements have E = 200 GPa and A = 500 mm².

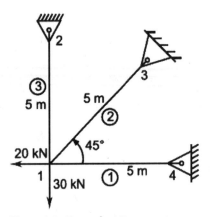

Figure 4.3. Example 4.2

Solution

$$E = 2 \times 10^5 \text{ N/mm}^2$$
$$A = 500 \text{ mm}^2.$$

(I) FEM by hand calculation

Elements	Node numbers		θ	$\cos \theta$	$\sin \theta$	L (mm)
	Local 1	Local 2				
1	1	4	0	1	0	5000
2	1	3	45	0.707	0.707	5000
3	1	2	90	0	1	5000

Stiffness matrices for elements 1, 2, and 3 are,

$$[k_1] = \frac{AE}{L}\begin{bmatrix} c^2 & cs & -c^2 & -cs \\ cs & s^2 & -cs & -s^2 \\ -c^2 & -cs & c^2 & cs \\ -cs & -s^2 & cs & s^2 \end{bmatrix} = \frac{500 \times 2 \times 10^5}{5000}\begin{bmatrix} 1 & 0 & -1 & 0 \\ 0 & 0 & 0 & 0 \\ -1 & 0 & 1 & 0 \\ 0 & 0 & 0 & 0 \end{bmatrix}$$

$$[k_1] = 20 \times 10^3 \begin{bmatrix} u_1 & v_1 & u_4 & v_4 \\ 1 & 0 & -1 & 0 \\ 0 & 0 & 0 & 0 \\ -1 & 0 & 1 & 0 \\ 0 & 0 & 0 & 0 \end{bmatrix}\begin{matrix} u_1 \\ v_1 \\ u_4 \\ v_4 \end{matrix}$$

$$[k_2] = 20 \times 10^3 \begin{bmatrix} u_1 & v_1 & u_3 & v_3 \\ 0.5 & 0.5 & -0.5 & -0.5 \\ 0.5 & 0.5 & -0.5 & -0.5 \\ -0.5 & -0.5 & 0.5 & 0.5 \\ -0.5 & -0.5 & 0.5 & 0.5 \end{bmatrix}\begin{matrix} u_1 \\ v_1 \\ u_3 \\ v_3 \end{matrix}$$

$$[k_3] = 20 \times 10^3 \begin{bmatrix} u_1 & v_1 & u_2 & v_2 \\ 0 & 0 & 0 & 0 \\ 0 & 1 & 0 & -1 \\ 0 & 0 & 0 & 0 \\ 0 & -1 & 0 & 1 \end{bmatrix}\begin{matrix} u_1 \\ v_1 \\ u_2 \\ v_2 \end{matrix}.$$

Global stiffness matrix is,

$$[K] = 20 \times 10^3 \begin{bmatrix} 1+0.5 & 0.5 & 0 & 0 & -0.5 & -0.5 & -1 & 0 \\ 0.5 & 0.5+1 & 0 & -1 & -0.5 & -0.5 & 0 & 0 \\ 0 & 0 & 0 & 0 & 0 & 0 & 0 & 0 \\ 0 & -1 & 0 & 1 & 0 & 0 & 0 & 0 \\ -0.5 & -0.5 & 0 & 0 & 0.5 & 0.5 & 0 & 0 \\ -0.5 & -0.5 & 0 & 0 & 0.5 & 0.5 & 0 & 0 \\ -1 & 0 & 0 & 0 & 0 & 0 & 1 & 0 \\ 0 & 0 & 0 & 0 & 0 & 0 & 0 & 0 \end{bmatrix} \begin{matrix} u_1 \\ v_1 \\ u_2 \\ v_2 \\ u_3 \\ v_3 \\ u_4 \\ v_4 \end{matrix}$$

with column headers $u_1\ v_1\ u_2\ v_2\ u_3\ v_3\ u_4\ v_4$.

$$[K] = 20 \times 10^3 \begin{bmatrix} 1.5 & 0.5 & 0 & 0 & -0.5 & -0.5 & -1 & 0 \\ 0.5 & 1.5 & 0 & -1 & -0.5 & -0.5 & 0 & 0 \\ 0 & 0 & 0 & 0 & 0 & 0 & 0 & 0 \\ 0 & -1 & 0 & 1 & 0 & 0 & 0 & 0 \\ -0.5 & -0.5 & 0 & 0 & 0.5 & 0.5 & 0 & 0 \\ -0.5 & -0.5 & 0 & 0 & 0.5 & 0.5 & 0 & 0 \\ -1 & 0 & 0 & 0 & 0 & 0 & 1 & 0 \\ 0 & 0 & 0 & 0 & 0 & 0 & 0 & 0 \end{bmatrix} \begin{matrix} u_1 \\ v_1 \\ u_2 \\ v_2 \\ u_3 \\ v_3 \\ u_4 \\ v_4 \end{matrix}.$$

Global equation is,

$$20 \times 10^3 \begin{bmatrix} 1+0.5 & 0.5 & 0 & 0 & -0.5 & -0.5 & -1 & 0 \\ 0.5 & 0.5+1 & 0 & -1 & -0.5 & -0.5 & 0 & 0 \\ 0 & 0 & 0 & 0 & 0 & 0 & 0 & 0 \\ 0 & -1 & 0 & 1 & 0 & 0 & 0 & 0 \\ -0.5 & -0.5 & 0 & 0 & 0.5 & 0.5 & 0 & 0 \\ -0.5 & -0.5 & 0 & 0 & 0.5 & 0.5 & 0 & 0 \\ -1 & 0 & 0 & 0 & 0 & 0 & 1 & 0 \\ 0 & 0 & 0 & 0 & 0 & 0 & 0 & 0 \end{bmatrix} \begin{Bmatrix} u_1 \\ v_1 \\ u_2 \\ v_2 \\ u_3 \\ v_3 \\ u_4 \\ v_4 \end{Bmatrix} = \begin{Bmatrix} -20 \\ -30 \\ R_{2x} \\ R_{2y} \\ R_{3x} \\ R_{3y} \\ R_{4x} \\ R_{4y} \end{Bmatrix} \times 10^3.$$

Using the elimination method for applying boundary conditions,

$$u_2 = v_2 = u_3 = v_3 = u_4 = v_4 = 0.$$

Then the above matrix reduces to,

$$20\begin{bmatrix} 1.5 & 0.5 \\ 0.5 & 1.5 \end{bmatrix}\begin{bmatrix} u_1 \\ v_1 \end{bmatrix} = \begin{bmatrix} -20 \\ -30 \end{bmatrix}.$$

Solving the above matrix and equations, we get,

$$u_1 = -0.375 \text{ mm}$$
$$v_1 = -0.875 \text{ mm}.$$

Stress calculation

Stress in element 1 is,

$$\sigma_1 = \frac{E}{L_1}[-\cos\theta \quad -\sin\theta \quad \cos\theta \quad \sin\theta]\begin{bmatrix} u_1 \\ v_1 \\ u_4 \\ v_4 \end{bmatrix} = \frac{2\times 10^5}{5000}[-c \quad -s \quad c \quad s]\begin{bmatrix} u_1 \\ v_1 \\ u_4 \\ v_4 \end{bmatrix}$$

$$\sigma_1 = \frac{2\times 10^5}{5000}[-1 \quad 0 \quad 1 \quad 0]\begin{bmatrix} -0.375 \\ -0.875 \\ 0 \\ 0 \end{bmatrix} = 15 \text{ MPa}.$$

Stress in element 2 is,

$$\sigma_2 = \frac{E}{L_2}[-\cos\theta \quad -\sin\theta \quad \cos\theta \quad \sin\theta]\begin{bmatrix} u_1 \\ v_1 \\ u_3 \\ v_3 \end{bmatrix} = \frac{2\times 10^5}{5000}[-c \quad -s \quad c \quad s]\begin{bmatrix} u_1 \\ v_1 \\ u_3 \\ v_3 \end{bmatrix}$$

$$\sigma_2 = \frac{2\times 10^5}{5000}[-0.707 \quad -0.707 \quad 0.707 \quad 0.707]\begin{bmatrix} -0.375 \\ -0.875 \\ 0 \\ 0 \end{bmatrix} = 35.352 \text{ MPa}.$$

Stress in element 3 is,

$$\sigma_3 = \frac{E}{L_3}[-\cos\theta \quad -\sin\theta \quad \cos\theta \quad \sin\theta]\begin{bmatrix} u_1 \\ v_1 \\ u_2 \\ v_2 \end{bmatrix} = \frac{2\times 10^5}{5000}[-c \quad -s \quad c \quad s]\begin{bmatrix} u_1 \\ v_1 \\ u_2 \\ v_2 \end{bmatrix}$$

$$\sigma_3 = \frac{2\times 10^5}{5000}[0 \quad -1 \quad 0 \quad 1]\begin{bmatrix} -0.375 \\ -0.875 \\ 0 \\ 0 \end{bmatrix} = 35 \text{ MPa}.$$

(II) Software results

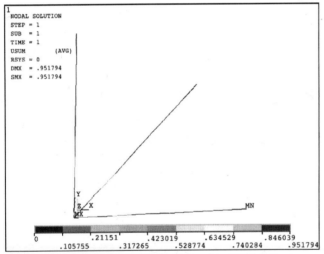

Figure 4.3(a). **Deflection pattern for a truss for Example 4.2 (refer to Appendix C for color figures).**

Deflection value at nodes

The following degree of freedom results are in global coordinates system

NODE	UX	UY	UZ	USUM
1	−0.37486	−0.87486	0.0000	0.95179
2	0.0000	0.0000	0.0000	0.0000
3	0.0000	0.0000	0.0000	0.0000
4	0.0000	0.0000	0.0000	0.0000

Finite Element Analysis Trusses

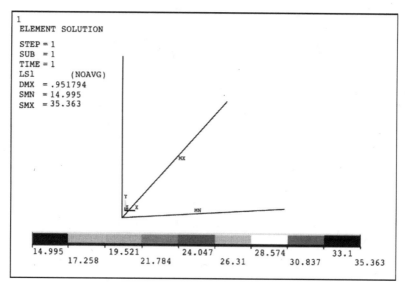

Figure 4.3(b). Stress pattern for a truss for Example 4.2 (refer to Appendix C for color figures).

Stress values of elements

STAT ELEM	CURRENT LS1
1	14.995
2	35.363
3	34.995

Answers for Example 4.2

Parameter	FEM-hand calculations	Software results
Displacement of node 1 in		
x-direction	−0.375 mm	−0.37486 mm
y-direction	−0.875 mm	−0.87486 mm
Stress in Element 1	15 MPa	14.995 MPa
Stress in Element 2	35.352 MPa	35.363 MPa
Stress in Element 3	35 MPa	34.995 MPa

Example 4.3

The bar truss shown in Figure 4.4, determine the displacement of node 1 and the axial stress in each member. Take E = 210 GPa and A = 600 mm². Solve the problem if node 1 settles an amount of δ = 25 mm in the negative y-direction.

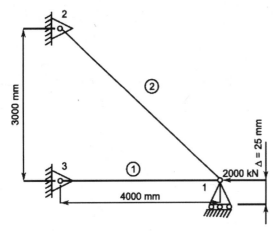

Figure 4.4. Example 4.3

Solution

(I) FEM by hand calculation

Elements	Node numbers		θ	$\cos \theta$	$\sin \theta$	L (mm)
	Local 1	Local 2				
1	3	1	0	1	0	4000
2	2	1	−36.87	0.8	−0.6	5000

Angle calculation

For 2nd element,

$$\sin \theta = \frac{3}{5} = 0.6 \Rightarrow \theta = -36.87°.$$

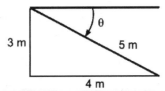

Figure 4.4(a). Angle calculation for 2nd element for Example 4.3.

Finite Element Analysis Trusses

Stiffness matrices for element 1 and 2 are,

$$[k_1] = \frac{AE}{L}\begin{bmatrix} c^2 & cs & -c^2 & -cs \\ cs & s^2 & -cs & -s^2 \\ -c^2 & -cs & c^2 & cs \\ -cs & -s^2 & cs & s^2 \end{bmatrix} = \frac{600 \times 210 \times 10^3}{4000}\begin{bmatrix} 1 & 0 & -1 & 0 \\ 0 & 0 & 0 & 0 \\ -1 & 0 & 1 & 0 \\ 0 & 0 & 0 & 0 \end{bmatrix}$$

$$[k_1] = 31.5 \times 10^3 \begin{bmatrix} u_3 & v_3 & u_1 & v_1 \\ 1 & 0 & -1 & 0 \\ 0 & 0 & 0 & 0 \\ -1 & 0 & 1 & 0 \\ 0 & 0 & 0 & 0 \end{bmatrix}\begin{matrix} u_3 \\ v_3 \\ u_1 \\ v_1 \end{matrix}$$

$$[k_2] = \frac{AE}{L}\begin{bmatrix} c^2 & cs & -c^2 & -cs \\ cs & s^2 & -cs & -s^2 \\ -c^2 & -cs & c^2 & cs \\ -cs & -s^2 & cs & s^2 \end{bmatrix} = \frac{600 \times 210 \times 10^3}{5000}\begin{bmatrix} 0.64 & -0.48 & -0.64 & 0.48 \\ -0.48 & 0.36 & 0.48 & -0.36 \\ -0.64 & 0.48 & 0.64 & -0.48 \\ 0.48 & -0.36 & -0.48 & 0.36 \end{bmatrix}$$

$$[k_2] = 25.2 \times 10^3 \begin{bmatrix} u_2 & v_2 & u_1 & v_1 \\ 0.64 & -0.48 & -0.64 & 0.48 \\ -0.48 & 0.36 & 0.48 & -0.36 \\ -0.64 & 0.48 & 0.64 & -0.48 \\ 0.48 & -0.36 & -0.48 & 0.36 \end{bmatrix}\begin{matrix} u_2 \\ v_2 \\ u_1 \\ v_1 \end{matrix}$$

Global stiffness matrix is,

$$[K] = 10^3 \begin{bmatrix} u_1 & v_1 & u_2 & v_2 & u_3 & v_3 \\ 31.5+16.13 & -12.1 & -16.13 & 12.1 & -31.5 & 0 \\ -12.1 & 9.1 & 12.1 & -9.1 & 0 & 0 \\ -16.13 & 12.1 & 16.13 & -12.1 & 0 & 0 \\ 12.1 & -9.1 & -12.1 & 9.1 & 0 & 0 \\ -31.5 & 0 & 0 & 0 & 31.5 & 0 \\ 0 & 0 & 0 & 0 & 0 & 0 \end{bmatrix}\begin{matrix} u_1 \\ v_1 \\ u_2 \\ v_2 \\ u_3 \\ v_3 \end{matrix}$$

$$[K] = 10^3 \begin{bmatrix} & u_1 & v_1 & u_2 & v_2 & u_3 & v_3 \\ 47.63 & -12.1 & -16.13 & 12.1 & -31.5 & 0 \\ -12.1 & 9.1 & 12.1 & -9.1 & 0 & 0 \\ -16.13 & 12.1 & 16.13 & -12.1 & 0 & 0 \\ 12.1 & -9.1 & -12.1 & 9.1 & 0 & 0 \\ -31.5 & 0 & 0 & 0 & 31.5 & 0 \\ 0 & 0 & 0 & 0 & 0 & 0 \end{bmatrix} \begin{matrix} u_1 \\ v_1 \\ u_2 \\ v_2 \\ u_3 \\ v_3 \end{matrix}$$

Global equation is,

$$[K] = 10^3 \begin{bmatrix} 47.63 & -12.1 & -16.13 & 12.1 & -31.5 & 0 \\ -12.1 & 9.1 & 12.1 & -9.1 & 0 & 0 \\ -16.13 & 12.1 & 16.13 & -12.1 & 0 & 0 \\ 12.1 & -9.1 & -12.1 & 9.1 & 0 & 0 \\ -31.5 & 0 & 0 & 0 & 31.5 & 0 \\ 0 & 0 & 0 & 0 & 0 & 0 \end{bmatrix} \begin{Bmatrix} u_1 \\ v_1 \\ u_2 \\ v_2 \\ u_3 \\ v_3 \end{Bmatrix} = \begin{Bmatrix} -2000 \\ R_{1y} \\ R_{2x} \\ R_{2y} \\ R_{3x} \\ R_{3y} \end{Bmatrix} 10^3.$$

Using the elimination method for applying boundary conditions,

i.e., $\quad u_2 = v_2 = u_3 = v_3 = 0.$

Then the above matrix reduces to,

$$\begin{bmatrix} 47.63 & -12.1 \\ -12.1 & 9.1 \end{bmatrix} \begin{bmatrix} u_1 \\ v_1 \end{bmatrix} = \begin{bmatrix} -2000 \\ R_{1y} \end{bmatrix}.$$

We know that $v_1 = -25$ mm, substitute this in the above matrix, then,

$$\begin{bmatrix} 47.63 & -12.1 \\ -12.1 & 9.1 \end{bmatrix} \begin{bmatrix} u_1 \\ -25 \end{bmatrix} = \begin{bmatrix} -2000 \\ R_{1y} \end{bmatrix}.$$

Solving the above matrix and equations we get,

$$u_1 = -48.34 \text{ mm}.$$

Stress calculation

Stress in element 1 is,

$$\sigma_1 = \frac{E}{L_1}[-\cos\theta \quad -\sin\theta \quad \cos\theta \quad \sin\theta]\begin{bmatrix}u_3\\v_3\\u_1\\v_1\end{bmatrix} = \frac{210\times 10^3}{4000}[-c \quad -s \quad c \quad s]\begin{bmatrix}u_3\\v_3\\u_1\\v_1\end{bmatrix}$$

$$\sigma_1 = \frac{210\times 10^3}{4000}[-1 \quad 0 \quad 1 \quad 0]\begin{bmatrix}0\\0\\-48.34\\-25\end{bmatrix} = -2537.85 \text{ MPa}.$$

Stress in element 2 is,

$$\sigma_2 = \frac{E}{L_2}[-\cos\theta \quad -\sin\theta \quad \cos\theta \quad \sin\theta]\begin{bmatrix}u_2\\v_2\\u_1\\v_1\end{bmatrix} = \frac{210\times 10^3}{5000}[-c \quad -s \quad c \quad s]\begin{bmatrix}u_2\\v_2\\u_1\\v_1\end{bmatrix}$$

$$\sigma_1 = \frac{210\times 10^3}{5000}[-0.8 \quad 0.6 \quad 0.8 \quad -0.6]\begin{bmatrix}0\\0\\-48.34\\-25\end{bmatrix} = -994.22 \text{ MPa}.$$

(II) Software results

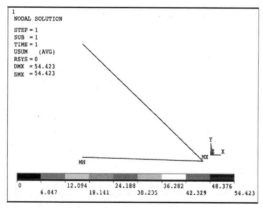

Figure 4.4(b). Deflection pattern for a truss for Example 4.3 (refer to Appendix C for color figures).

Deflection value at nodes

The following degree of freedom results are in global coordinates system

NODE	UX	UY	UZ	USUM
1	−48.341	−25.000	0.0000	54.423
2	0.0000	0.0000	0.0000	0.0000
3	0.0000	0.0000	0.0000	0.0000

Maximum absolute values

NODE	1	1	0	1
VALUE	−48.341	−25.000	0.0000	54.423

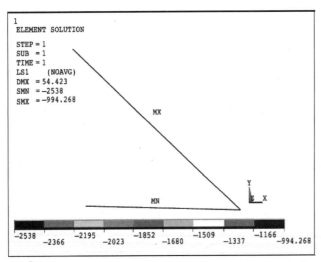

Figure 4.4(c). Stress pattern for a truss for Example 4.3 (refer to Appendix C for color figures).

Stress values of elements

STAT ELEM	CURRENT LS1
1	−2537.9
2	−994.27

Answers for Example 4.3

Parameter	FEM-hand calculations	Software results
Displacement of node 1 in		
x-direction	−48.34 mm	−48.341 mm
y-direction	−25 mm	−25 mm
Stress in Element 1	−2537.85 MPa	−2537.9 MPa
Stress in Element 2	−994.22 MPa	−994.27 MPa

Procedure for solving the problems using ANSYS® 12.0 academic teaching software
For Example 4.3

PREPROCESSING

1. **Main Menu > Preprocessor > Element Type > Add/Edit/Delete > Add > Structural Link > 2D spar 1 > OK > Close**

Figure 4.5. Element selection.

2. **Main Menu > Preprocessor > Real Constants > Add/Edit/Delete > Add > OK**

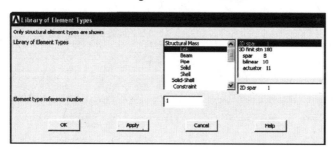

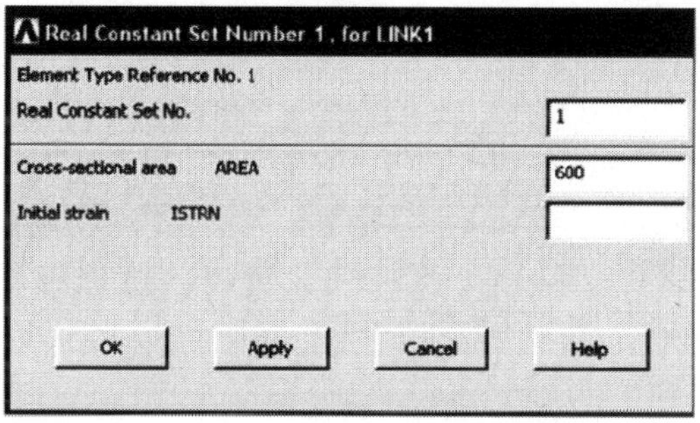

Figure 4.6. Enter the cross-sectional area.

Cross-sectional area AREA > **Enter 600 > OK > Close**

Enter the material properties.
3. **Main Menu > Preprocessor > Material Props > Material Models**
   Material Model Number 1, Click **Structural > Linear > Elastic > Isotropic**
   Enter **EX = 2.1E5 and PRXY = 0.3 > OK**
   (**Close** the Define Material Model Behavior window.)
   Create the nodes and elements as shown in the figure.
4. **Main Menu > Preprocessor > Modeling > Create > Nodes > In Active CS**
   Enter the coordinates of node 1 > **Apply** Enter the coordinates of node 2 > **Apply** Enter the coordinates of node 3 > **OK**

	Node locations	
Node number	X-coordinate	Y-coordinate
1	0	0
2	−4000	3000
3	−4000	0

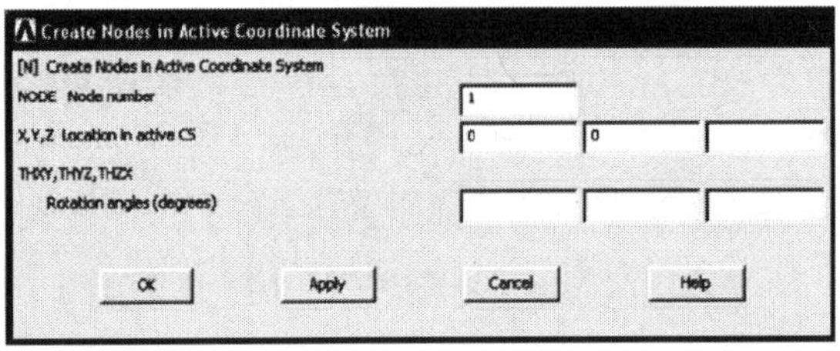

Figure 4.7. Enter the node coordinates.

5. **Main Menu > Preprocessor > Modeling > Create > Elements > Auto Numbered > Thru nodes** Pick the 1st and 2nd node > **Apply** Pick the 1st and 3rd node > **OK**

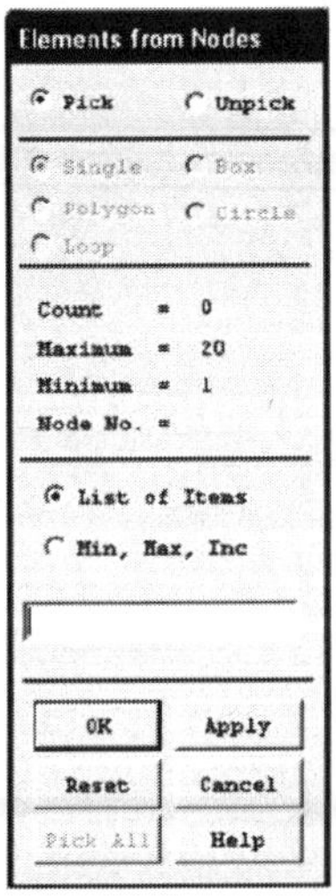

Figure 4.8. Pick the nodes to create elements.

Apply the displacement boundary conditions and loads.
6. **Main Menu > Preprocessor > Loads > Define Loads > Apply > Structural > Displacement > On Nodes** Pick the 2nd and 3rd node > **Apply > All DOF=0 > OK**
7. **Main Menu > Preprocessor > Loads > Define Loads > Apply > Structural > Displacement > On Nodes** Pick the 1st node > **Apply > UY=-25 > OK**
8. **Main Menu > Preprocessor > Loads > Define Loads > Apply > Structural > Force/Moment > On Nodes** Pick the 1st node > **OK > Force/Moment value= -2000e3 > OK**

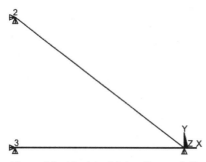

Figure 4.9. Model with loading and displacement boundary conditions.

The model-building step is now complete, and we can proceed to the solution. First to be safe, save the model.

Solution

The interactive solution proceeds.
9. **Main Menu > Solution > Solve > Current LS > OK**
The **/STATUS Command** window displays the problem parameters and **the Solve Current Load Step** window is shown. Check the solution options in the **/STATUS** window and if all is OK, select **File > Close.**
In the **Solve Current Load Step** window, select **OK,** and the solution is complete, **close** the '**Solution is Done!'** window.

POST-PROCESSING

We can now plot the results of this analysis and also list the computed values.
10. **Main Menu > General Postproc > Plot Results > Contour Plot > Nodal Solu > DOF Solution > Displacement vector sum > OK**
This result is shown in Figure 4.4(b).

To find the axial stress, the following procedure is followed.

11. MAIN Menu > General Postproc > Element Table > Define Table > Add

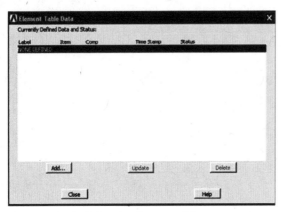

Figure 4.10. Defining the element table.

Select **By sequence num and LS** and type 1 **after LS** (as shown in Figure 4.11) >**OK**

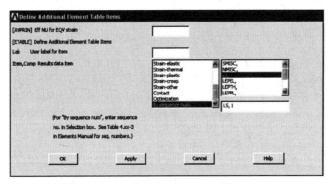

Figure 4.11. Selecting options in element table.

12. Main Menu > General Postproc > Plot Results > Contour Plot > Elem Table > Select LS1 > OK

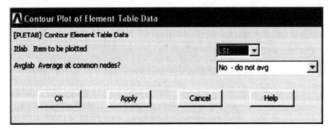

Figure 4.12. Selecting options for finding out axial stress.

This result is shown in Figure 4.4(c).

PROBLEMS

1. For a 5 bar truss shown in Figure 4.13, determine the following:
   (a) nodal displacements
   (b) stresses in each element
   (c) reaction forces.

   Take E = 200 GPa and Area A = 750 mm² for all elements.

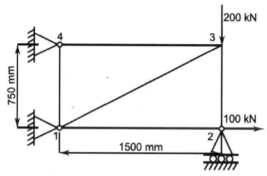

Figure 4.13. Problem 1

2. For the 3 bar truss shown in Figure 4.14, determine the displacement of node 1 and the stresses in elements. Take A = 300 mm² and E = 210 GPa.

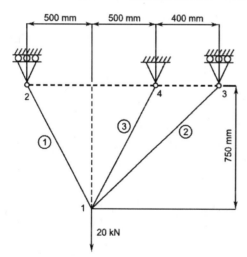

Figure 4.14. Problem 2

3. Consider the truss shown in Figure 4.15, determine the nodal displacements, element stresses, and reactions. Take E = 200 GPa. $A_1 = A_2 = A_3 = 500$ mm², $P_1 = 300$ kN, $P_2 = 200$ kN.

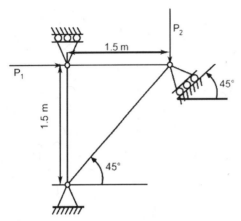

Figure 4.15. Problem 3

4. Consider the truss structure shown in Figure 4.16, determine the stresses of the truss structure. Take all members have elastic modulus (E) of 210 GPa and cross-sectional area (A) of 250 mm².

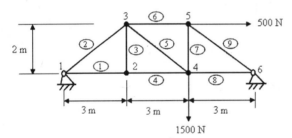

Figure 4.16. Problem 4

5. Consider the truss structure shown in Figure 4.17, derive the finite element matrix equations using 2 elements. Determine the displacements and the stresses in the member. Assume all members have elastic modulus (E) of 200 GPa and cross-sectional area (A) of 300 mm².

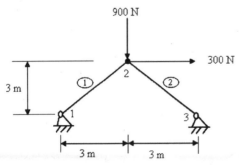

Figure 4.17. Problem 5

6. Consider the truss structure shown in Figure 4.18, determine the nodal displacement and the element forces assuming that all elements have the same AE.

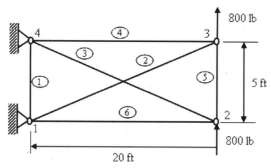

Figure 4.18. Problem 6

7. Determine the nodal displacements, element stresses, and support reactions for the 3 member truss shown in Figure 4.19. Take $A_1 = 10$ in², $A_2 = 15$ in², $A_3 = 10$ in² and $E = 20$ msi for all members.

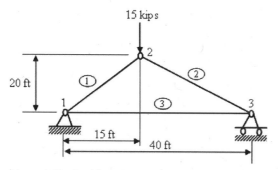

Figure 4.19. Problem 7

8. Determine the nodal displacements, element stresses and support reactions for the three member truss shown in Figure 4.20. Take $A_1 = 1$ in², $A_2 = 2$ in², $A_3 = 3$ in², and $E = 30$ Mlb/in² for all members.

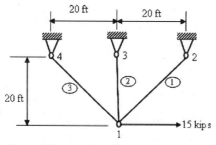

Figure 4.20. Problem 8

9. Determine the nodal displacements, element stresses, and support reactions for the 3 member truss shown in Figure 4.21. Take $A_1 = 6$ cm², $A_2 = 8$ cm², $A_2 = 8$ cm², and $E = 20$ MN/cm² for all members.

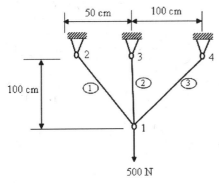

Figure 4.21. Problem 9

REFERENCES

1. Y. W. Hwon and H. Bang, "The Finite Element Method Using MATLAB, Second Edition," CRC Press, 2000.
2. D. L. Logan, "A First Course in the Finite Element Method, Fifth Edition," Cengage Learning, 2012.
3. S. Moaveni, "Finite Element Analysis: Theory and Application with ANSYS, Third Edition," Prentice Hall, 2008.
4. J. N. Reddy, "An Introduction to the Finite Element Method, Third Edition," McGraw Hill Higher Education, 2004.
5. C. T. F. Ross, "Finite Element Method in Structural Mechanics," Ellis Horwood Limited Publishers, 1985.
6. F. L. Stasa, "Applied Finite Element Analysis for Engineering," Holt, Rinehart and Winston, 1985.
7. L. J. Segerlind, "Applied Finite Element Analysis, Second Edition," John Wiley and Sons, 1984.

Chapter 5
FINITE ELEMENT ANALYSIS OF BEAMS

5.1 INTRODUCTION

Beam is very common structure in many engineering applications because of its efficient load carrying capability. Beam by definition is a transversely loaded structural member. Beam element is used in the analysis of beams.

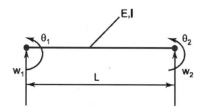

Figure 5.1. Beam element.

This element has 2 end nodes each having 2 degrees of freedom, namely transverse displacement and slope. Beam element gives accurate results if acted upon by nodal forces and moments. A greater number of small elements will be necessary in the case of a beam acted upon by distributed loads in order to get good results. The interpolation equation and element stiffness matrix for beam element are given by

$$w = \begin{bmatrix} N_1 & N_2 & N_3 & N_4 \end{bmatrix} \begin{Bmatrix} w_1 \\ \theta_1 \\ w_2 \\ \theta_2 \end{Bmatrix} \tag{5.1}$$

175

$$[K] = \frac{EI}{L^3} \begin{bmatrix} 12 & 6L & -12 & 6L \\ 6L & 4L^2 & -6L & 2L^2 \\ -12 & -6L & 12 & -6L \\ 6L & 2L^2 & -6L & 4L^2 \end{bmatrix}. \tag{5.2}$$

5.2 SIMPLY SUPPORTED BEAMS

Example 5.1

For the beam shown in Figure 5.2, determine the nodal displacements, slope, and reactions. Take $E = 210$ GPa and $I = 4 \times 10^{-4}$ m^4.

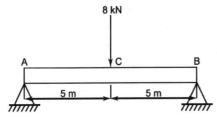

Figure 5.2. The beam for Example 5.1.

Solution

(I) Analytical method [refer to Figure 5.2]

$$L = 10 \text{ m}$$
$$P = 8 \text{ kN}.$$

Deflection,

$$\delta_C = -\frac{PL^3}{48EI} = -\frac{8 \times 10^3 \times (10)^3}{48 \times 210 \times 10^9 \times 4 \times 10^{-4}} = -1.98 \times 10^{-3} \text{ m} = -1.98 \text{ mm}$$

$$|\theta_C| = |\theta_B| = \frac{PL^2}{16EI} = \frac{8 \times 10^3 \times (10)^2}{16 \times 210 \times 10^9 \times 4 \times 10^{-4}} = 5.95 \times 10^{-4} \text{ rad}$$

$\theta_C = 0$, by symmetry.
Reaction,

$$R_A = R_B = \frac{8}{2} = 4 \text{ kN}.$$

(II) FEM by hand calculations [refer to Figure 5.2(a)]

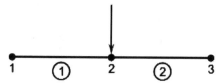

Figure 5.2(a). Finite element model for Example 5.1.

Element stiffness matrices are,

$$[k] = \frac{EI}{L^3}\begin{bmatrix} 12 & 6L & -12 & 6L \\ 6L & 4L^2 & -6L & 2L^2 \\ -12 & -6L & 12 & -6L \\ 6L & 2L^2 & -6L & 4L^2 \end{bmatrix}$$

$$[k_1] = \frac{210 \times 10^9 \times 4 \times 10^{-4}}{(5)^3}\begin{bmatrix} 12 & 6(5) & -12 & 6(5) \\ 6(5) & 4(5)^2 & -6(5) & 2(5)^2 \\ -12 & -6(5) & 12 & -6(5) \\ 6(5) & 2(5)^2 & -6(5) & 4(5)^2 \end{bmatrix}$$

$$[k_1] = 672 \times 10^3 \begin{bmatrix} 12 & 30 & -12 & 30 \\ 30 & 100 & -30 & 50 \\ -12 & -30 & 12 & -30 \\ 30 & 50 & -30 & 100 \end{bmatrix}\begin{matrix} w_1 \\ \theta_1 \\ w_2 \\ \theta_2 \end{matrix}$$

Due to symmetry,

$$[k_1] = [k_2]$$

$$[k_2] = 672 \times 10^3 \begin{bmatrix} 12 & 30 & -12 & 30 \\ 30 & 100 & -30 & 50 \\ -12 & -30 & 12 & -30 \\ 30 & 50 & -30 & 100 \end{bmatrix}\begin{matrix} w_2 \\ \theta_2 \\ w_3 \\ \theta_3 \end{matrix}$$

Global equation is,

$$[K]\{r\} = \{R\} \qquad (5.3)$$

$$672 \times 10^3 \begin{bmatrix} \overset{w_1}{12} & \overset{\theta_1}{30} & \overset{w_2}{-12} & \overset{\theta_2}{30} & \overset{w_3}{0} & \overset{\theta_3}{0} \\ 30 & 100 & -30 & 50 & 0 & 0 \\ -12 & -30 & 12+12 & -30+30 & -12 & 30 \\ 30 & 50 & -30+30 & 100+100 & -30 & 50 \\ 0 & 0 & -12 & -30 & 12 & -30 \\ 0 & 0 & 30 & 50 & -30 & 100 \end{bmatrix} \begin{Bmatrix} w_1 \\ \theta_1 \\ w_2 \\ \theta_2 \\ w_3 \\ \theta_3 \end{Bmatrix} = \begin{Bmatrix} R_1 \\ 0 \\ -8 \times 10^3 \\ 0 \\ R_3 \\ 0 \end{Bmatrix}.$$

Using the elimination method for applying boundary conditions,

$$w_1 = w_3 = 0.$$

The above matrix reduces to

$$672 \times 10^3 \begin{bmatrix} \overset{\theta_1}{100} & \overset{w_2}{-30} & \overset{\theta_2}{50} & \overset{\theta_3}{0} \\ -30 & 24 & 0 & 30 \\ 50 & 0 & 200 & 50 \\ 0 & 30 & 50 & 100 \end{bmatrix} \begin{Bmatrix} \theta_1 \\ w_2 \\ \theta_2 \\ \theta_3 \end{Bmatrix} = \begin{Bmatrix} 0 \\ -8 \times 10^3 \\ 0 \\ 0 \end{Bmatrix}.$$

By solving the above equations, we get,

$$w_2 = -0.002 \text{ m} = -2 \text{ mm},$$

$\theta_1 = -0.0006$ rad, $\theta_2 = 0$ rad, and $\theta_3 = 0.0006$ rad.

Reaction calculation

$$672 \times 10^3 (30 \times \theta_1 - 12 \times w_2) = R_1$$

$$672 \times 10^3 (30 \times (-0.0006) - 12 \times (-0.002)) = R_1$$

$$R_1 = 4.032 \text{ kN}$$

$$672 \times 10^3 (-12 \times w_2 - 30 \times \theta_3) = R_3$$

$$672 \times 10^3 (-12 \times (-0.002) - 30 \times (0.0006)) = R_3$$

$$R_3 = 4.032 \text{ kN}.$$

Finite Element Analysis of Beams

(III) Software results

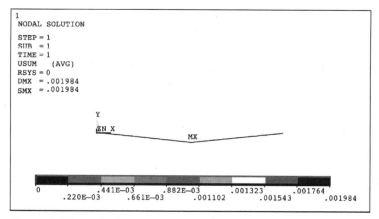

Figure 5.2(b). Deflection pattern for a simply supported beam (refer to Appendix C for color figures).

Deflection values at nodes (in meters)

The following degree of freedom results are in global coordinates

NODE	UX	UY	UX	USUM
1	0.0000	0.0000	0.0000	0.0000
2	0.0000	−0.19841E-02	0.0000	−0.19841E-02
3	0.0000	0.0000	0.0000	0.0000

The following degree of freedom results are in global coordinates

NODE	ROTZ
1	−0.59524E-03
2	0.0000
3	0.59524E-03

Reaction values

The following X, Y, Z solutions are in global coordinates

NODE	FX	FY	MZ
1	0.0000	4000.0	
3	0.0000	4000.0	

Answers for Example 5.1

Parameter	Analytical method	FEM-hand calculations	Software results
Displacement at node 2	−1.98 mm	−2 mm	−1.9841 mm
Slope at node			
1	−5.95 × 10⁻⁴ rad	−0.0006 rad	−0.59524 × 10⁻³ rad
2	0	0	0
3	−5.95 × 10⁻⁴ rad	−0.0006 rad	0.59524 × 10⁻³ rad
Reaction at node			
1	4 kN	4.032 kN	4 kN
3	4 kN	4.032 kN	4 kN

Example 5.2

For the beam shown in Figure 5.3, determine displacements, slopes, reactions, maximum bending moment, shear force, and maximum bending stress. Take $E = 210$ GPa and $I = 2 \times 10^{-4}$ m⁴. The beam has rectangular cross-section of depth $h = 1$ m.

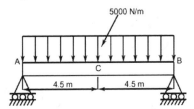

Figure 5.3. The beam for Example 5.2.

Solution

(I) Analytical method [refer to Figure 5.3]

Reaction,

$$R_A = R_B = \frac{5000 \times 9}{2} = 22500 \text{ N} = 22.5 \text{ kN}$$

$$\delta_C = -\frac{5PL^4}{388EI} = -\frac{5 \times 5000 \times (9)^4}{384 \times 210 \times 10^9 \times 2 \times 10^{-4}} = -0.0102 \text{ m} = -10.2 \text{ mm}$$

$$|\theta_A| = |\theta_B| = \frac{PL^3}{24EI} = \frac{5000 \times (9)^3}{24 \times 210 \times 10^9 \times 2 \times 10^{-4}} = 3.62 \times 10^{-3} \text{ rad}$$

$\theta_C = 0$, by symmetry.

Maximum bending moment,

$$M_{max} = \frac{PL^2}{8} = \frac{5000 \times (9)^2}{8} = 50625 \text{ N-m.}$$

Shear force,

$$SF = \frac{PL}{2} = \frac{5000 \times 9}{2} = 22500 \text{ N.}$$

Maximum bending stress,

$$f_{max} = \frac{M_{max}}{I} \times y_{max} \tag{5.4}$$

$$y_{max} = \frac{h}{2} = \frac{1}{2} = 0.5 \text{ m}$$

$$f_{max} = \frac{50625}{2 \times 10^{-4}} \times 0.5 = 126.56 \text{ MPa.}$$

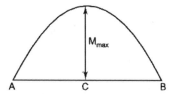

Figure 5.3(a). Bending moment diagram.

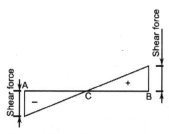

Figure 5.3(b). Shear force diagram.

(II) FEM by hand calculations [refer to Figure 5.3(c)]

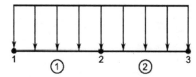

Figure 5.3(c). Finite element model for Example 5.2.

Stiffness matrices are,

$$[k]=\frac{EI}{L^3}\begin{bmatrix} 12 & 6L & -12 & 6L \\ 6L & 4L^2 & -6L & 2L^2 \\ -12 & -6L & 12 & -6L \\ 6L & 2L^2 & -6L & 4L^2 \end{bmatrix}$$

$$[k_1]=\frac{210\times10^9\times2\times10^{-4}}{(4.5)^3}\begin{bmatrix} 12 & 6(4.5) & -12 & 6(4.5) \\ 6(4.5) & 4(4.5)^2 & -6(4.5) & 2(4.5)^2 \\ -12 & -6(4.5) & 12 & -6(4.5) \\ 6(4.5) & 2(4.5)^2 & -6(4.5) & 4(4.5)^2 \end{bmatrix}$$

$$\begin{matrix} & w_1 & \theta_1 & w_2 & \theta_2 & \end{matrix}$$

$$[k_1]=460905.35\begin{bmatrix} 12 & 27 & -12 & 27 \\ 27 & 81 & -27 & 40.5 \\ -12 & -27 & 12 & -27 \\ 27 & 40.5 & -27 & 81 \end{bmatrix}\begin{matrix} w_1 \\ \theta_1 \\ w_2 \\ \theta_2 \end{matrix}.$$

Due to symmetry,

$$[k_1]=[k_2]$$

$$\begin{matrix} & w_2 & \theta_2 & w_3 & \theta_3 & \end{matrix}$$

$$[k_2]=460905.35\begin{bmatrix} 12 & 27 & -12 & 27 \\ 27 & 81 & -27 & 40.5 \\ -12 & -27 & 12 & -27 \\ 27 & 40.5 & -27 & 81 \end{bmatrix}\begin{matrix} w_2 \\ \theta_2 \\ w_3 \\ \theta_3 \end{matrix}.$$

Nodal force calculation

For element 1,

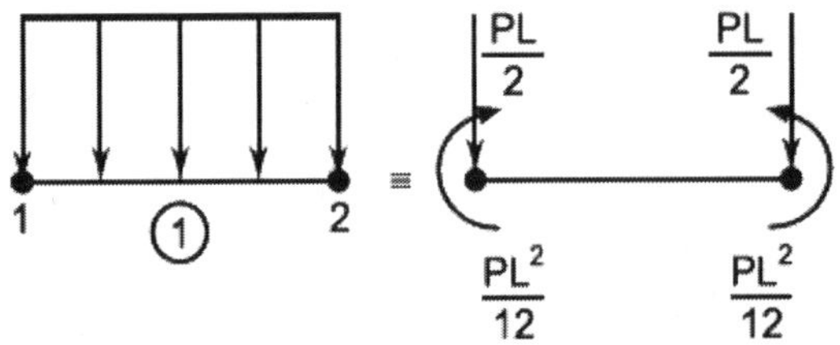

Figure 5.3(d). **Nodal force calculation for element 1 in Example 5.2.**

Nodal forces and moments for element 1 is,

$$\{F_1\} = \begin{Bmatrix} -\dfrac{PL}{2} \\ -\dfrac{PL^2}{12} \\ -\dfrac{PL}{2} \\ \dfrac{PL^2}{12} \end{Bmatrix} = \begin{Bmatrix} -\dfrac{5000 \times 4.5}{2} \\ -\dfrac{5000 \times (4.5)^2}{12} \\ -\dfrac{5000 \times 4.5}{2} \\ \dfrac{5000 \times (4.5)^2}{12} \end{Bmatrix} = \begin{Bmatrix} -11250 \\ -8437.5 \\ -11250 \\ 8437.5 \end{Bmatrix} \begin{matrix} f_1 \\ m_1 \\ f_2 \\ m_2 \end{matrix}.$$

For element 2,

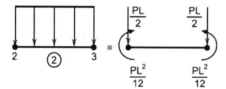

Figure 5.3(e). Nodal force calculation for element 2 in Example 5.2.

Due to symmetry,

$$\{F_1\} = \{F_2\}$$

$$\{F_2\} = \begin{Bmatrix} -11250 \\ -8437.5 \\ -11250 \\ 8437.5 \end{Bmatrix} \begin{matrix} f_2 \\ m_2 \\ f_3 \\ m_3 \end{matrix}.$$

Global equation is,

$$[K]\{r\} = \{R\} \qquad (5.5)$$

$$460905.35 \begin{bmatrix} 12 & 27 & -12 & 27 & 0 & 0 \\ 27 & 81 & -27 & 40.5 & 0 & 0 \\ -12 & -27 & 12+12 & -27+27 & -12 & 27 \\ 27 & 40.5 & -27+27 & 81+81 & -27 & 40.5 \\ 0 & 0 & -12 & -27 & 12 & -27 \\ 0 & 0 & 27 & 40.5 & -27 & 81 \end{bmatrix} \begin{Bmatrix} w_1 \\ \theta_1 \\ w_2 \\ \theta_2 \\ w_3 \\ \theta_3 \end{Bmatrix} = \begin{Bmatrix} -11250 + R_{1y} \\ -8437.5 \\ -11250 - 11250 \\ 8437.5 - 8437.5 \\ 11250 + R_{3y} \\ 8437.5 \end{Bmatrix}.$$

Using the elimination method for applying boundary conditions,
$$w_1 = w_3 = 0.$$
The above matrix reduces to

$$460905.35 \begin{bmatrix} \theta_1 & w_2 & \theta_2 & \theta_3 \\ 81 & -27 & 40.5 & 0 \\ -27 & 24 & 0 & 27 \\ 40.5 & 0 & 162 & 40.5 \\ 0 & 27 & 40.5 & 81 \end{bmatrix} \begin{Bmatrix} \theta_1 \\ w_2 \\ \theta_2 \\ \theta_3 \end{Bmatrix} = \begin{Bmatrix} -8437.5 \\ -22500 \\ 0 \\ 8437.5 \end{Bmatrix}.$$

By solving the above equations, we get,
$$w_2 = -0.0102 \text{ m},$$
$$\theta_1 = -0.0036 \text{ rad},\ \theta_2 = 0 \text{ rad, and } \theta_3 = 0.0036 \text{ rad}.$$

Reactions are calculated from 1st and 5th rows of global matrix.

$$460905.35 \begin{bmatrix} 12 & 27 & -12 & 27 & 0 & 0 \end{bmatrix} \begin{Bmatrix} w_1 \\ \theta_1 \\ w_2 \\ \theta_2 \\ w_3 \\ \theta_3 \end{Bmatrix} = -11250 + R_{1y}$$

$$\therefore \quad 11615 = -11250 + R_{1y}$$

$$\therefore \quad R_{1y} = 22865 \text{ N} = 22.865 \text{ kN}.$$

Similarly from 5th row
$$R_{3y} = 22.865 \text{ kN}.$$

(III) Software results

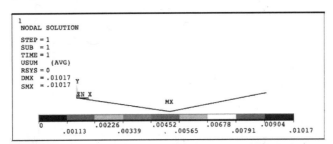

Figure 5.3(f). Deflection pattern for a simply supported beam (refer to Appendix C for color figures).

Deflection values at nodes (in meters)

The following degree of freedom results are in global coordinates

NODE	UX	UY	UX	USUM
1	0.0000	0.0000	0.0000	0.0000
2	0.0000	−0.10170E−01	0.0000	−0.10170E−01
3	0.0000	0.0000	0.0000	0.0000

Slope values at nodes

The following degree of freedom results are in global coordinates

NODE	ROTZ
1	−0.36161E−02
2	0.0000
3	0.36161E−02

Reaction values

The following X, Y, Z solutions are in global coordinates

NODE	FX	FY	MZ
1	0.0000	22500	
3	0.0000	22500	

Total values

VALUE	0.0000	45000	0.0000

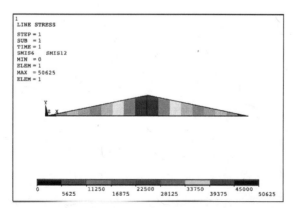

Figure 5.3(g). Bending moment diagram for a simply supported beam (refer to Appendix C for color figures).

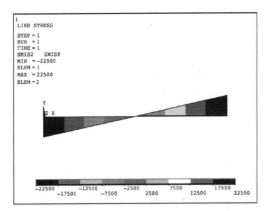

Figure 5.3(h). Shear force diagram for a simply supported beam (refer to Appendix C for color figures).

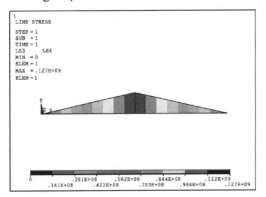

Figure 5.3(i). Bending stress for a simply supported beam (refer to Appendix C for color figures).

Answers for Example 5.2

Parameter	Analytical method	FEM-hand calculations	Software results
Displacement at node 2	−0.0102 m	−0.0102 m	−0.01017 m
Slope at node			
1	-3.62×10^{-3} rad	0.0036 rad	-0.36161×10^{-2} rad
2	0	0	0
3	3.62×10^{-3} rad	0.0036 rad	0.36161×10^{-2} rad
Reaction at node			
1	22500 N	22865 N	22500 N
3	22500 N	22865 N	22500 N
Maximum bending moment	50625 N-m	……	50625 N-m
Shear force	22500 N	…….	22500 N
Maximum bending stress	126.56 MPa	…….	127 MPa

Example 5.3

For the beam shown in Figure 5.4, determine displacements, slopes, and reactions. Take $E = 200$ GPa and $I = 6.25 \times 10^{-4}$ m^4.

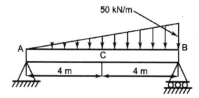

Figure 5.4. The beam for Example 5.3.

Solution

(I) Analytical method [refer to Figure 5.4]

Reaction,

$$R_A = \frac{PL}{6} = \frac{50 \times 10^3 \times 8}{6} = 66666.67 \text{ N} = 66.67 \text{ kN}$$

$$R_B = \frac{PL}{3} = \frac{50 \times 10^3 \times 8}{3} = 133333.33 \text{ N} = 133.33 \text{ kN}$$

$$\theta_A = -\frac{7PL^3}{360EI} = \frac{7 \times 50 \times 10^3 \times (8)^3}{360 \times 200 \times 10^9 \times 6.25 \times 10^{-4}} = -0.00398 \text{ rad}$$

$$\theta_B = -\frac{PL^3}{45EI} = \frac{50 \times 10^3 \times (8)^3}{45 \times 200 \times 10^9 \times 6.25 \times 10^{-4}} = 0.00455 \text{ rad}$$

$$\delta_C = \frac{1}{EI}\left(\frac{PL}{36}x^3 - \frac{P}{120 \times L}x^5 - \frac{7PL^3}{360}x\right)_{x=\frac{L}{2}}$$

$$\delta_C = \frac{1}{EI}\left(\frac{PL}{36}\times\left(\frac{L}{2}\right)^3 - \frac{P}{120 \times L}\times\left(\frac{L}{2}\right)^5 - \frac{7PL^3}{360}\times\left(\frac{L}{2}\right)\right) = \frac{1}{EI}\left(\frac{PL^4}{288} - \frac{PL^4}{3840} - \frac{7PL^4}{720}\right)$$

$$\delta_C = \frac{PL^4}{EI}\left(\frac{40 - 3 - 112}{11520}\right)$$

$$\therefore \delta_C = -\frac{75PL^4}{11520EI} = -\frac{75 \times 50 \times 10^3 \times (8)^4}{11520 \times 200 \times 10^9 \times 6.25 \times 10^{-4}} = -0.01067 \text{ m}$$

$$\theta_C = \frac{1}{EI}\left(\frac{PL}{12}x^2 - \frac{P}{24 \times L}x^4 - \frac{7PL^3}{360}x\right)_{x=\frac{L}{2}}$$

$$\theta_C = \frac{1}{EI}\left(\frac{PL}{12}\times\left(\frac{L}{2}\right)^2 - \frac{P}{24\times L}\times\left(\frac{L}{2}\right)^4 - \frac{7PL^3}{360}\times\left(\frac{L}{2}\right)\right) = \frac{PL^3}{EI}\left(\frac{1}{48} - \frac{1}{384} - \frac{7}{360}\right)$$

$$\therefore \quad \theta_C = -\frac{1.2153\times10^{-3}\times PL^3}{EI} = -\frac{1.2153\times10^{-3}\times 50\times(8)^3}{200\times10^9 \times 6.25\times10^{-4}} = 2.4889\times10^{-4}\ \text{rad}.$$

(II) FEM by hand calculations [refer to Figure 5.4(a)]

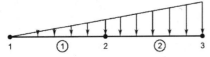

Figure 5.4(a). Finite element model for Example 5.3.

Stiffness matrices are,

$$[k_1] = \frac{EI}{L_1^3}\begin{bmatrix} 12 & 6L & -12 & 6L \\ 6L & 4L^2 & -6L & 2L^2 \\ -12 & -6L & 12 & -6L \\ 6L & 2L^2 & -6L & 4L^2 \end{bmatrix} = \frac{200\times10^9 \times 6.25\times10^{-4}}{(4)^3}\begin{bmatrix} 12 & 6(4) & -12 & 6(4) \\ 6(4) & 4(4)^2 & -6(4) & 2(4)^2 \\ -12 & -6(4) & 12 & -6(4) \\ 6(4) & 2(4)^2 & -6(4) & 4(4)^2 \end{bmatrix}$$

$$[k_1] = 195.3125\times10^4 \begin{bmatrix} 12 & 24 & -12 & 24 \\ 24 & 64 & -24 & 32 \\ -12 & -24 & 12 & -24 \\ 24 & 32 & -24 & 64 \end{bmatrix}\begin{matrix} w_1 \\ \theta_1 \\ w_2 \\ \theta_2 \end{matrix}$$

Due to symmetry,

$$[k_1] = [k_2]$$

$$[k_2] = \frac{EI}{L_2^3}\begin{bmatrix} 12 & 6L & -12 & 6L \\ 6L & 4L^2 & -6L & 2L^2 \\ -12 & -6L & 12 & -6L \\ 6L & 2L^2 & -6L & 4L^2 \end{bmatrix} = \frac{200\times10^9 \times 6.25\times10^{-4}}{(4)^3}\begin{bmatrix} 12 & 6(4) & -12 & 6(4) \\ 6(4) & 4(4)^2 & -6(4) & 2(4)^2 \\ -12 & -6(4) & 12 & -6(4) \\ 6(4) & 2(4)^2 & -6(4) & 4(4)^2 \end{bmatrix}$$

$$[k_2] = 195.3125\times10^4 \begin{bmatrix} 12 & 24 & -12 & 24 \\ 24 & 64 & -24 & 32 \\ -12 & -24 & 12 & -24 \\ 24 & 32 & -24 & 64 \end{bmatrix}\begin{matrix} w_2 \\ \theta_2 \\ w_3 \\ \theta_3 \end{matrix}$$

Finite Element Analysis of Beams

Global stiffness matrix is,

$$[K] = 195.3125 \times 10^4 \begin{bmatrix} 12 & 24 & -12 & 24 & 0 & 0 \\ 24 & 64 & -24 & 32 & 0 & 0 \\ -12 & -24 & 12+12 & -24+24 & -12 & 24 \\ 24 & 32 & -24+24 & 64+64 & -24 & 32 \\ 0 & 0 & -12 & -24 & 12 & -24 \\ 0 & 0 & 24 & 32 & -24 & 64 \end{bmatrix} \begin{matrix} w_1 \\ \theta_1 \\ w_2 \\ \theta_2 \\ w_3 \\ \theta_3 \end{matrix}$$

with columns labeled $w_1, \theta_1, w_2, \theta_2, w_3, \theta_3$.

Load vector,

$$\{F\} = \frac{L}{20} \begin{Bmatrix} 7P_1 + 3P_2 \\ \dfrac{L}{3}(3P_1 + 2P_2) \\ 3P_1 + 7P_2 \\ -\dfrac{L}{3}(2P_1 + 3P_2) \end{Bmatrix}.$$

For element 1,

$$P_1 = 0,\ P_2 = -25\ \text{kN/m},\ L = 4\ \text{m}$$

$$\{F_1\} = \frac{4}{20} \begin{Bmatrix} -75 \\ \dfrac{4}{3}(-50) \\ -175 \\ -\dfrac{4}{3}(-75) \end{Bmatrix} = \begin{Bmatrix} -15\ \text{kN} \\ -13.33\ \text{kN-m} \\ -35\ \text{kN} \\ 20\ \text{kN-m} \end{Bmatrix} = \begin{Bmatrix} -15000\ \text{N} \\ -1333\ \text{N-m} \\ -35000\ \text{N} \\ 2000\ \text{N-m} \end{Bmatrix}.$$

For element 2,

$$P_1 = -25\ \text{kN/m},\ P_2 = -50\ \text{kN/m},\ L = 4\ \text{m}$$

$$\{F_2\} = \frac{4}{20} \begin{Bmatrix} -175 - 150 \\ \dfrac{4}{3}(-75 - 100) \\ -75 - 350 \\ -\dfrac{4}{3}(-50 - 150) \end{Bmatrix} = \begin{Bmatrix} -65000\ \text{N} \\ -46667\ \text{N-m} \\ -85000\ \text{N} \\ 53333\ \text{N-m} \end{Bmatrix}.$$

Global load vector is,

$$\{F\} = \begin{Bmatrix} -15000 \\ -13333 \\ -100000 \\ -26667 \\ -85000 \\ 53333 \end{Bmatrix}.$$

Global equation is,

$$[K]\{r\} = \{R\} \tag{5.6}$$

$$195.3125 \times 10^4 \begin{bmatrix} \overset{w_1}{12} & \overset{\theta_1}{24} & \overset{w_2}{-12} & \overset{\theta_2}{24} & \overset{w_3}{0} & \overset{\theta_3}{0} \\ 24 & 64 & -24 & 32 & 0 & 0 \\ -12 & -24 & 12+12 & -24+24 & -12 & 24 \\ 24 & 32 & -24+24 & 64+64 & -24 & 32 \\ 0 & 0 & -12 & -24 & 12 & -24 \\ 0 & 0 & 24 & 32 & -24 & 64 \end{bmatrix} \begin{Bmatrix} w_1 \\ \theta_1 \\ w_2 \\ \theta_2 \\ w_3 \\ \theta_3 \end{Bmatrix} = \begin{Bmatrix} -15000 + R_1 \\ -13333 \\ -100000 \\ -26667 \\ -85000 + R_3 \\ 53333 \end{Bmatrix}.$$

Using the elimination method for applying boundary conditions,

$$w_1 = w_3 = 0.$$

The above matrix reduces to

$$195.3125 \times 10^4 \begin{bmatrix} \overset{\theta_1}{64} & \overset{w_2}{-24} & \overset{\theta_2}{32} & \overset{\theta_3}{0} \\ -24 & 24 & 0 & 24 \\ 32 & 0 & 128 & 32 \\ 0 & 24 & 32 & 64 \end{bmatrix} \begin{Bmatrix} \theta_1 \\ w_2 \\ \theta_2 \\ \theta_3 \end{Bmatrix} = \begin{Bmatrix} -13333 \\ -100000 \\ -26667 \\ 53333 \end{Bmatrix}.$$

By solving the above equations, we get,

$$\begin{Bmatrix} \theta_1 \\ w_2 \\ \theta_2 \\ \theta_3 \end{Bmatrix} = \begin{Bmatrix} -0.00398 \text{ rad} \\ -0.01067 \text{ m} \\ -0.00025 \text{ rad} \\ 0.00455 \text{ rad} \end{Bmatrix}.$$

Reaction calculation

$$1953125(12 \times w_1 + 24 \times \theta_1 - 12 \times w_2 + 24 \times \theta_2) = -15000 + R_1$$
$$R_1 = 66796.875 \text{ N} = 66.79 \text{ kN}$$

$$1953125(-12(-0.01067) - 24 \times (-0.00025) - 24 \times (0.00455)) = -85000 + R_3$$
$$R_3 = 133515.63 \text{ N} = 133.52 \text{ kN}.$$

(III) Software results

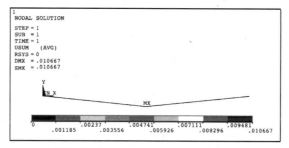

Figure 5.4(b). Deflection pattern for a simply supported beam (refer to Appendix C for color figures).

Deflection values at nodes (in meters)

The following degree of freedom results are in global coordinates

NODE	UX	UY	UX	USUM
1	0.0000	0.0000	0.0000	0.0000
2	0.0000	−0.10667E-01	0.0000	0.10667E-01
3	0.0000	0.0000	0.0000	0.0000

Maximum absolute values

NODE	0	2	0	2
VALUE	0.0000	−0.10667E-01	0.0000	0.10667E-01

Slope values at nodes

The following degree of freedom results are in global coordinates

NODE	ROTZ
1	−0.39822E-02
2	−0.24889E-03
3	0.45511E-02

Reaction values

The following X, Y, Z solutions are in global coordinates

NODE	FX	FY	MZ
1	0.0000	66667	
3	0.0000	0.13333E +06	

Total values

VALUE	0.0000	0.20000E + 06	0.0000

Answers for Example 5.3

Parameter	Analytical method	FEM-hand calculations	Software results
Displacement at node 2	−0.01067 m	−0.01067 m	−0.010667 m
Slope at node			
1	−0.00398 rad	−0.00398 rad	−0.0039822 rad
2	−2.4889 × 10⁻⁴ rad	−0.00025 rad	−0.00024889 rad
3	0.00455 rad	0.00455 rad	0.0045511 rad
Reaction at node			
1	66.67 kN	66.79 kN	66.667 kN
3	133.33 kN	133.52 kN	133.33 kN

Example 5.4

Calculate the maximum deflection in the beam shown in Figure 5.5. Take E = 200 GPa.

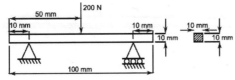

Figure 5.5. The beam for Example 5.4.

Solution

(I) Analytical method [refer Figure 5.5(a)]

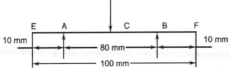

Figure 5.5(a). Analytical method for Example 5.4.

Finite Element Analysis of Beams

$$I = \frac{bh^3}{12}$$

$$I = \frac{0.01 \times (0.01)^3}{12} = 8.333 \times 10^{-10} \text{ m}^4$$

$$\delta_C = \frac{PL^2}{48EI} = -\frac{200 \times (0.08)^3}{48EI} = -\frac{2.133 \times 10^{-3}}{EI} = -\frac{2.133 \times 10^{-3}}{200 \times 10^9 \times 8.33 \times 10^{-10}}$$

$$\delta_C = -1.2803 \times 10^{-5} \text{ m} = -0.0128 \text{ mm}$$

$$\theta_C = 0$$

$$|\theta_A| = |\theta_B| = \frac{PL^2}{16EI} = -\frac{200 \times (0.08)^2}{16EI} = \frac{0.08}{EI} = \frac{0.08}{200 \times 10^9 \times 8.33 \times 10^{-10}} = 4.802 \times 10^{-4} \text{ rad}$$

$$\delta_E = \delta_F = \theta_B \times BF$$

$$\delta_E = \delta_F = 4.802 \times 10^{-4} \times 10 = 4.802 \times 10^{-3} \text{ mm.}$$

(II) FEM by hand calculations [refer to Figure 5.5(b)]

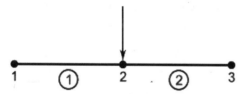

Figure 5.5(b). Finite element model for Example 5.4.

For beam,

$$I = \frac{bh^3}{12}$$

$$I = \frac{10 \times (10)^3}{12} = 833.34 \text{ mm}^4.$$

For element 1 and 2, L = 40 mm

$$[k_1] = \frac{EI}{L}\begin{bmatrix} 12 & 6L & -12 & 6L \\ 6L & 4L^2 & -6L & 2L^2 \\ -12 & -6L & 12 & -6L \\ 6L & 2L^2 & -6L & 4L^2 \end{bmatrix} = \frac{200 \times 10^3 \times 833.34}{(40)^3}\begin{bmatrix} 12 & 6(40) & -12 & 6(40) \\ 6(40) & 4(40)^2 & -6(4) & 2(40)^2 \\ -12 & -6(40) & 12 & -6(40) \\ 6(40) & 2(40)^2 & -6(40) & 4(40)^2 \end{bmatrix}$$

$$[k_1] = 2604.1875 \begin{bmatrix} w_1 & \theta_1 & w_2 & \theta_2 \\ 12 & 240 & -12 & 240 \\ 240 & 6400 & -240 & 3200 \\ -12 & -240 & 12 & -240 \\ 240 & 3200 & -240 & 6400 \end{bmatrix} \begin{matrix} w_1 \\ \theta_1 \\ w_2 \\ \theta_2 \end{matrix}.$$

Due to symmetry,

$$[k_1] = [k_2]$$

$$[k_2] = \frac{EI}{L} \begin{bmatrix} 12 & 6L & -12 & 6L \\ 6L & 4L^2 & -6L & 2L^2 \\ -12 & -6L & 12 & -6L \\ 6L & 2L^2 & -6L & 4L^2 \end{bmatrix} = \frac{200 \times 10^3 \times 833.34}{(40)^3} \begin{bmatrix} 12 & 6(40) & -12 & 6(40) \\ 6(40) & 4(40)^2 & -6(4) & 2(40)^2 \\ -12 & -6(40) & 12 & -6(40) \\ 6(40) & 2(40)^2 & -6(40) & 4(40)^2 \end{bmatrix}$$

$$[k_2] = 2604.1875 \begin{bmatrix} w_2 & \theta_2 & w_3 & \theta_3 \\ 12 & 240 & -12 & 240 \\ 240 & 6400 & -240 & 3200 \\ -12 & -240 & 12 & -240 \\ 240 & 3200 & -240 & 6400 \end{bmatrix} \begin{matrix} w_2 \\ \theta_2 \\ w_3 \\ \theta_3 \end{matrix}.$$

Global stiffness matrix is,

$$[K] = 2604.1875 \begin{bmatrix} w_1 & \theta_1 & w_2 & \theta_2 & w_3 & \theta_3 \\ 12 & 240 & -12 & 240 & 0 & 0 \\ 240 & 6400 & -24 & 3200 & 0 & 0 \\ -12 & -240 & 12+12 & -240+240 & -12 & 240 \\ 240 & 3200 & -240+240 & 6400+6400 & -240 & 3200 \\ 0 & 0 & -12 & -240 & 12 & -240 \\ 0 & 0 & 240 & 3200 & -240 & 6400 \end{bmatrix} \begin{matrix} w_1 \\ \theta_1 \\ w_2 \\ \theta_2 \\ w_3 \\ \theta_3 \end{matrix}.$$

Global load vector is,

$$\{F\} = \begin{Bmatrix} R_1 \\ 0 \\ -200 \\ 0 \\ R_3 \\ 0 \end{Bmatrix}.$$

Finite Element Analysis of Beams

Global equation is,

$$[K]\{r\} = \{R\} \qquad (5.7)$$

$$2604.1875 \begin{bmatrix} 12 & 240 & -12 & 240 & 0 & 0 \\ 240 & 6400 & -240 & 3200 & 0 & 0 \\ -12 & -240 & 24 & 0 & -12 & 240 \\ 240 & 3200 & 0 & 12800 & -240 & 3200 \\ 0 & 0 & -12 & -240 & 12 & -240 \\ 0 & 0 & 240 & 3200 & -240 & 6400 \end{bmatrix} \begin{Bmatrix} w_1 \\ \theta_1 \\ w_2 \\ \theta_2 \\ w_3 \\ \theta_3 \end{Bmatrix} = \begin{Bmatrix} R_1 \\ 0 \\ -200 \\ 0 \\ R_3 \\ 0 \end{Bmatrix}.$$

Using the elimination method for applying boundary conditions,
$$w_1 = w_3 = 0.$$

The above matrix reduces to

$$2604.1875 \begin{bmatrix} 6400 & -240 & 3200 & 0 \\ -240 & 24 & 0 & 240 \\ 3200 & 0 & 12800 & 3200 \\ 0 & 240 & 3200 & 6400 \end{bmatrix} \begin{Bmatrix} \theta_1 \\ w_2 \\ \theta_2 \\ \theta_3 \end{Bmatrix} = \begin{Bmatrix} 0 \\ -200 \\ 0 \\ 0 \end{Bmatrix}.$$

By solving the above equations, we get,

$$\begin{Bmatrix} \theta_1 \\ w_2 \\ \theta_2 \\ \theta_3 \end{Bmatrix} = \begin{Bmatrix} -4.8 \times 10^{-4} \text{ rad} \\ -0.0128 \text{ mm} \\ 0 \\ 4.8 \times 10^{-4} \text{ rad} \end{Bmatrix}.$$

At $\theta_2 = 0$, max deflection between supports is 0.0128 mm.
Deflection at ends (overhang) = $4.8 \times 10^{-4} \times 10 = 4.8 \times 10^{-3}$ mm.

(III) Software results

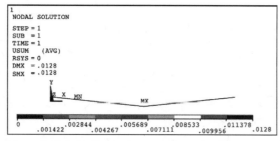

Figure 5.5(c). Deflection pattern for a simply supported beam (refer to Appendix C for color figures).

Deflection values at nodes (in mm)

The following degree of freedom results are in global coordinates

NODE	UX	UY	UX	USUM
1	0.0000	0.48000E-02	0.0000	0.48000E-02
2	0.0000	0.0000	0.0000	0.0000
3	0.0000	−0.12800E-01	0.0000	0.12800E-01
4	0.0000	0.0000	0.0000	0.0000
5	0.0000	0.48000E-02	0.0000	0.48000E-02

Maximum absolute values

NODE	0	3	0	3
VALUE	0.0000	−0.12800E-01	0.0000	0.12800E-01

Slope values at nodes

The following degree of freedom results are in global coordinates

NODE	ROTZ
1	−0.48000E-03
2	−0.48000E-03
3	0.0000
4	0.48000E-03
5	0.48000E-03

Answers for Example 5.4

Parameter	Analytical method	FEM-hand calculations	Software results
Deflection at applied load	−0.0128 mm	−0.0128 mm	−0.0128 mm
Deflection at ends (overhang)	4.802×10^{-3} mm	4.8×10^{-3} mm	4.8×10^{-3} mm
Slope at hinged support	-4.802×10^{-4} rad	-4.8×10^{-4} rad	-4.8×10^{-4} rad
Slope at roller support	4.802×10^{-4} rad	4.8×10^{-4} rad	4.8×10^{-4} rad

Procedure for solving the problems using ANSYS® 11.0 academic teaching software.
For Example 5.2

PREPROCESSING

1. **Main Menu > Preprocessor > Element Type > Add/Edit/Delete > Add > Beam > 2D elastic 3 > OK > Close**

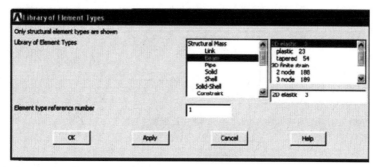

Figure 5.6. Element selection.

2. **Main Menu > Preprocessor > Real Constants > Add/Edit/Delete > Add > OK**

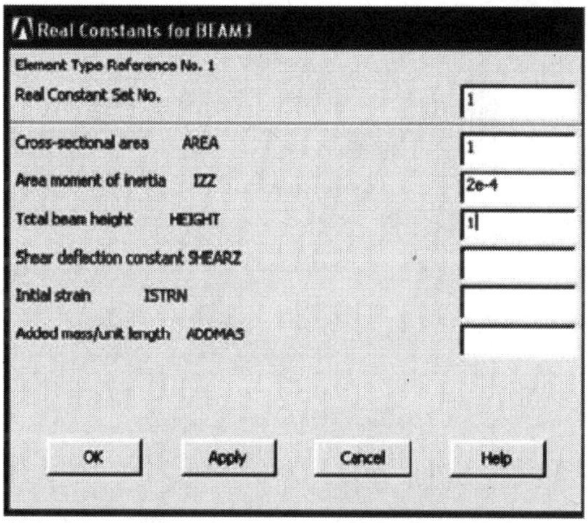

Figure 5.7. Enter the area, moment of inertia, and height of beam.

Cross-sectional area AREA > **Enter 1**
Area moment of inertia IZZ > **Enter 2e-4**
Total beam height HEIGHT > **Enter 1 > OK > Close**
Enter the material properties.

3. **Main Menu > Preprocessor > MATERIAL Props > Material Models**
Material Model Number 1, Click **Structural > Linear > Elastic > Isotropic**
Enter **EX = 210E9 and PRXY = 0.3 > OK**
(**Close** the define material model behavior window.)
Create the nodes and elements as shown in the table below and Figure 5.8.

4. **Main Menu > Preprocessor > Modeling > Create > Nodes > In Active CS**
Enter the coordinates of node 1 > **Apply** Enter the coordinates of node 2 > **Apply** Enter the coordinates of node 3 > **OK**.

Node locations		
Node number	X-coordinate	Y-coordinate
1	0	0
2	4.5	0
3	9	0

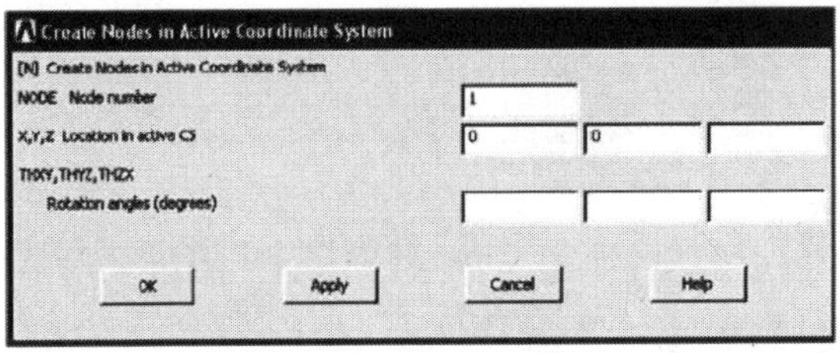

Figure 5.8. Enter the node coordinates.

5. **Main Menu > Preprocessor > Modeling > Create > Elements > Auto Numbered > Thru nodes** Pick the 1st and 2nd node > **Apply** Pick the 2nd and 3rd node > **OK**

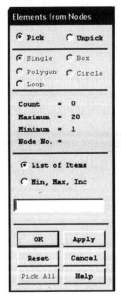

Figure 5.9. Pick the nodes to create elements.

Apply the displacement boundary conditions and loads.

6. **Main Menu > Preprocessor > Loads > Define Loads > Apply > Structural > Displacement > On Nodes** Pick the 1st node and 3rd node > **Apply** > Select UX and UY and Enter displacement value = 0 > **OK**

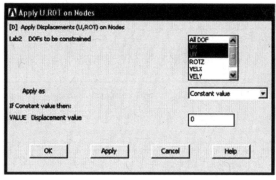

Figure 5.10. Apply the displacement constraint.

7. **Main Menu > Preprocessor > Loads > Define Loads > Apply > Structural > Pressure > On Beams** Pick the 1st element > **OK** > Enter Pressure values at node I = 5000 > **OK**

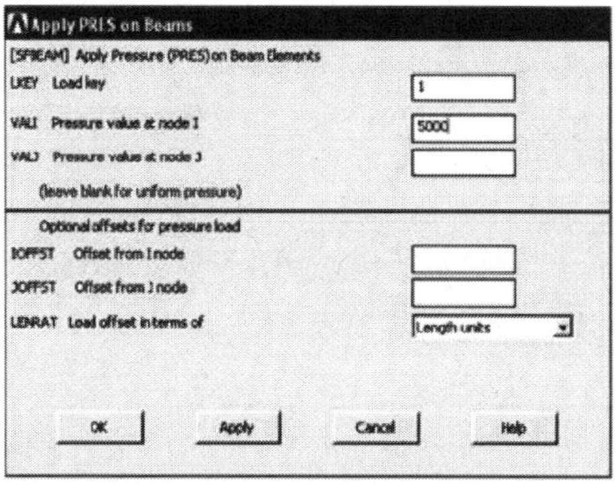

Figure 5.11. Applying loads on element 1.

8. **Main Menu > Preprocessor > Loads > Define Loads > Apply > Structural > Pressure > On Beams** Pick the 2nd element > **OK** > Enter Pressure value at node I = 5000 > **OK**

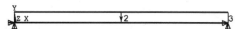

Figure 5.12. Model with loading and displacement boundary conditions.

The model-building step is now complete, and we can proceed to the solution. First, to be safe, save the model.

Solution

The interactive solution proceeds.

9. **Main Menu > Solution > Solve > Current LS > OK**

   The **/STATUS Command** window displays the problem parameters and **the Solve Current Load Step** window is shown. Check the solution options in the **/STATUS** window and if all is OK, select **File > Close**.

   In the **Solve Current Load Step** window, select **OK**, and when the solution is complete, **close** the 'Solution is Done!' window.

POST-PROCESSING

We can now plot the results of this analysis and also list the computed values.

10. **Main Menu > General Postproc > Plot Results > Contour Plot > Nodal Solu > DOF Solution > Displacement vector sum > OK**

    This result is shown in figure 5.3(f).

11. **Main Menu > General Postproc > List Results > Nodal Solution > Select Rolation vector sum > OK**

12. **Main Menu > General Postproc > List Results > Reaction Solu > OK**

    To find the **bending moment diagram,** the following procedure is followed.

13. **Main Menu > General Postproc > Element Table > Define Table > Add** as shown in Figure 5.13.

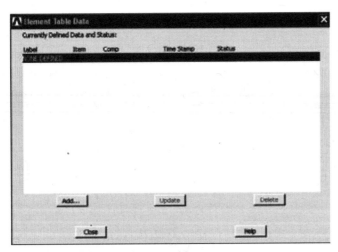

Figure 5.13. Define the element table.

Select **By sequence num and SMISC** and type **6 after SMISC** (as shown in Figure 5.14) > **APPLY**

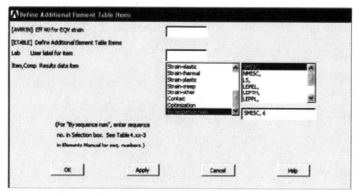

Figure 5.14. Selecting options in element table.

Then again select **By sequence num and SMISC** and type **12 after SMISC > OK**

14. Main Menu > General Postproc > Plot Results > Contour Plot > Line Elem Res > Select **SMIS 6 and SMIS 12 in the rows of LabI and LabJ** respectively as shown in Figure 5.15 > **OK**

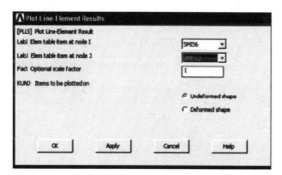

Figure 5.15. Selecting options for finding out bending moment.

This result is shown in Figure 5.3(g).

To find the **shear force diagram** the following procedure is followed.

15. Main Menu > General Postproc > Element Table > Define Table > Add
    Select **By sequence num and SMISC** and type **2 after SMISC > APPLY**
    Then again select **By sequence num and SMISC** and type **8 after SMISC > OK**

16. Main Menu > General Postproc > Plot Results > Contour Plot > Lone Elem Res > Select **SMIS 2 and SMIS 8 > OK**
    This result is shown in Figure 5.3(h).
    To find the **bending stress,** the following procedure is followed.

17. **Main Menu > General Postproc > Element Table > Define Table > Add**
    Select **By sequence num** and **LS** and type **3 after LS > APPLY**
    Then again select **By sequence num and LS** and type **6 after LS > OK**
18. **Main Menu > General Postproc > Plot Results > Contour Plot > Line Elem Res > Select LS 3 and LS 6 > OK**
    This result is shown is Figure 5.3(i).

5.3 CANTILEVER BEAMS

Example 5.5

Beam subjected to concentrated load. For the beam shown in Figure 5.16, determine the deflections and reactions. Let E = 210 GPa and $I = 2 \times 10^{-4}$ m⁴. Take 2 elements.

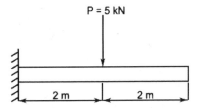

Figure 5.16. **Beam subjected to concentrated load for Example 5.5.**

Solution

(I) Analytical method [refer to Figure 5.16(a)]

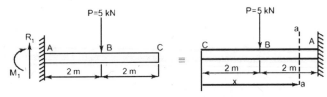

Figure 5.16(a). **Analytical method for Example 5.5.**

The solution is obtained by Macaulay's method. The number within the brackets <> is to be neglected whenever it is less than zero.
At section a-a

$$EIy'' = -P<x-2>$$

$$EIy' = \frac{-P<x-2>^2}{2} + C_1$$

$$EIy = \frac{-P<x-2>^3}{6} + C_1 x + C_2$$

At $x = 4, y' = 0 \Rightarrow C_1 = 2P$

At $x = 4, y = 0 \Rightarrow C_2 = \frac{-20P}{3}$

$$y' = \frac{1}{EI}\left(\frac{-P<x-2>^2}{2} + 2P\right)$$

$$y = \frac{1}{EI}\left(\frac{-P<x-2>^3}{6} + 2Px - \frac{20P}{3}\right)$$

$$y'_B = y'_{x=2} = \frac{1}{EI}(0+2P) = \frac{2\times 5000}{210\times 10^9 \times 2\times 10^{-4}} = 2.381\times 10^{-4} \text{ rad}$$

$$y'_C = y'_{x=0} = \frac{1}{EI}(0+2P) = \frac{2\times 5000}{210\times 10^9 \times 2\times 10^{-4}} = 2.381\times 10^{-4} \text{ rad.}$$

Similarly,

$$y_B = y_{x=2} = \frac{1}{EI}\left(4P - \frac{20P}{3}\right) = \frac{2\times 5000}{210\times 10^9 \times 2\times 10^{-4}}\left(4\times 5000 - \frac{20\times 5000}{3}\right)$$

$$y_B = y_{x=2} = -3.1746\times 10^{-4} \text{ m}$$

$$y_C = y_{x=0} = \frac{1}{EI}\left(-\frac{20P}{3}\right) = \frac{2\times 5000}{210\times 10^9 \times 2\times 10^{-4}}\left(-\frac{20\times 5000}{3}\right)$$

$$y_C = y_{x=0} = -7.9365\times 10^{-4} \text{ m}$$

$$\sum F_y = 0 \Rightarrow R_1 = 5 \text{ kN}$$

$$\sum M = 0 \Rightarrow M_1 = 10 \text{ kN-m.}$$

(II) FEM by hand calculations

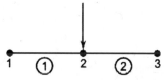

Figure 5.16(b). Finite element model.

Element stiffness matrix for element 1 is,

$$[k_1] = \frac{EI}{L^3}\begin{bmatrix} 12 & 6L & -12 & 6L \\ 6L & 4L^2 & -6L & 2L^2 \\ -12 & -6L & 12 & -6L \\ 6L & 2L^2 & -6L & 4L^2 \end{bmatrix} = \frac{210\times10^9 \times 2\times10^{-4}}{(2)^3}\begin{bmatrix} 12 & 6(2) & -12 & 6(2) \\ 6(2) & 4(2)^2 & -6(2) & 2(2)^2 \\ -12 & -6(2) & 12 & -6(2) \\ 6(2) & 2(2)^2 & -6(2) & 4(2)^2 \end{bmatrix}$$

$$[k_1] = 5.25\times10^6 \begin{bmatrix} w_1 & \theta_1 & w_2 & \theta_2 \\ 12 & 12 & -12 & 12 \\ 12 & 16 & -12 & 8 \\ -12 & -12 & 12 & -12 \\ 12 & 8 & -12 & 16 \end{bmatrix}\begin{matrix} w_1 \\ \theta_1 \\ w_2 \\ \theta_2 \end{matrix}$$

Element stiffness matrix for element 2 is,

$$[k_2] = \frac{EI}{L^3}\begin{bmatrix} 12 & 6L & -12 & 6L \\ 6L & 4L^2 & -6L & 2L^2 \\ -12 & -6L & 12 & -6L \\ 6L & 2L^2 & -6L & 4L^2 \end{bmatrix} = \frac{210\times10^9 \times 2\times10^{-4}}{(2)^3}\begin{bmatrix} 12 & 6(2) & -12 & 6(2) \\ 6(2) & 4(2)^2 & -6(2) & 2(2)^2 \\ -12 & -6(2) & 12 & -6(2) \\ 6(2) & 2(2)^2 & -6(2) & 4(2)^2 \end{bmatrix}$$

$$[k_2] = 5.25\times10^6 \begin{bmatrix} w_2 & \theta_2 & w_3 & \theta_3 \\ 12 & 12 & -12 & 12 \\ 12 & 16 & -12 & 8 \\ -12 & -12 & 12 & -12 \\ 12 & 8 & -12 & 16 \end{bmatrix}\begin{matrix} w_2 \\ \theta_2 \\ w_3 \\ \theta_3 \end{matrix}$$

Global stiffness matrix is,

$$[K] = 5.25\times10^6 \begin{bmatrix} w_1 & \theta_1 & w_2 & \theta_2 & w_3 & \theta_3 \\ 12 & 12 & -12 & 12 & 0 & 0 \\ 12 & 16 & -12 & 8 & 0 & 0 \\ -12 & -12 & 12+12 & -12+12 & -12 & 12 \\ 12 & 8 & -12+12 & 16+16 & -12 & 8 \\ 0 & 0 & -12 & -12 & 12 & -12 \\ 0 & 0 & 12 & 8 & -12 & 16 \end{bmatrix}\begin{matrix} w_1 \\ \theta_1 \\ w_2 \\ \theta_2 \\ w_3 \\ \theta_3 \end{matrix}$$

The global equations are,

$$[K]\{r\} = \{R\} \tag{5.8}$$

$$5.25 \times 10^6 \begin{bmatrix} \overset{w_1}{12} & \overset{\theta_1}{12} & \overset{w_2}{-12} & \overset{\theta_2}{12} & \overset{w_3}{0} & \overset{\theta_3}{0} \\ 12 & 16 & -12 & 8 & 0 & 0 \\ -12 & -12 & 24 & 0 & -12 & 12 \\ 12 & 8 & 0 & 32 & -12 & 8 \\ 0 & 0 & -12 & -12 & 12 & -12 \\ 0 & 0 & 12 & 8 & -12 & 16 \end{bmatrix} \begin{Bmatrix} w_1 \\ \theta_1 \\ w_2 \\ \theta_2 \\ w_3 \\ \theta_3 \end{Bmatrix} = \begin{Bmatrix} 0+R_1 \\ 0+M_1 \\ -5 \times 10^3 \\ 0 \\ 0 \\ 0 \end{Bmatrix}.$$

By using the elimination method for applying boundary conditions,

$$w_1 = \theta_1 = 0.$$

The above matrix reduces to,

$$5.25 \times 10^6 \begin{bmatrix} \overset{w_2}{24} & \overset{\theta_2}{0} & \overset{w_3}{-12} & \overset{\theta_3}{12} \\ 0 & 32 & -12 & 8 \\ -12 & -12 & 12 & -12 \\ 12 & 8 & -12 & 16 \end{bmatrix} \begin{Bmatrix} w_2 \\ \theta_2 \\ w_3 \\ \theta_3 \end{Bmatrix} = \begin{Bmatrix} -5 \times 10^3 \\ 0 \\ 0 \\ 0 \end{Bmatrix}.$$

By solving the above matrix and equations, we get,
Deflections and slopes as

$$w_2 = -0.3175 \times 10^{-3} \text{ m}$$
$$\theta_2 = -0.2381 \times 10^{-3} \text{ rad}$$
$$w_3 = -0.7937 \times 10^{-3} \text{ m}$$
$$\theta_3 = -0.2381 \times 10^{-3} \text{ rad}$$

Reaction calculation

$$5.25 \times 10^6 (12w_1 + 12\theta_1 - 12w_2 + 12\theta_2) = R_1$$
$$5.25 \times 10^6 \left(12 \times 0 + 12 \times 0 - 12\left(-0.3175 \times 10^{-3}\right) + 12\left(-0.2381 \times 10^{-3}\right)\right) = R_1$$
$$\therefore \quad R_1 = 5002.2 \text{ N} \approx 5 \text{ kN}$$

$$5.25 \times 10^6 \left(-12 w_2 + 8\theta_2\right) = M_1$$
$$5.25 \times 10^6 \left(-12\left(-0.3175 \times 10^{-3}\right) + 8\left(-0.238 \times 10^{-3}\right)\right) = M_1$$
$$\therefore \ M_1 = 10002.3 \ \text{N-m} \approx 10 \ \text{kN-m}.$$

(III) Software results

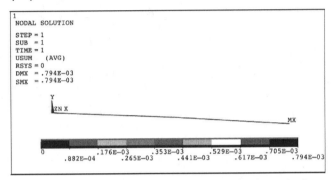

Figure 5.16(c). Deflection pattern for a cantilever beam (refer to Appendix C for color figures).

Deflection values at nodes (in meters)

The following degree of freedom results are in global coordinates

NODE	UX	UY	UZ	USUM
1	0.0000	0.0000	0.0000	0.0000
2	0.0000	−0.31746E-03	0.0000	0.31746E-03
3	0.0000	−0.79365E-03	0.0000	0.79365E-03

Maximum absolute values

NODE	0	3	0	3
VALUE	0.0000	−0.79365E-03	0.0000	0.79365E-03

Rotational deflection values at nodes

The following degree of freedom results are in global coordinates

NODE	ROTZ
1	0.0000
2	−0.23810E-03
3	−0.23810E-03

Finite Element Analysis of Beams

Reaction values

The following X, Y, Z solutions are in global coordinates

NODE	FX	FY	MZ
1	0.0000	5000.0	10000.

Answers for Example 5.5

Parameter	Analytical method	FEM-hand calculations	Software results
Deflection at node			
2	-3.1746×10^{-4} m	-0.3175×10^{-3} m	-0.31746×10^{-3} m
3	-7.9365×10^{-4} m	-0.7937×10^{-3} m	-0.79365×10^{-3} m
Rotational deflection at node			
2	-2.381×10^{-4} rad	-0.2381×10^{-3} rad	-0.2381×10^{-3} rad
3	-2.381×10^{-4} rad	-0.2381×10^{-3} rad	-0.2381×10^{-3} rad
Reaction force at node 1	5 kN	5 kN	5 kN
Reaction moment at node 1	10 kN-m	10 kN-m	10 kN-m

Example 5.6

Propped cantilever beam with distributed load. Find nodal displacements and support reactions for the beam shown in Figure 5.17. Let $E = 70$ GPa and $I = 6 \times 10^{-4}$ m^4.

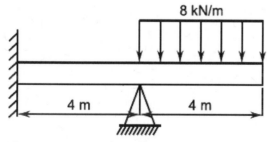

Figure 5.17. Propped cantilever beam with distributed load for Example 5.6.

Solution

(I) Analytical method [refer to Figure 5.17(a)]

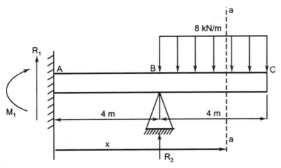

Figure 5.17(a). Analytical method for Example 5.6.

The solution is obtained by Macaulay's method. The number within the brackets <> is to be neglected whenever it is less than zero.

$$\sum F_y = 0 \Rightarrow R_1 = -R_2 + 8 \times 10^3 \times 4 = 32000 - R_2$$

$$\sum M = 0 \Rightarrow M_1 = 4R_2 - (8 \times 10^3 \times 4) \times 6 = (4R_2 - 192000) \text{ N-m}.$$

At section a-a

$$M_x = M_1 + R_1 x + R_2 <x-4> - \frac{8 \times 10^3}{2} <x-4>^2$$

$$EIy'' = (4R_2 - 192000) + (32000 - R_2)x + R_2 <x-4> - 4 \times 10^3 <x-4>^2$$

$$EIy' = 4R_2 x - 192000 x + \left(32000 \times \frac{x^2}{2}\right) - \left(R_2 \times \frac{x^2}{2}\right)$$

$$+ \frac{R_2}{2} <x-4>^2 - \frac{4 \times 10^3 <x-4>^3}{3} + C_1 \qquad (5.9)$$

$$EIy = \left(4R_2 \times \frac{x^2}{2}\right) - \left(192000 \times \frac{x^2}{2}\right) + \left(32000 \times \frac{x^3}{6}\right) - \left(R_2 \times \frac{x^3}{6}\right)$$

$$+ \frac{R_2}{6} <x-4>^3 - \frac{4 \times 10^3 <x-4>^4}{12} + C_1 x + C_2. \qquad (5.10)$$

Boundary conditions are,

At $x = 0, y = 0 \Rightarrow C_2 = 0$

At $x = 0, y' = 0 \Rightarrow C_1 = 0$

At $x = 4, y = 0 \Rightarrow R_2 = 56008$ N

$\therefore \quad R_1 = 32000 - R_2 = -24008$ N

$M_1 = 4R_2 - 192000 = 32032$ N-m (Clockwise), (negative).

Substituting in equations (5.9) and (5.10)

$$EIy' = 4 \times 56008x - 192000x + \left(32000 \times \frac{x^2}{2}\right) - \left(56008 \times \frac{x^2}{2}\right)$$

$$+ \frac{56008}{2}<x-4>^2 - \frac{4 \times 10^3 <x-4>^3}{3}$$

$$y'|_{x=4} = -0.00152 \text{ rad}$$
$$y'|_{x=8} = -0.00355 \text{ rad}$$

and

$$EIy = \left(4 \times 56008 \times \frac{x^2}{2}\right) - \left(192000 \times \frac{x^2}{2}\right) + \left(32000 \times \frac{x^3}{6}\right) - \left(56008 \times \frac{x^3}{6}\right)$$

$$+ \frac{56008}{6}<x-4>^3 - \frac{4 \times 10^3 <x-4>^4}{12}$$

$$y|_{x=8} = y_C = y_3 = -0.0122 \text{ m}$$

$y_A = y_1 = 0$ and $y_B = y_2 = 0$ (Given boundary conditions).

(II) FEM by hand calculations

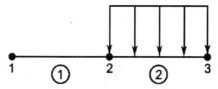

Figure 5.17(b). Finite element model for Example 5.6.

$$E = 70 \times 10^3 \text{ N/mm}^2 \text{ and } I = 6 \times 10^8 \text{ mm}^4$$

Stiffness matrix for element 1 is,

$$[k] = \frac{EI}{L^3} \begin{bmatrix} 12 & 6L & -12 & 6L \\ 6L & 4L^2 & -6L & 2L^2 \\ -12 & -6L & 12 & -6L \\ 6L & 2L^2 & -6L & 4L^2 \end{bmatrix}$$

$$[k_1] = \frac{70 \times 10^3 \times 6 \times 10^8}{(4000)^3} \begin{bmatrix} 12 & 6(4000) & -12 & 6(4000) \\ 6(4000) & 4(4000)^2 & -6(4000) & 2(4000)^2 \\ -12 & -6(4000) & 12 & -6(4000) \\ 6(4000) & 2(4000)^2 & -6(4000) & 4(4000)^2 \end{bmatrix}$$

$$[k_1] = 656.25 \begin{array}{c} \\ \end{array} \begin{matrix} w_1 & \theta_1 & w_2 & \theta_2 \end{matrix}$$
$$[k_1] = 656.25 \begin{bmatrix} 12 & 24000 & -12 & 24000 \\ 24000 & 64 \times 10^6 & -24000 & 32 \times 10^6 \\ -12 & -24000 & 12 & -24000 \\ 24000 & 32 \times 10^6 & -24000 & 64 \times 10^6 \end{bmatrix} \begin{matrix} w_1 \\ \theta_1 \\ w_2 \\ \theta_2 \end{matrix}$$

Due to symmetry,

$$[k_1] = [k_2]$$

$$[k_2] = 656.25 \begin{bmatrix} 12 & 24000 & -12 & 24000 \\ 24000 & 64 \times 10^6 & -24000 & 32 \times 10^6 \\ -12 & -24000 & 12 & -24000 \\ 24000 & 32 \times 10^6 & -24000 & 64 \times 10^6 \end{bmatrix} \begin{matrix} w_2 \\ \theta_2 \\ w_3 \\ \theta_3 \end{matrix}$$

Nodal force calculation
For element 2,

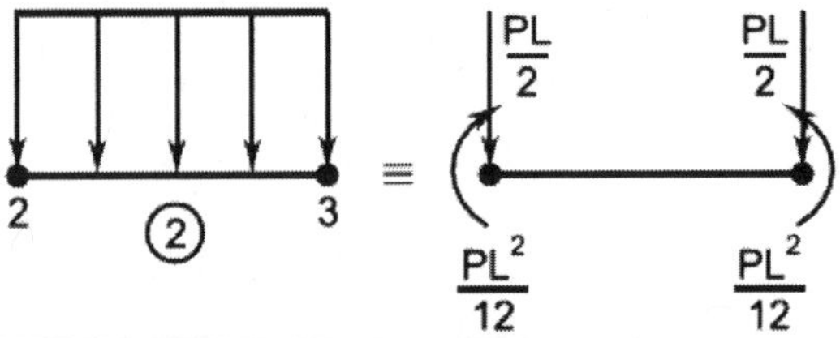

Figure 5.17(c). Nodal force calculation for element 2 in Example 5.6.

Finite Element Analysis of Beams

$$P = 8 \text{ N/mm}$$
$$L = 4000 \text{ mm}$$
$$\frac{PL}{2} = \frac{8 \times 4000}{2} = 16000 \text{ N}$$
$$\frac{PL^2}{12} = \frac{8 \times (4000)^2}{12} = 10.667 \times 10^6 \text{ N-mm.}$$

The nodal forces and moments for element 2 is,

$$[F_2] = \begin{Bmatrix} -\dfrac{PL}{2} \\ -\dfrac{PL^2}{12} \\ -\dfrac{PL}{2} \\ -\dfrac{PL^2}{12} \end{Bmatrix} = \begin{Bmatrix} -1600 \\ -10.667 \times 10^6 \\ -16000 \\ -10.667 \times 10^6 \end{Bmatrix} \begin{matrix} f_1 \\ m_1 \\ f_2 \\ m_2 \end{matrix}.$$

The global equations are,

$$[K]\{r\} = \{R\} \quad (5.11)$$

$$656.25 \begin{bmatrix} 12 & 24000 & -12 & 24000 & 0 & 0 \\ 24000 & 64 \times 10^6 & -24000 & 32 \times 10^6 & 0 & 0 \\ -12 & -24000 & 12+12 & -24000+24000 & -12 & 24000 \\ 24000 & 32 \times 10^6 & -24000+24000 & 64 \times 10^6 + 64 \times 10^6 & -24000 & 32 \times 10^6 \\ 0 & 0 & -12 & -24000 & 12 & -24000 \\ 0 & 0 & 24000 & 32 \times 10^6 & -24000 & 64 \times 10 \end{bmatrix} \begin{matrix} w_1 \\ \theta_1 \\ w_2 \\ \theta_2 \\ w_3 \\ \theta_3 \end{matrix}$$

$$\times \begin{Bmatrix} w_1 \\ \theta_1 \\ w_2 \\ \theta_2 \\ w_3 \\ \theta_3 \end{Bmatrix} = \begin{Bmatrix} 0 + R_1 \\ 0 + M_1 \\ 16000 + R_2 \\ -10.667 \times 10^6 \\ -16000 \\ 10.667 \times 10^6 \end{Bmatrix}$$

By using the elimination method for applying boundary conditions,

$$w_1 = \theta_1 = w_2 = 0.$$

The above matrix reduces to

$$656.25 \begin{bmatrix} 128\times10^6 & -24000 & 32\times10^6 \\ -24000 & 12 & -24000 \\ 32\times10^6 & -24000 & 64\times10^6 \end{bmatrix} \begin{Bmatrix} \theta_2 \\ w_3 \\ \theta_3 \end{Bmatrix} = \begin{Bmatrix} -10.667\times10^6 \\ -16000 \\ 10.667\times10^6 \end{Bmatrix}.$$

Solving the above matrix and equations, we get,

$$w_2 = -12.19 \text{ mm} = -0.01219 \text{ m}$$
$$\theta_3 = -0.00355 \text{ rad}$$
$$\theta_2 = -0.00152 \text{ rad}.$$

Reaction calculation

$$656.25(12w_1 + 24000\times\theta_1 - 12w_2 + 24000\times\theta_2) = R_1$$
$$656.25(12\times0 + 24000\times0 - 12\times0 + 24000\times(-0.00152)) = R_1$$
$$\therefore \quad R_1 = -23.94 \text{ kN} \approx -24 \text{ kN}$$

$$656.25(24000w_1 + 64\times10^6\times\theta_1 - 24000w_2 + 32\times10^6\times\theta_2) = M_1$$
$$656.25(24000\times0 + 64\times10^6\times0 - 24000\times0 + 32\times10^6\times(-0.00152)) = M_1$$
$$\therefore \quad M_1 = -31.92 \text{ kN-m} \approx -32 \text{ kN-m}$$

$$656.25(-12w_1 - 24000\times\theta_1 + 24w_2 + 0\times\theta_2 - 12w_3 + 24000\times\theta_3) = R_2 - 16000$$
$$656.25(-12\times0 - 24000\times0 + 24\times0 + 0\times\theta_2 - 12(-12.19) + 24000\times(-0.00355))$$
$$= R_2 - 16000.$$
$$\therefore \quad R_2 = 56.08 \text{ kN} \approx 56 \text{ kN}$$

(III) Software results

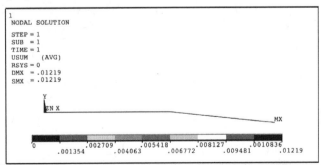

Figure 5.17(d). Deflection pattern for a cantilever beam (refer to Appendix C for color figures).

Deflection values at nodes (in meters)
The following degree of freedom results are in global coordinates

NODE	UX	UY	UZ	USUM
1	0.0000	0.0000	0.0000	0.0000
2	0.0000	0.0000	0.0000	0.31746E-03
3	0.0000	−0.12190E-01	0.0000	0.12190E-01

Rotational deflection values at nodes
The following degree of freedom results are in global coordinates

NODE	ROTZ
1	0.0000
2	−0.15238E-02
3	−0.35556E-02

Reaction values
The following X, Y, Z solutions are in global coordinates

NODE	FX	FY	MZ
1	0.0000	−24000	−32000
2	0.0000	56000	

Answer for Example 5.6

Parameter	Analytical method	FEM-hand calculations	Software results
Deflection at node 3	−0.0122 m	−0.01219 m	−0.01219 m
Rotational deflection at node			
2	−0.00152 rad	−0.00152 rad	−0.001524 rad
3	−0.00355 rad	−0.00355 rad	−0.00355 rad
Reaction force at			
1	−24 kN	−24 kN	−24 kN
2	56 kN	56 kN	56 kN
Reaction moment at node 1	−32 kN-m	−32 kN-m	−32 kN-m

Example 5.7

Propped cantilever beam with varying load. For the beam shown in Figure 5.18, determine the nodal displacements, slopes, reactions, maximum bending moment, shear force, and maximum bending stress. Take E = 200 GPa.

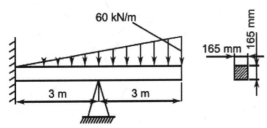

Figure 5.18. Propped cantilever beam with varying load for Example 5.7.

Solution

(I) FEM hand calculations

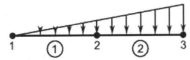

Figure 5.18(a). Finite element model for Example 5.7.

$$I = \frac{bh^3}{12} = \frac{165 \times (165)^3}{12} = 61766718.75 \text{ mm}^4 = 6.18 \times 10^{-5} \text{ m}^4.$$

Stiffness matrices for element 1 and 2 are,

$$[k] = \frac{EI}{L^3} \begin{bmatrix} 12 & 6L & -12 & 6L \\ 6L & 4L^2 & -6L & 2L^2 \\ -12 & -6L & 12 & -6L \\ 6L & 2L^2 & -6L & 4L^2 \end{bmatrix}$$

$$[k_1] = \frac{200 \times 10^9 \times 6.18 \times 10^{-5}}{(3)^3} \begin{bmatrix} 12 & 6(3) & -12 & 6(3) \\ 6(3) & 4(3)^2 & -6(3) & 2(3)^2 \\ -12 & -6(3) & 12 & -6(3) \\ 6(3) & 2(3)^2 & -6(3) & 4(3)^2 \end{bmatrix}$$

$$[k_1] = \frac{200 \times 10^9 \times 6.18 \times 10^{-5}}{(3)^3} \begin{bmatrix} 12 & 18 & -12 & 18 \\ 18 & 36 & -18 & 18 \\ -12 & -18 & 12 & -18 \\ 18 & 18 & -18 & 36 \end{bmatrix}$$

$$[k_1] = 457.78 \times 10^3 \begin{bmatrix} w_1 & \theta_1 & w_2 & \theta_2 \\ 12 & 18 & -12 & 18 \\ 18 & 36 & -18 & 18 \\ -12 & -18 & 12 & -18 \\ 18 & 18 & -18 & 36 \end{bmatrix} \begin{matrix} w_1 \\ \theta_1 \\ w_2 \\ \theta_2 \end{matrix}.$$

Due to symmetry,

$$[k_1] = [k_2]$$

$$[k_1] = 457.78 \times 10^3 \begin{bmatrix} w_2 & \theta_2 & w_3 & \theta_3 \\ 12 & 18 & -12 & 18 \\ 18 & 36 & -18 & 18 \\ -12 & -18 & 12 & -18 \\ 18 & 18 & -18 & 36 \end{bmatrix} \begin{matrix} w_2 \\ \theta_2 \\ w_3 \\ \theta_3 \end{matrix}.$$

Nodal force calculation

For element 1,

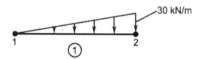

Figure 5.18(b). Nodal force calculation for element 1 for Example 5.7.

$$P_1 = 0 \text{ and } P_2 = 30 \text{ kN/m}$$
$$L = 3 \text{ m}.$$

The nodal forces and moments for element 1 is,

$$[F_1] = \frac{L}{20} \begin{Bmatrix} -(7P_1 + 3P_2) \\ -\frac{L}{3}(3P_1 + 2P_2) \\ -(3P_1 + 7P_2) \\ \frac{L}{3}(2P_1 + 3P_2) \end{Bmatrix} = \frac{3}{20} \begin{Bmatrix} -(7 \times 0 + 3 \times 30 \times 10^3) \\ -\frac{3}{3}(3 \times 0 + 2 \times 30 \times 10^3) \\ -(3 \times 0 + 7 \times 30 \times 10^3) \\ \frac{3}{3}(2 \times 0 + 3 \times 30 \times 10^3) \end{Bmatrix} = \begin{Bmatrix} -13.5 \times 10^3 \\ -9 \times 10^3 \\ -31.5 \times 10^3 \\ 13.5 \times 10^3 \end{Bmatrix} \begin{matrix} f_1 \\ m_1 \\ f_2 \\ m_2 \end{matrix}.$$

For element 2,

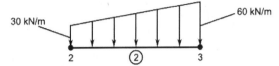

Figure 5.18(c). Nodal force calculation for element 2 for Example 5.7.

$$P_1 = 30 \times 10^3 \text{ N/m and } P_2 = 60 \times 10^3 \text{ N/m}$$
$$L = 3 \text{ m}.$$

The nodal forces and moments for element 2 is,

$$[F_2] = \frac{L}{20} \begin{Bmatrix} -(7P_1+3P_2) \\ -\dfrac{L}{3}(3P_1+2P_2) \\ -(3P_1+7P_2) \\ -\dfrac{L}{3}(2P_1+3P_2) \end{Bmatrix} = \frac{3}{20} \begin{Bmatrix} -(7 \times 30 \times 10^3 + 3 \times 60 \times 10^3) \\ -\dfrac{3}{3}(3 \times 30 \times 10^3 + 2 \times 60 \times 10^3) \\ -(3 \times 30 \times 10^3 + 7 \times 60 \times 10^3) \\ \dfrac{3}{3}(2 \times 30 \times 10^3 + 3 \times 60 \times 10^3) \end{Bmatrix} = \begin{Bmatrix} -58.5 \times 10^3 \\ -31.5 \times 10^3 \\ -76.5 \times 10^3 \\ 36 \times 10^3 \end{Bmatrix} \begin{matrix} f_2 \\ m_2 \\ f_3 \\ m_3 \end{matrix}.$$

The combined nodal forces and moments matrix is,

$$[F] = \begin{Bmatrix} -13.5 \times 10^3 \\ -9 \times 10^3 \\ (-31.5-58.5) \times 10^3 \\ (13.5-31.5) \times 10^3 \\ -76.5 \times 10^3 \\ 36 \times 10^3 \end{Bmatrix} = \begin{Bmatrix} -13.5 \times 10^3 \\ -9 \times 10^3 \\ -90 \times 10^3 \\ -18 \times 10^3 \\ -76.5 \times 10^3 \\ 36 \times 10^3 \end{Bmatrix} \begin{matrix} f_1 \\ m_1 \\ f_2 \\ m_2 \\ f_3 \\ m_3 \end{matrix}.$$

The global equations are,

$$457.78 \times 10^3 \begin{bmatrix} w_1 & \theta_1 & w_2 & \theta_2 & w_3 & \theta_3 \\ 12 & 18 & -12 & 18 & 0 & 0 \\ 18 & 36 & -18 & 18 & 0 & 0 \\ -12 & -18 & 12+12 & -18+18 & -12 & 18 \\ 18 & 18 & -18+18 & 36+36 & -18 & 18 \\ 0 & 0 & -12 & -18 & 12 & -18 \\ 0 & 0 & 18 & 18 & -18 & 36 \end{bmatrix} \begin{Bmatrix} w_1 \\ \theta_1 \\ w_2 \\ \theta_2 \\ w_3 \\ \theta_3 \end{Bmatrix} = 10^3 \times \begin{Bmatrix} -13.5+R_1 \\ -9+M_1 \\ -90+R_2 \\ -18 \\ -76.5 \\ 36 \end{Bmatrix}.$$

By using the elimination method for applying boundary conditions, $w_1 = \theta_1 = w_2 = 0$. The above matrix reduces to,

$$457.78 \begin{bmatrix} 72 & -18 & 18 \\ -18 & 12 & -18 \\ 18 & -18 & 36 \end{bmatrix} \begin{Bmatrix} \theta_2 \\ w_3 \\ \theta_3 \end{Bmatrix} = \begin{Bmatrix} -18 \\ -76.5 \\ 36 \end{Bmatrix}.$$

By solving the above matrix and equations, we get

$\theta_2 = -0.0128$ rad, $w_3 = -0.0811$ m, and $\theta_3 = -0.0319$ rad.

Reaction calculation

$457.78 \times 10^3 (12w_1 + 18\theta_1 - 12w_2 + 18\theta_2) = -13.5 \times 10^3 + R_1$

$457.78 \times 10^3 (12 \times 0 + 18 \times 0 - 12 \times 0 + 18 \times (-0.0128)) = -13.5 \times 10^3 + R_1$

$\therefore \quad R_1 = -91972.5$ N

$457.78 \times 10^3 (18w_1 + 36\theta_1 - 18w_2 + 18\theta_2) = -9 \times 10^3 + M_1$

$457.78 \times 10^3 (18 \times 0 + 36 \times 0 - 18 \times 0 + 18 \times (-0.0128)) = -9 \times 10^3 + M_1$

$\therefore \quad M_1 = -96472.5$ N-m

$457.78 \times 10^3 (-12w_1 - 18\theta_1 + 24w_2 + 0 \times \theta_2 - 12w_3 + 18\theta_3) = -90 \times 10^3 + R_2$

$457.78 \times 10^3 (-12 \times 0 - 18 \times 0 + 24 \times 0 + 0 \times \theta_2 - 12 \times (-0.0811) + 18 \times (-0.0319))$

$= -90 \times 10^3 + R_2.$

$\therefore \quad R_2 = 272654.22$ N

(II) Software results

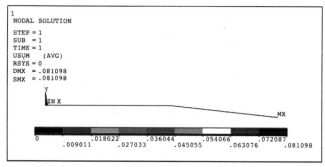

Figure 5.18(d). Deflection pattern for a cantilever beam (refer to Appendix C for color figures).

Deflection values at nodes (in meters)

The following degree of freedom results are in global coordinates

NODE	UX	UY	UZ	USUM
1	0.0000	0.0000	0.0000	0.0000
2	0.0000	0.0000	0.0000	0.0000
3	0.0000	−0.81098E-01	0.0000	0.810898E-01

Maximum absolute values

NODE	0	3	0	3
VALUE	0.0000	−0081098E-01	0.0000	0.81098E-01

Rotational deflection values at nodes

The following degree of freedom results are in global coordinates

NODE	ROTZ
1	0.0000
2	−0.12834E-01
3	−0.31948E-01

Reaction values

The following X, Y, Z solutions are in global coordinates

NODE	FX	FY	MZ
1	0.0000	−92250	−96750
2	0.0000	0.27225E + 06	

Total values

VALUE	0.0000	0.18000E + 06	−96750

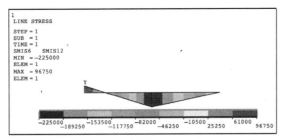

Figure 5.18(e). Bending moment diagram for a propped cantilever beam (refer to Appendix C for color figures).

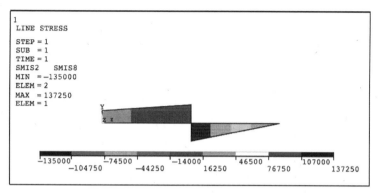

Figure 5.18(f). Shear force diagram for a propped cantilever beam (refer to Appendix C for color figures).

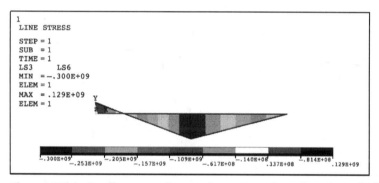

Figure 5.18(g). Bending stress diagram for a propped cantilever beam (refer to Appendix C for color figures).

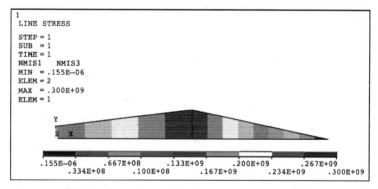

Figure 5.18(h). Maximum stress diagram for a propped cantilever beam (refer to Appendix C for color figures).

Answers for Example 5.7

Parameter	FEM-hand calculations	Software results
Deflection at node 3	−0.0811 m	−0.081098 m
Rotational deflection at node		
2	−0.0128 rad	−0.012834 rad
3	−0.0319 rad	−0.031948 rad
Reaction force at		
1	−91.97 kN	−92.25 kN
2	272.65 kN	272.25 kN
Reaction moment at node 1	−96.47 kN-m	−96.75 kN-m
Maximum bending moment	…..	96750 N-m
Shear force	…..	137250 N
Maximum bending stress	…..	129 MPa
Maximum stress (bending stress + direct stress)	…..	300 MPa

Example 5.8

Propped cantilever beam with stepped loading. Analyze the beam in Figure 5.19 by finite element method and determine the reactions. Also, determine the deflections.

Given $\quad E = 200$ GPa and $I = 5 \times 10^{-4}$ m^4.

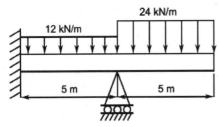

Figure 5.19. Propped cantilever beam with stepped loading for Example 5.8.

Solution

(I) FEM by hand calculations

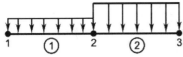

Figure 5.19(a). Finite element model for Example 5.8.

Stiffness matrix for element 1 and 2 are,

$$[k]=\frac{EI}{L^3}\begin{bmatrix} 12 & 6L & -12 & 6L \\ 6L & 4L^2 & -6L & 2L^2 \\ -12 & -6L & 12 & -6L \\ 6L & 2L^2 & -6L & 4L^2 \end{bmatrix}$$

$$[k_1]=\frac{200\times10^9\times5\times10^{-4}}{(5)^3}\begin{bmatrix} 12 & 6(5) & -12 & 6(5) \\ 6(5) & 4(5)^2 & -6(5) & 2(5)^2 \\ -12 & -6(5) & 12 & -6(5) \\ 6(5) & 2(5)^2 & -6(5) & 4(5)^2 \end{bmatrix}$$

$$\begin{matrix} & w_1 & \theta_1 & w_2 & \theta_2 \end{matrix}$$

$$[k_1]=800\times10^3\begin{bmatrix} 12 & 30 & -12 & 30 \\ 30 & 100 & -30 & 50 \\ -12 & -30 & 12 & -30 \\ 30 & 50 & -30 & 100 \end{bmatrix}\begin{matrix} w_1 \\ \theta_1 \\ w_2 \\ \theta_2 \end{matrix}$$

$$\begin{matrix} & w_2 & \theta_2 & w_3 & \theta_3 \end{matrix}$$

$$[k_1]=800\times10^3\begin{bmatrix} 12 & 30 & -12 & 30 \\ 30 & 100 & -30 & 50 \\ -12 & -30 & 12 & -30 \\ 30 & 50 & -30 & 100 \end{bmatrix}\begin{matrix} w_2 \\ \theta_2 \\ w_3 \\ \theta_3 \end{matrix}.$$

Nodal force calculation

For element 1,

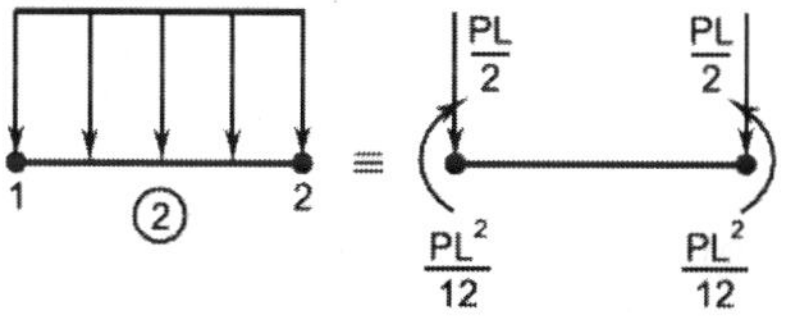

Figure 5.19(b). **Nodal force calculation for element 1 in Example 5.8.**

$$P=12\text{ kN/m}=12\times10^3\text{ N/m}$$

$$\text{L}=5\text{ m}$$

$$\frac{\text{PL}}{2}=\frac{12\times10^3\times5}{2}=30\times10^3\text{ N}$$

$$\frac{\text{PL}^2}{12}=\frac{12\times10^3\times(5)^2}{2}=25\times10^3\text{ N-m.}$$

The nodal forces and moments for element 1 is,

$$[F_1] = \begin{Bmatrix} -\dfrac{PL}{2} \\ -\dfrac{PL^2}{12} \\ -\dfrac{PL}{2} \\ \dfrac{PL^2}{12} \end{Bmatrix} = \begin{Bmatrix} -30\times 10^3 \\ -25\times 10^3 \\ -30\times 10^3 \\ 25\times 10^3 \end{Bmatrix} \begin{matrix} f_1 \\ m_1 \\ f_2 \\ m_2 \end{matrix}.$$

For element 2,

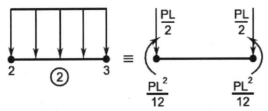

Figure 5.19(c). Nodal force calculation for element 2 in Example 5.8.

$P = 24 \text{ kN/m} = 24\times 10^3 \text{ N/m}$

$L = 5 \text{ m}$

$\dfrac{PL}{2} = \dfrac{24\times 10^3 \times 5}{2} = 60\times 10^3 \text{ N}$

$\dfrac{PL^2}{12} = \dfrac{24\times 10^3 \times (5)^2}{2} = 50\times 10^3 \text{ N-m}.$

The nodal forces and moments for element 2 is,

$$[F_2] = \begin{Bmatrix} -\dfrac{PL}{2} \\ -\dfrac{PL^2}{12} \\ -\dfrac{PL}{2} \\ \dfrac{PL^2}{12} \end{Bmatrix} = \begin{Bmatrix} -60\times 10^3 \\ -50\times 10^3 \\ -60\times 10^3 \\ 50\times 10^3 \end{Bmatrix} \begin{matrix} f_2 \\ m_2 \\ f_3 \\ m_3 \end{matrix}.$$

Finite Element Analysis of Beams

The combined nodal forces and moments is,

$$[F] = \begin{Bmatrix} -30 \times 10^3 \\ -25 \times 10^3 \\ -30 \times 10^3 - 60 \times 10^3 \\ 25 \times 10^3 - 50 \times 10^3 \\ -60 \times 10^3 \\ 50 \times 10^3 \end{Bmatrix} = \begin{Bmatrix} -30 \times 10^3 \\ -25 \times 10^3 \\ -90 \times 10^3 \\ -25 \times 10^3 \\ -60 \times 10^3 \\ 50 \times 10^3 \end{Bmatrix} \begin{matrix} f_1 \\ m_1 \\ f_2 \\ m_2 \\ f_3 \\ m_3 \end{matrix}.$$

The global equations are,

$$[K]\{r\} = \{R\} \tag{5.12}$$

$$800 \times 10^3 \begin{bmatrix} 12 & 30 & -12 & 30 & 0 & 0 \\ 30 & 100 & -30 & 50 & 0 & 0 \\ -12 & -30 & 12+12 & -30+30 & -12 & 30 \\ 30 & 50 & -30+30 & 100+100 & -30 & 50 \\ 0 & 0 & -12 & -30 & 12 & -30 \\ 0 & 0 & 30 & 50 & -30 & 100 \end{bmatrix} \begin{Bmatrix} w_1 \\ \theta_1 \\ w_2 \\ \theta_2 \\ w_3 \\ \theta_3 \end{Bmatrix} = \begin{Bmatrix} -30 \times 10^3 + R_1 \\ -25 \times 10^3 + M_1 \\ -90 \times 10^3 + R_2 \\ -25 \times 10^3 \\ -60 \times 10^3 \\ 50 \times 10^3 \end{Bmatrix}.$$

By using the elimination method for applying boundary conditions,

$$w_1 = \theta_1 = w_2 = 0,$$

the above matrix reduces to

$$800 \times 10^3 \begin{bmatrix} 200 & -30 & 50 \\ -30 & 12 & -30 \\ 50 & -30 & 100 \end{bmatrix} \begin{Bmatrix} \theta_2 \\ w_3 \\ \theta_3 \end{Bmatrix} = \begin{Bmatrix} -25 \times 10^3 \\ -60 \times 10^3 \\ 50 \times 10^3 \end{Bmatrix}.$$

By solving the above matrix and equations, we get

Reaction calculation

$$800 \times 10^3 (12 w_1 + 30 \theta_1 - 12 w_2 + 30 \theta_2) = -30 \times 10^3 + R_1$$

$$800 \times 10^3 (12 \times 0 + 30 \times 0 - 12 \times 0 + 30 \times (-0.003438)) = -30 \times 10^3 + R_1$$

$$\therefore \quad R_1 = -52512 \text{ N}$$

$$800 \times 10^3 \left(30w_1 + 100\theta_1 - 30w_2 + 50\theta_2\right) = -25 \times 10^3 + M_1$$
$$800 \times 10^3 \left(30 \times 0 + 100 \times 0 - 30 \times 0 + 50 \times (-0.003438)\right) = -25 \times 10^3 + M_1$$
$$\therefore \quad M_1 = -112520 \text{ N-m}$$

$$800 \times 10^3 \left(-12w_1 - 30\theta_1 + 24w_2 + 0 \times \theta_2 - 12w_3 + 30\theta_3\right) = -90 \times 10^3 + R_2$$
$$457.78 \times 10^3 \left(-12 \times 0 - 30 \times 0 + 24 \times 0 + 0 \times \theta_2 - 12 \times (-0.0035938) + 30 \times (-0.008438)\right)$$
$$= -90 \times 10^3 + R_2.$$
$$\therefore \quad R_2 = 232492.8 \text{ N}$$

(II) Software results

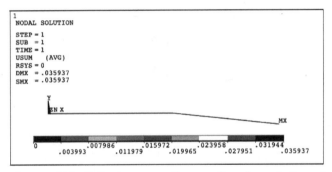

Figure 5.19(d). Deflection pattern for a cantilever beam (refer to Appendix C for color figures).

Deflection values at nodes (in meters)

The following degree of freedom results are in global coordinates

NODE	UX	UY	UZ	USUM
1	0.0000	0.0000	0.0000	0.0000
2	0.0000	0.0000	0.0000	0.0000
3	0.0000	−0.35938E-01	0.0000	0.35938E-01

Maximum absolute values

NODE	0	3	0	3
VALUE	0.0000	−0.35938E-01	0.0000	0.35938E-01

Rotational deflection values at nodes
The following degree of freedom results are in global coordinates

NODE	ROTZ
1	0.0000
2	−0.34375E−02
3	−0.84375E−02

Reaction values
The following X, Y, Z solutions are in global coordinates

NODE	FX	FY	MZ
1	0.0000	−52500	−0.11250E+06
2	0.0000	0.23250E+06	

Answers for Example 5.8

Parameter	FEM-hand calculations	Software results
Deflection at node 3	−0.035938 m	−0.035938 m
Rotational deflection at node		
2	−0.003438 rad	−0.0034375 rad
3	−0.003438 rad	−0.0034375 rad
Reaction force at		
1	−52.512 kN	−52.5 kN
2	232.493 kN	232.25 kN
Reaction moment at node 1	−112.52 kN-m	−112.5 kN-m

Procedure for solving the problems using ANSYS® 11.0 academic teaching software.
For Example 5.7

PREPROCESSING

1. **Main Menu > Preprocessor > Element Type > Add/Edit/Delete > Add > Beam > 2D elastic 3 > OK > Close**

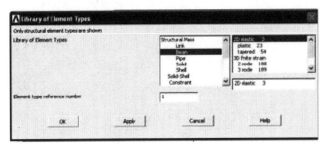

Figure 5.20. Element selection.

2. **Main Menu > Preprocessor > Sections > Beam > Common sections,** following dialog box appears

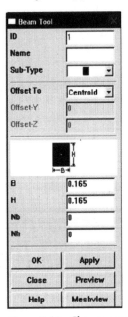

Figure 5.21. Choose cross-section of the beam.

In that dialog box, select **Sub-Type,** choose **Square Cross-Section,** then Enter value of B = 0.165 and H = 0.165 as shown in Figure 5.21.
Click on **Preview > OK**
The following figure appears on the screen.

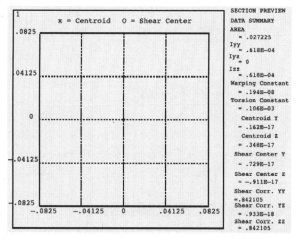

Figure 5.22. Details of geometrical properties of the beam.

From Figure 5.22, note down the values of Area A = 0.027225 m² and moment of inertia $I_{zz} = 0.681 \times 10^{-4}$ m⁴.

3. **Main Menu > Preprocessor > Real Constants > Add/Edit/Delete > Add > OK**

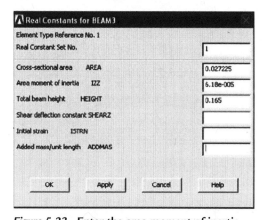

Figure 5.23. Enter the area moment of inertia.

Cross-sectional area AREA > **Enter 0.027225**
Area moment of inertia I_{ZZ} > **Enter 0.618e-4**
Total beam height HEIGHT > **Enter 0.165 > OK > Close**
Enter the material properties.

4. **Main Menu > Preprocessor > Material Props > Material Models**
   Material Model Number 1, click **Structural > Linear > Elastic > Isotropic**
   Enter **EX = 200E9 and RRXY = 0.3 > OK**
   (**Close** the Define Material Model Behavior window.)
   Create the nodes and elements as shown in the figure.

5. **Main Menu > Preprocessor > Modeling > Create > Nodes > In Active CS**
   Enter the coordinated of node 1 > **Apply** Enter the coordinates of node 2 > **Apply** Enter the coordinate of node 3 > **OK**.

Node locations		
Node number	X-coordinate	Y-coordinate
1	0	0
2	3	0
3	6	0

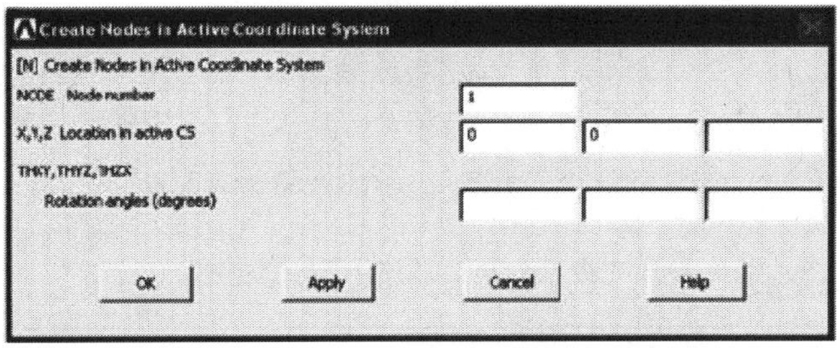

   Figure 5.24. Enter the node coordinate.

6. **Main Menu > Preprocessor > Modeling > Create > Elements > Auto Numbered > Thru nodes** Pick the 1st and 2nd node > **Apply** Pick the 2nd and 3rd node > **OK**

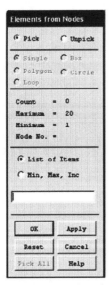

Figure 5.25. Pick the nodes to create elements.

Apply the displacement boundary conditions and loads.

7. **Main Menu > Preprocessor > Loads > Define Loads > Apply > Structural > Displacement > On Nodes** Pick the 1st node > **Apply > All DOF = 0 > OK**
8. **Main Menu > Preprocessor > loads > Define Loads > Apply > Structural > Displacement > On Nodes** Pick the 2nd node > **Apply > Select UX and UY = 0 > OK**

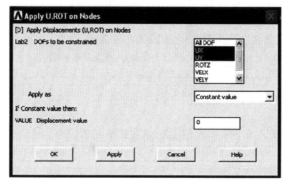

Figure 5.26. Applying boundary conditions on node 2.

9. **Main Menu > Preprocessor > Loads > Define Loads > Apply > Structural > Pressure > On Beams** Pick the 1st element > **OK > Enter Pressure value at node I = 0 and Pressure value at node J = 30e3 > OK**

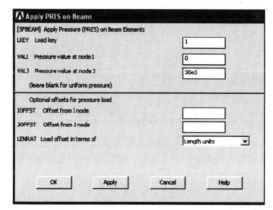

Figure 5.27. Applying loads on element 1.

10. **Main Menu > Preprocessor > Loads > Define Loads > Apply > Structural > Pressure > On Beams** Pick the 2nd element > **OK > Enter Pressure value at node I = 30e3 and Pressure value at node J = 60e3 > OK**

Figure 5.28. Model with loading and displacement boundary conditions.

The model-building step is now complete, and we can proceed to the solution. First to be safe, save the model.

Solution

The interactive solution proceeds.

11. **Main Menu > Solution > Solve > Current LS > OK**
    The **/STATUS Command** window displays the problem parameters and **the Solve Control Load Step** window is shown. Click the solution options in the **/STATUS** window and if all is OK, select **File > Close**.
    In the **Solve Current Load Step** window, select **OK,** and when the solution is complete, **close** the **"Solution is Done!'** window.

POST-PROCESSING

We can now plot the results of this analysis and also list the computed values.

12. **Main Menu > General Postproc > Plot Results > Contour Plot > Nodal Solu > DOF Solution > Displacement vector sum > OK**
    This result is shown in Figure 5.18(d).
13. **Main Menu > General Postproc > List Results > Nodal Solu > Select Roatation vector sum > OK**
14. **Main Menu > General Postproc > List Results > Reaction Solu > PL**
    To find the **bending moment diagram** following procedure is followed.
15. **Main Menu > General Postproc > Element Table > Define Table > Add**

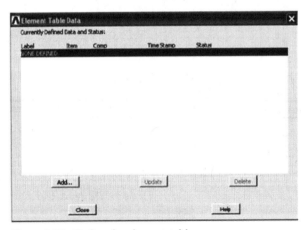

Figure 5.29. Define the element table.

Select **By sequence num and SMISC** and type **6 after SMISC** (as shown in Figure 5.30) > **APPLY**

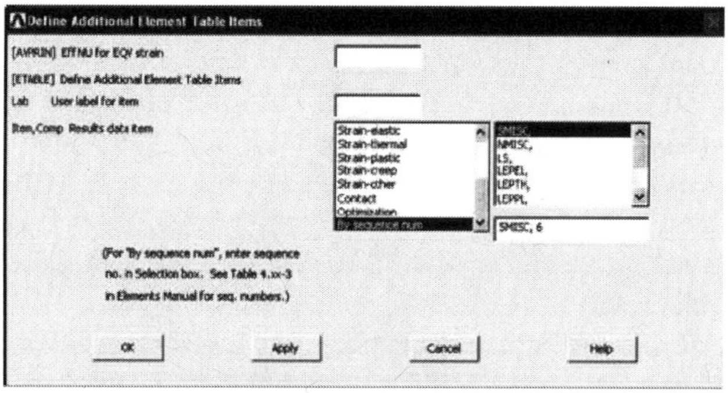

Finite Element Analysis of Beams

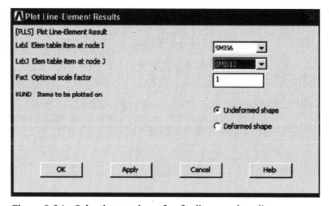

Figure 5.30. Selecting options in element table.

Then again select **By sequence num and SMISC** and type **12 after SMISC > OK**

16. **Main Menu > General Postproc > Plot Results > Contour Plot > Line Elem Res >** Select **SMIS 6 and SMIS 12 in the rows of LabI and LabJ,** respectively as shown in Figure 5.31 > **OK**

Figure 5.31. Selecting options for finding out bending moment.

This result is shown in Figure 5.18(e).

To find the **shear force diagram** following procedure is followed.

17. **Main Menu > General Postproc > Element Table > Define Table > Add**
Select **By sequence num and SMISC** and type **2 after SMISC > APPLY**
Then again select **By sequence num and SMISC** and type **8 after SMISC > OK > Close**

18. **Main Menu > General Postproc > Plot Results > Contour Plot > Line Elem Res >** Select **SMIS 2 and SMIS 8 > OK**
This result is shown in Figure 5.18(f).

To find the **bending stress** the following procedure is followed.

19. **Main Menu > General Postproc > Element Table > Define Table > Add**
    Select **By sequence num and LS** and type **3 after LS > APPLY**
    Then again select **By sequence num and LS** and type **6 after LS > OK**
20. **Main Menu > General Postproc > Plot Results > Contour Plot > Line Elem Res >**
    Select **LS 3 and LS 6 > OK**
    This result is shown in Figure 5.18(g).
    To find the **maximum stress (direct stress + bending stress)** following procedure is followed.
21. **Main Menu > General Postproc > Element Table > Define Table > Add**
    Select **By sequence num and NMISC** and **type 1 after NMISC > APPLY**
    Then again select **By sequence num and NMISC and type 3 after NMISC > OK**
22. **Main Menu > General Postproc > Plot Results > Contour Plot > Line Elem Res >** Select **NMISC 1 and NIMS 3 > OK**

PROBLEMS

1. For the bean shown in Figure 5.32, determine the deflection, slopes, reactions, maximum bending moment, shear force, and maximum bending stress. Take E = 210 GPa and $I = 7 \times 10^{-4}$ m^4.

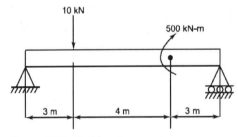

Figure 5.32. Problem 1

2. Find the deflection, slopes, reactions, maximum bending moment, shear force, and maximum bending stress for the aluminum beam shown in Figure 5.33. Take E = 200 GPa and $I = 3 \times 10^{-4}$ m^4.

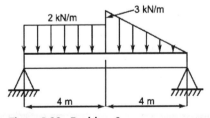

Figure 5.33. Problem 2

3. Find the deflection at the load and the slopes at the end for the shaft shown in Figure 5.34. Also find the maximum bending moment, maximum bending stress, and reactions developed in the bearings. Consider the shaft to be simply supported at bearings A and B. Take E = 200 GPa.

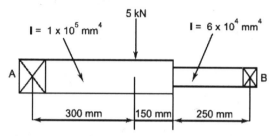

Figure 5.34. Problem 3

4. Find the deflection of the bean shown in Figure 5.35 under self-weight. Take E = 200 GPa and mass density ρ = 7800 kg/m³.

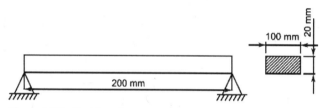

Figure 5.35. Problem 4

5. Find the deflection and bending stress distribution for the cantilever beam shown in Figure 5.36 under combined loading. Take E = 200 GPa.

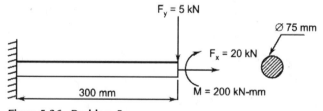

Figure 5.36. Problem 5

 $\left[I = \dfrac{\pi d^4}{64} \right]$

6. For the beam shown in Figure 5.37, determine the deflection at nodes and reaction. Also, plot the bending moment diagram, shear force diagram and find the bending stress. Take E = 200 GPa and $I = 8 \times 10^{-4}$ m^4.

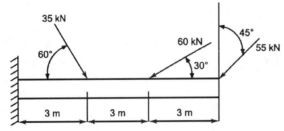

Figure 5.37. Problem 6

7. A cantilever beam is shown in Figure 5.38. Using 2 beam elements determine the nodal deflection and reaction. Take $E = 0.25 \times 10^5$ N/mm^2 and $I = 8 \times 10^{-4}$ m^4.

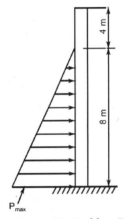

Figure 5.38. Problem 7

$[P_{max} = \rho \times g \times h]$

8. Determine the deflection, reaction, and bending stress for the beam shown in Figure 5.39. Also, plot the bending moment and shear force diagram. Take E = 207 GPa, W = 150 N/mm, h = 800 mm, b = 400 mm, t_1 = 40 mm, t_2 = 40 mm, and t_3 = 50 mm.

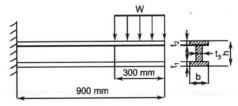

Figure 5.39. Problem 8

9. Figure 5.40 presents a beam fixed at one end, supported by a cable at the other end, subjected to a uniformly distributed load of 70 lb/in. Take $E = 30 \times 10^6$ psi, Beam cross-section = 4 in × 4 in, and cable cross-section = 1 in². Determine the finite element equilibrium equations of the system by using one finite element for the beam and one finite element for the cable, the displacement of nodes 1 and 2, and the stress distribution in the beam and in the cable.

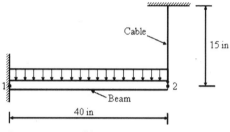

Figure 5.40. Problem 9

REFERENCES

1. Y. W. Hwon and H. Bang, "The Finite Element Method Using MATLAB, Second Edition," CRC Press, 2000.
2. D. L. Logan, "A First Course in the Finite Element Method, Fifth Edition," Cengage Learning, 2012.
3. S. Moaveni, "Finite Element Analysis: Theory and Application with ANSYS, Third Edition," Prentice Hall, 2008.
4. J. N. Reddy, "An Introduction to the Finite Element Method, Third Edition," McGraw Hill Higher Education, 2004.
5. C. T. F. Ross, "Finite Element Method in Structural Mechanics," Ellis Horwood Limited Publishers, 1985.
6. F. L. Stasa, "Applied Finite Element Analysis for Engineering," Holt, Rinehart and Winston, 1985.
7. L. J. Segerlind, "Applied Finite Element Analysis, Second Edition," John Wiley and Sons, 1984.
8. S. S. Rao, "The Finite Element Method in Engineering, Fifth Edition," Butterworth-Heinemann, 2011.
9. N. Willems and W. M. Lucas, Jr., "Structural Analysis for Engineering," McGraw-Hill, 1978.
10. R. C. Hibbleer, "Mechanics of Materials, Second Edition," Macmillan, 1994.

Chapter 6

Stress Analysis of a Rectangular Plate with a Circular Hole

6.1 INTRODUCTION

Two dimensional problems in structural analysis are dealt with in this chapter. Hand calculations, even with 2 elements, become too long and hence are not given for these problems: only analytical method solutions and software solutions using ANSYS have been provided.

Two dimensional problems can either be plane stress or plane strain problems. Method of analysis is the same for both, except that stress strain matrix is different in 2 cases.

Plane bodies that are flat and of constant thickness that are subjected to in-plane loading fall under the category of plane stress problems. Stress components σ_z, τ_{xz}, and τ_{yz} assume zero values in these problems.

Some of the elements used in the analysis of 2 dimensional problems are constant strain triangles (CST), linear strain triangle (LST), linear quadrilateral, isoparametric quadrilateral, etc. Each of these elements has 2 degrees of freedom per node namely the translation in x and y directions.

Stress within the element may be calculated using the equation,

$$\{\sigma\} = [D][B]\{q\}. \tag{6.1}$$

6.2 A RECTANGULAR PLATE WITH A CIRCULAR HOLE

The stress analysis of a rectangular plate with a circular hole problem is assumed as a 2 dimensional plane stress problem. Plane stress is defined as a state of stress in which the normal stress and the shear stress directed perpendicular to the plane are assumed to be negligible.

The above problem can be categorized into 3 sub cases.

Sub Case 1

A rectangular plate with a very small circular hole at the center with one vertical edge fixed and the other vertical edge is acted upon by a horizontal tensile load in the form of pressure.

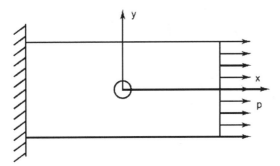

Figure 6.1. Rectangular plate with a very small circular hole subjected to tensile load at one edge.

Sub Case 2

A rectangular plate with a small circular hole at the center and a horizontal tensile force in the form of pressure is acting on both the vertical edges of the plate.

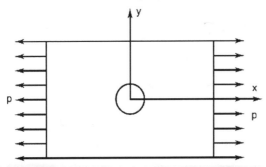

Figure 6.2. Rectangular plate with a hole subjected to tensile load at both the edges.

The above problem is solved by exploiting the symmetric geometry and symmetric loading boundary conditions. Now we can draw the above Figure 6.2 as below for the analysis purpose (refer Figure 6.3).

Place the origin of *x–y* coordinates at the center of the hole and pull on both ends of the plate. Then points on the centerlines will not move perpendicular to them but move along the centerlines. This indicated the appropriate displacement conditions to use as shown in Figure 6.3.

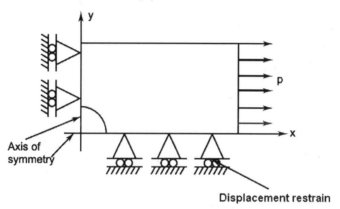

Figure 6.3. Finite element model of one-quarter of the plate.

Sub Case 3

A rectangular plate with a large circular hole at the center and a uniform pressure acts on the boundary of the hole.

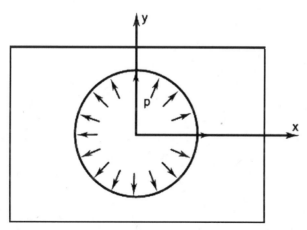

Figure 6.4. Rectangular plate with a hole subjected to uniform pressure at the boundary of the hole.

The above problem can be solved considering one quarter of the plate and by exploiting the symmetric geometry and loading conditions. The finite element model is shown below.

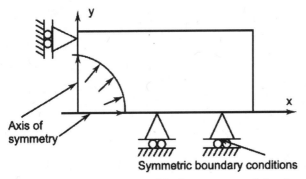

Figure 6.5. Finite element model of one-quarter of the plate.

Example 6.1

A rectangular plate of size 1000 mm × 500 mm is subjected to uniform pressure as shown in Figure 6.6. The plate has a thickness of 10 mm and has a central hole 50 mm in diameter. The material of the plate is steel with Young's modulus E = 210 GPa and Poisson's ratio, $v = 0.3$. Assume a case of plane stress. Plot the Von Mises stress distribution and compare result with analytical method.

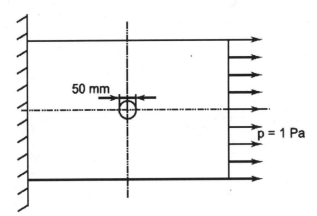

Figure 6.6. Rectangular plate with very small circular hole at the center of the plate.

Solution

(1) Analytical method

Comparing the above case with the infinite plate with a very small circular hole, for this, the stress concentration factor is (SCF) = 3.

$$\text{SCF} = \frac{\text{Maximum stress}}{\text{Nominal stress}}. \quad (6.2)$$

Hence,

$$\text{Tensile force} = \text{Pressure} \times \text{cross-sectional area} \quad (6.3)$$
$$\text{Tensile force} = 1 \times 0.5 \times 0.01 = 0.005 \text{ N}$$

$$\text{Nominal stress} = \frac{\text{Tensile force}}{\text{Cross-sectional area}} \quad (6.4)$$

$$\text{Nominal stress} = \frac{0.005}{0.5 \times 0.01} = 1 \text{ N/m}^2$$

Maximum stress = SCF × Nominal stress = 3 × 1 = 3 Pa.

(II) Software results

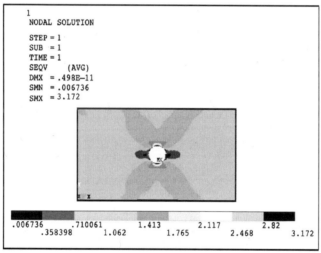

Figure 6.6(a). Von Mises stress distribution pattern (refer to Appendix C for color figures).

From the software, we got, Maximum stress (Von Mises stress) = 3.172 Pa.

Answers for Example 6.1

Parameter	Analytical method	Software results	Percentage of error
Maximum stress	3 Pa	3.172 Pa	5.42

Example 6.2

A rectangular plate with hole at the center is subjected to uniform pressure as shown in Figure 6.7. The plate is under plane stress. Find the maximum deflection and maximum stress distribution. Also find the deformed shape of the hole. Assume plate thickness, $t = 25$ mm, $E = 207$ GPa, and $v = 0.3$.

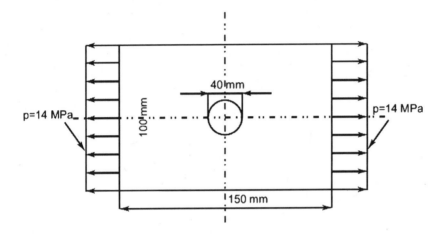

Figure 6.7. **Rectangular plate with a hole with symmetrical loading.**

Solution

(I) Analytical method

$$\text{Geometric factor} = \frac{\text{Diameter of hole}}{\text{Width of plate}} = \frac{d}{w} \qquad (6.5)$$

$$\text{Geometric factor} = \frac{40}{100} = 0.4.$$

From the design data handbook,

for $\frac{d}{w}$ of 0.4 the stress concentration factor (SCF) = 2.25

$$\text{SCF} = \frac{\text{Maximum stress}}{\text{Nominal stress}}.$$

Hence,

Tensile force = Pressure × Cross-sectional area = $14 \times 100 \times 25 = 35000$ N

$$\text{Nominal stress} = \frac{\text{Tensile force}}{\text{Cross-sectional area}} = \frac{\text{Tensile force}}{(w-d)t} = \frac{35000}{(100-40)25}$$
$$= 23.33 \text{ MPa}$$

Maximum stress = SCF × Nominal stress = 2.25 × 23.333 = 52.5 MPa.

(II) Software results

For the analysis using software, one quarter of the plate is modeled and analyzed.

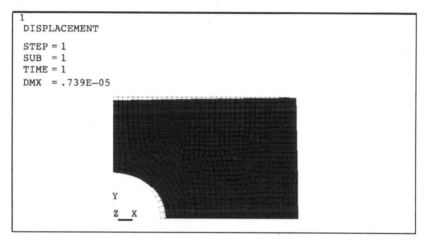

Figure 6.7(a). Deformed shape of the hole (refer to Appendix C for color figures).

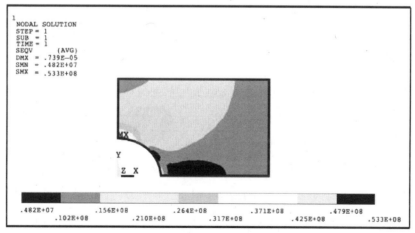

Figure 6.7(b). Von Mises stress distribution pattern (refer to Appendix C for color figures).

From the software, we got, maximum stress = 53.3 MPa.

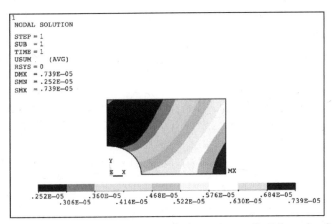

Figure 6.7(c). Deflection pattern (refer to Appendix C for color figures).

Answers for Example 6.2

Parameter	Analytical method	Software results	Percentage of error
Maximum stress	52.5 MPa	53.3 MPa	1.5
Maximum deflection	……	7.39×10^3 mm	…..

Example 6.3

Determine the stress distribution and displacement for a rectangular plate with a hole at the center of the plate with uniform thickness of 10 mm. A uniform pressure of p = 10 MPa acts on the boundary of the hole as shown in Figure 6.8. Assume Young's modulus E = 120 GPa and the Poisson's ratio is 0.28. Assume plane stress condition.

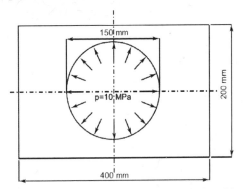

Figure 6.8. Rectangular plate with a hole subjected to uniform pressure at the boundary of the hole.

Solution

(1) Software results

For the analysis using software, one quarter of the plate is modeled and analyzed.

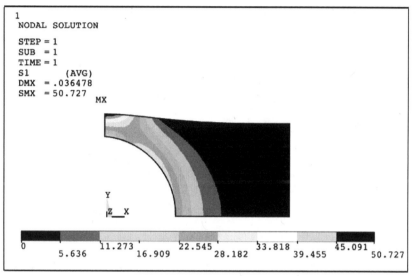

Figure 6.8(a). First principal stress distribution pattern (refer to Appendix C for color figures).

From the software, we got maximum stress = 50.727 MPa.

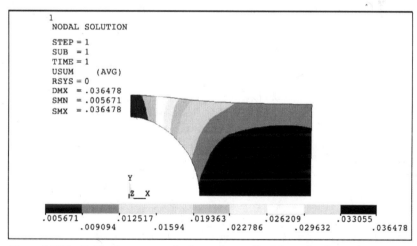

Figure 6.8(b). Deflection pattern (refer to Appendix C for color figures).

Validation of the Results

The reactions at the supports must balance the applied forces. Therefore, from the software, the total reaction force in the x-direction is -7500 N.

Applied force = (pressure) × (projected distance in x-direction of the
line along which the constant pressure acts) × (thickness)
$$= p \times r \times t \qquad (6.6)$$

Applied force = $10 \times 75 \times 10 = 7500$ N in positive x-direction.
So the reaction cancels out the applied force in the x-direction.

Answers for Example 6.3

Parameter	Software results
Maximum stress	50.727 MPa
Maximum deflection	0.036478 mm

Procedure for solving the problem using ANSYS® 11.0 academic teaching software.
For Example 6.2

PREPROCESSING

1. Main Menu > Preprocessor > Element Type > Add/Edit/Delete > Add > Structural Solid > Quad 4 node 42 > OK

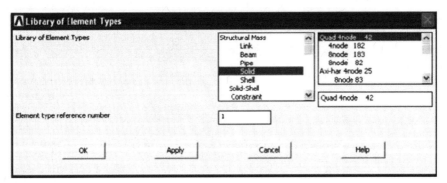

Figure 6.9. Element selection.

Select the option where you define the plate thickness.

2. **Options (Element behavior K3) > Plane strs w/thk > OK > Close**

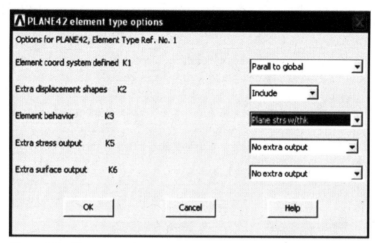

Figure 6.10. Element options.

3. **Main Menu > Preprocessor > Real Constants > Add/Edit/Delete > Add > OK**

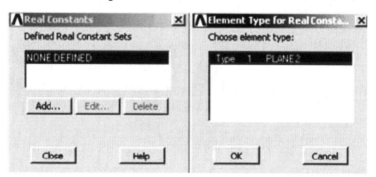

Figure 6.11. Real constants.

(Enter the plate thickness of 0.025 m) > **Enter 0.025 > OK > Close**

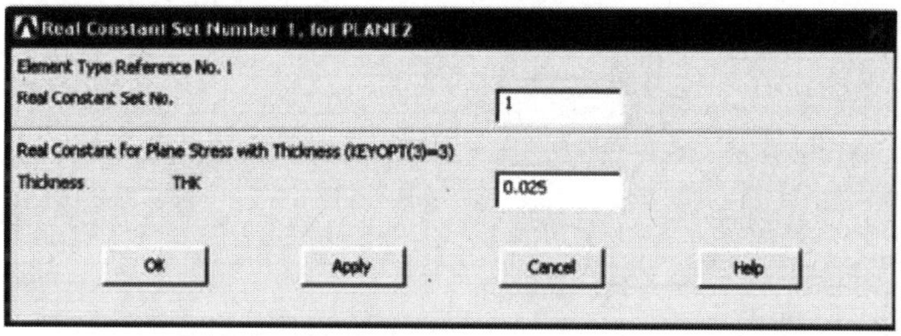

Figure 6.12. Enter the plate thickness.

Enter the material properties.

4. **Main Menu > Preprocessor > Material Props > Material Models**
   Material Model Number 1, click **Structural > Linear > Elastic > Isotropic**
   Enter **EX = 2.07E11** and **PRXY = 0.3 > OK** (**Close** the Define Material Model Behavior window.)
   Create the geometry for the upper-right quadrant of the plate by subtracting a 0.04 m diameter circle from a 0.075 × 0.05 m rectangle. Generate the rectangle first.
5. **Main Menu > Preprocessor > Modeling > Create > Areas > Rectangle > By 2 Corners**
   Enter (lower left corner) **WP X = 0.0, WP Y = 0.0 and Width = 0.075, Height = 0.05 > OK**
6. **Main Menu > Preprocessor > Modeling > Create > Areas > Circle > Solid Circle** Enter **WP X = 0.0, WP Y = 0.0 and Radius = 0.02 > OK**

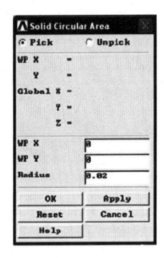

Figure 6.13. Create areas.

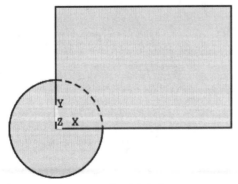

Figure 6.14. Rectangle and circle.

Now **subtract** the circle form the rectangle. (Read the messages in the window at the bottom of the screen as necessary.)

7. **Main Menu > Preprocessor > Modeling > Operate > Booleans > Subtract > Areas**

   Pick the rectangle > **OK,** then pick the circle > **OK**

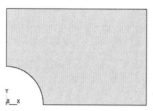

Figure 6.15. Geometry for quadrant of plate.

Create a mesh triangular element over the quadrant area.

8. **Main Menu > Preprocessor > Meshing > Mesh Tool**

   The **Mesh Tool** dialog box appears. In that dialog box, click on the **Smart Size** and move the slider available below the **Smart Size to 2** (i.e., towards **Fine** side). Then close the Mesh Tool box.

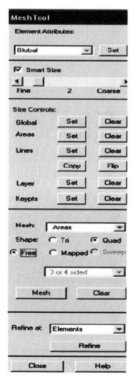

Figure 6.16. Mesh tool box.

9. **Main Menu > Preprocessor > Meshing > Mesh > Areas > Free** Pick the quadrant > **OK**

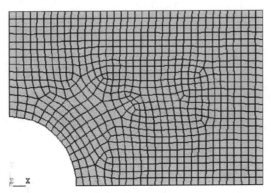

Figure 6.17. Quad element mesh.

Apply the displacement boundary conditions and loads.

10. **Main Menu > Preprocessor > Loads > Define Loads > Apply > Structural > Displacement > On Lines** Pick the left edge of the quadrant > **OK > UX = 0 > OK**

11. **Main Menu > Preprocessor > Loads > Define Loads > Apply > Structural > Displacement > On Lines** pick the bottom edge of the quadrant > **OK > UY = 0 > OK**

12. **Main Menu > Preprocessor > Loads > Define Loads > Apply > Structural > Displacement > On Lines.** Pick the right edge of the quadrant > **OK > Pressure = −14E6 > OK**

    (A positive pressure would be a compressive load, so we use a negative pressure. The pressure is shown as a single arrow.)

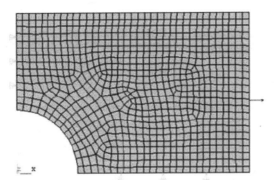

Figure 6.18. Model with loading and displacement boundary conditions.

The model-building step is now complete, and we can proceed to the solution. First to be safe, save the model.

Solution

The interactive solution proceeds

13. **Main Menu > Solution > Solve > Current LS > OK**
    The **/STATUS Command** window displays the problem parameters and **the Solve Current Load Step** window is shown. Check the solution options in the **/STATUS** window and if all is OK, select **File > Close**.
    In the **Solve Current Load Step** window, select **OK,** and when the solution is complete close the **'Solution is Done!'** window.

POST-PROCESSING

We can now plot the results of this analysis and also list the computed values. First examine the deformed shape.

14. **Main Menu > General Posrproc > Plot Results > Deformed Shape > Def. + Undeformed > OK**
    This result is shown in Figure 6.7(a).

15. **Main Menu > General Posrproc > Plot Results > Contour Plot > Nodal Solu > Stress > Von Mises stress > OK**
    This result is shown in Figure 6.7(b).

16. **Main Menu > General Posrproc > Plot Results > Contour Plot > Nodal Solu > DOF Solution > Displacement vector sum > OK**
    This result is shown in Figure 6.7(c).

PROBLEMS

1. Find the maximum stress in the aluminum plate shown in Figure 6.19. Consider an aluminum plate 10 mm thick with a hole at the center of the plate. Assume plane stress condition. Take E = 70 GPa and v = 0.35. Also, calculate the maximum stress by analytical method and compare the results.

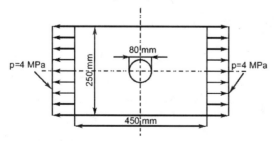

Figure 6.19. Problem 1

2. Find the maximum stress for the plate shown in Figure 6.20 if the hole is located halfway between the center line and the top edge as shown. Take E = 70 GPa and v = 0.35. Assume plane stress condition.

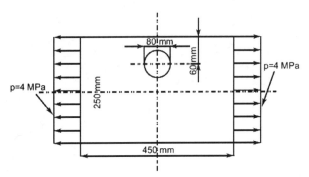

Figure 6.20. Problem 2

[*Model half of the plate by taking symmetry about y-axis.*]

3. For the plate shown in Figure 6.21, find the maximum stress. Take Young's modulus E = 210 GPa, Poisson's ratio = 0.3. Assume plane stress condition. Thickness of the plate = 10 mm with hole at the center of the plate.

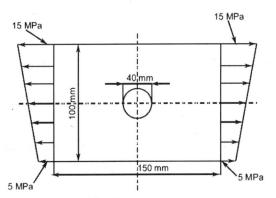

Figure 6.21. Problem 3

4. For the plate shown in Figure 6.22, find the maximum stress. Plate is made up of two materials.
For Material 1, E = 210 GPa and v = 0.3.
For Material 2, E = 70 GPa and v = 0.35.
Assume plane stress condition.
Thickness of the plate = 10 mm with a hole at the center of the plate.

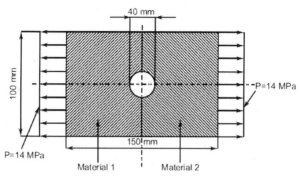

Figure 6.22. Problem 4

5. For the plate with a hole at the center shown in Figure 6.23, find the maximum stress. Take E = 210 GPa and v = 0.3, thickness of plate t = 10 mm. Assume plane stress condition.

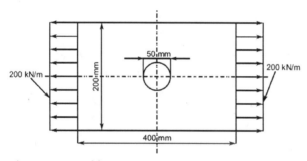

Figure 6.23. Problem 5

[*To find the pressure, divide distributed load by thickness of plate.*]

6. Determine the stresses in the plate with the round hole subjected to the tensile stresses in Figure 6.24. Find the maximum stress. Take E = 210 GPa and v = 0.25, thickness of plate t = 10 mm. Assume plane stress condition.

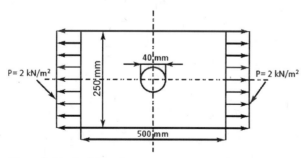

Figure 6.24. Problem 6

7. For the plate with a hole at the center shown in Figure 6.25, find the maximum stress. Take E = 210 GPa and ν = 0.3, thickness of plate t = 0.375 in. Assume plane stress condition.

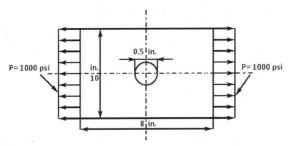

Figure 6.25. Problem 7

8. For the plate with a hole at the center shown in Figure 6.26, find the maximum stress. Take E = 30 × 10⁶ psi and ν = 0.25, thickness of plate t = 0.1 in. Assume plane stress condition.

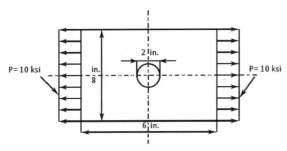

Figure 6.26. Problem 8

9. Find the maximum stress for the plate shown in Figure 6.27 if the hole is located halfway between the center line and the top edge as shown. Take E = 20 × 10⁶ N/cm² and ν = 0.25. Assume plane stress condition.

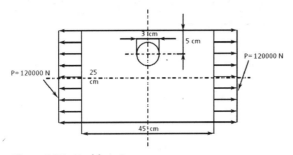

Figure 6.27. Problem 9

[*Model half of the plate by taking symmetry about y-axis.*]

REFERENCES

1. C. S. Desai and J. F. Abel, "Introduction to the Finite Element Method," Van Nostrad Reinhold, 1972.
2. L. J. Segerlind, "Applied Finite Element Analysis, Second Edition," John Wiley and Sons, 1984.
3. M. Asghar Bhatti, "Fundamental Finite Element Analysis and Applications with Mathematica and MATLAB Computations," John Wiley and Sons, 2005.
4. H. Grandin, Jr., "Fundamentals of the Finite Element Method," Macmillan Publishing Company, 1986.
5. D. L. Logan, "A First Course in the Finite Element Method, Fifth Edition," Cengage Learning, 2012.
6. F. L. Stasa, "Applied Finite Element Analysis for Engineering," Holt, Rinehart, and Winston, 1985.
7. N. Troyani, C. Gomes, and G. Sterlacci, "Theoretical Stress Concentration Factors for Short Rectangular Plates with Central Circular Holes," Journal of Mechanical Design, ASME, Vol. 124, pp. 126-128, 2002.
8. T. Hayashi, "Stress Analysis of a Rectangular Plate with a Circular Hole under Uniaxial Loading," Journal of Thermoplastic Composite Materials, Vol. 2, 1989.

Chapter 7

THERMAL ANALYSIS

7.1 INTRODUCTION

The computation of temperature distribution within a body will be used in this chapter due to its importance in many engineering applications. Conduction (q) is the transfer of heat through materials without any net motion of the mass of the material. The rate of heat flow in x-direction by conduction (q) is given by

$$q = kA \frac{\partial T}{\partial x} \qquad (7.1)$$

where
k is the thermal conductivity of the material, A is the area normal to x-direction through which heat flows, T is the temperature, and x is the length parameter.

Convection is the process by which thermal energy is transferred between a solid and a fluid surrounding it. The rate of heat flow by convection (q) is given by

$$q = hA(T - T_\infty) \qquad (7.2)$$

where
h is the heat transfer coefficient, A is the surface area of the body through which heat flows, T is the temperature of the surface of the body, and T_∞ is the temperature of the surrounding medium.

Thermal analysis is one of the scalar field problems. These problems have only 1 degree of freedom per node namely temperature. In this chapter, one-dimensional and two-dimensional heat conduction problems are dealt with. In these problems, a bar element with 2 end nodes each having temperature (T) as sole degree of freedom is useful. Nodal heat flow rates (Q) or heat fluxes are analogous quantities to nodal forces, in structural bar element.

The governing equation for this element is given by,

$$\frac{Ak}{L}\begin{bmatrix} 1 & -1 \\ -1 & 1 \end{bmatrix}\begin{Bmatrix} T_1 \\ T_2 \end{Bmatrix} = \frac{q}{2}\begin{Bmatrix} L \\ L \end{Bmatrix} + \begin{Bmatrix} Q_1 \\ Q_2 \end{Bmatrix} \qquad (7.3)$$

where,
q = heat generation rate per unit length

$$\frac{Ak}{L}\begin{bmatrix} 1 & -1 \\ -1 & 1 \end{bmatrix} = \text{element heat conductivity matrix.}$$

7.2 PROCEDURE OF FINITE ELEMENT ANALYSIS (RELATED TO THERMAL PROBLEMS)

Step 1. Select element type.
Step 2. Select temperature distribution function.
Step 3. Define the temperature gradient/temperature and heat flux/temperature gradient relationships.
Step 4. Derive the element conduction matrix and heat flux matrix.
Step 5. Assemble the element equations to obtain the global equations and introduce boundary conditions.
Step 6. Solve for the nodal temperatures.
Step 7. Solve for the element temperature gradients and heat fluxes.

7.3 ONE-DIMENSIONAL HEAT CONDUCTION

Example 7.1

A composite wall consists of 3 materials. The outer temperature is $T_0 = 20°C$. Convection heat transfer takes place on the inner surface of the wall with $T_\infty = 800°C$ and $h = 25$ W/m²°C. Determine the temperature distribution in the wall. Take $k_1 = 30$ W/m°C, $k_2 = 50$ W/m°C, $k_3 = 20$ W/m°C.

Thermal Analysis

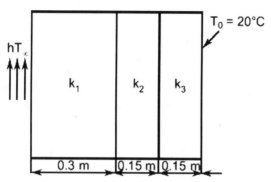

Figure 7.1. A composite wall consists of 3 materials for Example 7.1.

Solution

(1) Analytical method

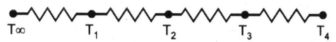

Figure 7.1(a). Analytical method for Example 7.1.

Heat flow rate per unit area,

$$Q = \frac{T_\infty - T_4}{\frac{1}{h} + \frac{L_1}{k_1} + \frac{L_2}{k_2} + \frac{L_3}{k_3}} = \frac{800 - 20}{\frac{1}{25} + \frac{0.3}{30} + \frac{0.15}{50} + \frac{0.15}{20}} = 12892.6 \text{ W/m}^2.$$

Now,

$$Q = h(T_\infty - T_1) = \frac{k_1(T_1 - T_2)}{L_1} = \frac{k_2(T_2 - T_3)}{L_2} = \frac{k_3(T_3 - T_4)}{L_3}$$

$$12892.6 = 25(800 - T_1) = \frac{30(T_1 - T_2)}{0.3} = \frac{50(T_2 - T_3)}{0.15} = \frac{20(T_3 - 20)}{0.15}.$$

By solving the above, we get

$$T_1 = 284.3°C$$
$$T_2 = 155.37°C$$
$$T_3 = 116.7°C.$$

(II) FEM by calculations [refer to Figure 7.1(b)]

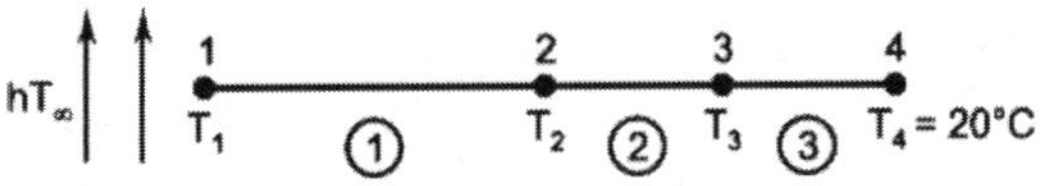

Figure 7.1(b). Finite element model for Example 7.1.

Governing equation is,

$$\frac{k}{L}\begin{bmatrix} 1 & -1 \\ -1 & 1 \end{bmatrix} \begin{Bmatrix} T_1 \\ T_2 \end{Bmatrix} = q \begin{Bmatrix} \frac{L}{2} \\ \frac{L}{2} \end{Bmatrix} + \begin{Bmatrix} -Q_1 \\ +Q_2 \end{Bmatrix}.$$

Since there is no heat generation specified, $q = 0$.
For element 1,

$$\frac{k_1}{L_1}\begin{bmatrix} 1 & -1 \\ -1 & 1 \end{bmatrix} \begin{Bmatrix} T_1 \\ T_2 \end{Bmatrix} = q \begin{Bmatrix} \frac{L}{2} \\ \frac{L}{2} \end{Bmatrix} + \begin{Bmatrix} -Q_1 \\ +Q_2 \end{Bmatrix}$$

$$\frac{30}{0.3}\begin{bmatrix} 1 & -1 \\ -1 & 1 \end{bmatrix} \begin{Bmatrix} T_1 \\ T_2 \end{Bmatrix} = 0 + \begin{Bmatrix} -Q_1 \\ +Q_2 \end{Bmatrix}.$$

For element 2,

$$\frac{k_2}{L_2}\begin{bmatrix} 1 & -1 \\ -1 & 1 \end{bmatrix} \begin{Bmatrix} T_2 \\ T_3 \end{Bmatrix} = q \begin{Bmatrix} \frac{L}{2} \\ \frac{L}{2} \end{Bmatrix} + \begin{Bmatrix} -Q_2 \\ +Q_3 \end{Bmatrix}$$

$$\frac{50}{0.15}\begin{bmatrix} 1 & -1 \\ -1 & 1 \end{bmatrix} \begin{Bmatrix} T_2 \\ T_3 \end{Bmatrix} = 0 + \begin{Bmatrix} -Q_2 \\ +Q_3 \end{Bmatrix}.$$

For element 3,

$$\frac{k_3}{L_3}\begin{bmatrix} 1 & -1 \\ -1 & 1 \end{bmatrix} \begin{Bmatrix} T_3 \\ T_4 \end{Bmatrix} = q \begin{Bmatrix} \frac{L}{2} \\ \frac{L}{2} \end{Bmatrix} + \begin{Bmatrix} -Q_3 \\ +Q_4 \end{Bmatrix}$$

$$\frac{20}{0.15}\begin{bmatrix} 1 & -1 \\ -1 & 1 \end{bmatrix} \begin{Bmatrix} T_3 \\ T_4 \end{Bmatrix} = 0 + \begin{Bmatrix} -Q_3 \\ +Q_4 \end{Bmatrix}.$$

Global equation after assembly,

$$\begin{bmatrix} 100 & -100 & 0 & 0 \\ -100 & 100+333.33 & -333.33 & 0 \\ 0 & -333.33 & 333.33+133.33 & -133.33 \\ 0 & 0 & -133.33 & 133.33 \end{bmatrix} \begin{Bmatrix} T_1 \\ T_2 \\ T_3 \\ T_4 \end{Bmatrix} = \begin{Bmatrix} -Q_1 \\ 0 \\ 0 \\ +Q_4 \end{Bmatrix}.$$

Boundary conditions are $T_4 = 20°C$ and

$$Q_1 = -h(T_\infty - T_1)$$
$$-Q_1 = 25(800 - T_1) = 20000 - 25T_1.$$

So modified equation,

$$\begin{bmatrix} 100+25 & -100 & 0 \\ -100 & 433.33 & -333.33 \\ 0 & -333.33 & 466.66 \end{bmatrix} \begin{Bmatrix} T_1 \\ T_2 \\ T_3 \end{Bmatrix} = \begin{Bmatrix} 20000 \\ 0 \\ 0 \end{Bmatrix} + \begin{Bmatrix} 0 \\ 0 \\ 20 \times 133.33 \end{Bmatrix}$$

$$\begin{bmatrix} 125 & -100 & 0 \\ -100 & 433.33 & -333.33 \\ 0 & -333.33 & 466.66 \end{bmatrix} \begin{Bmatrix} T_1 \\ T_2 \\ T_3 \end{Bmatrix} = \begin{Bmatrix} 20000 \\ 0 \\ 0 \end{Bmatrix} + \begin{Bmatrix} 0 \\ 0 \\ 2666.6 \end{Bmatrix}.$$

After solving the matrix and simultaneous equations, we get,

$$T_1 = 284.3°C$$
$$T_2 = 155.37°C$$
$$T_3 = 116.7°C.$$

(III) Software results
Temperature values

NODE	TEMP
1	284.30
2	155.37
3	116.69
4	20.000
5	800.00

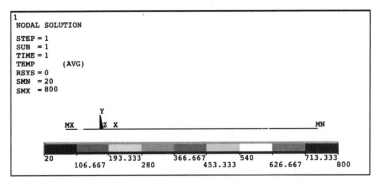

Figure 7.1(c). Temperature distribution in a composite wall (refer to Appendix C for color figures).

Answers for Example 7.1

Parameter	Analytical method	FEM-hand calculation	Software results
Temperature			
at node 1	284.3°C	284.3°C	284.3°C
at node 2	155.37°C	155.37°C	155.37°C
at node 3	116.7°C	116.7°C	116.69°C

Procedure for solving the problem using ANSYS® 11.0 academic teaching software.
For Example 7.1

PREPROCESSING

1. Main Menu > Preferences, then select Thermal > OK

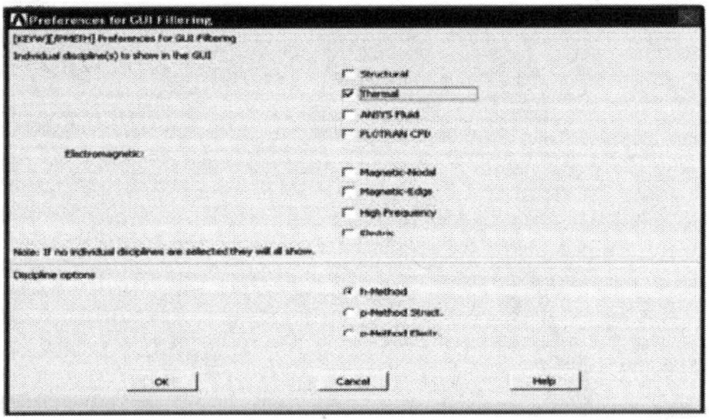

Figure 7.2. Selecting the preferences.

2. Main Menu > Preprocessor > Element Type > Add/Edit/Delete > Add > Click on Link > then on 2d conduction 32 > OK > Add > Click on Link > then on 3D convection 34 > OK > Close

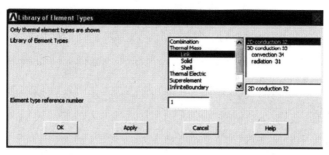

Figure 7.3. Selecting the element for conduction.

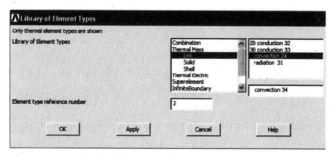

Figure 7.4. Selecting the element for convection.

3. Main Menu > Preprocessor > Real Constants > Add/Edit/Delete > Add > Click on Link 32 > OK

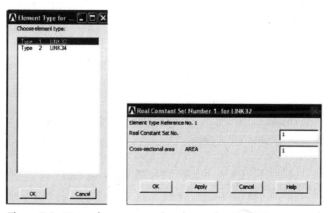

Figure 7.5. Enter the cross-sectional area for Link 32.

Enter cross-sectional area AREA > **Enter 1** > **OK**

Add > Click on Link 34 > OK
Enter cross-sectional area AREA > **Enter 1 > OK > Close**

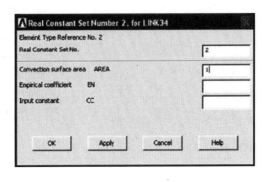

Figure 7.6. Enter the cross-sectional area for Link 34.

Enter the material properties.

4. **Main Menu > Preprocessor > Material Props > Material Models**
   Material Model Number 1,
   click **Thermal > Conductivity > Isotropic**
   Enter **KXX = 30 > OK**
   Then in the material model window, click on **Material menu > New Model > OK**
   Material Model Number 2,
   click **Thermal > Conductivity > Isotropic**
   Enter **KXX = 50 > OK**
   Then in the material model window, click on **Material menu > New Model > OK**
   Material Model Number 3,
   click **Thermal > Conductivity > Isotropic**
   Enter **KXX = 20 > OK**
   Then in the material model window, click on **Material menu > New Model > OK**
   Material Model Number 4,
   click **Thermal > Convection or Film Coef.**
   Enter **HF = 25 > OK**
   (**Close** the Define Material Model Behavior window.)
   Create the nodes and elements.
5. **Main Menu > Preprocessor > Modeling > Create > Nodes > In Active CS** Enter the coordinates of node 1 > **Apply** Enter the coordinates of node 2 > **Apply**

Thermal Analysis

Enter the coordinates of node 3 > **Apply** Enter the coordinates of node 4 > **Apply** Enter the coordinates of node 5 > **OK**

Node Locations		
Node number	X coordinates	Y coordinates
1	0	0
2	0.3	0
3	0.45	0
4	0.6	0
5	−0.1	0

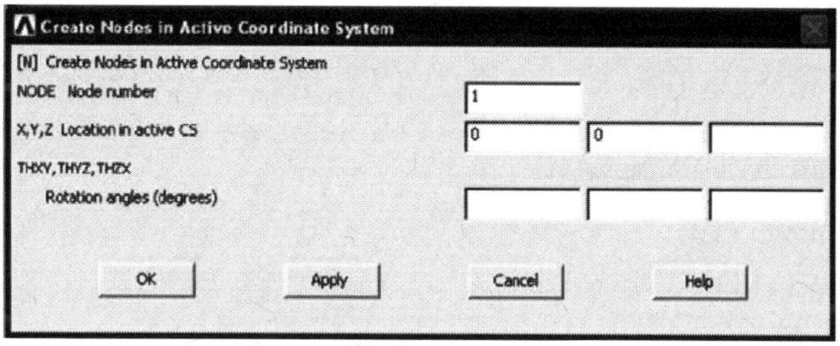

Figure 7.7. Enter the node coordinates.

6. Main Menu > Preprocessor > Modeling > Create > Elements > Elem Attributes > OK > Auto Numbered > Thru nodes Pick the 1st and 2nd node > OK

Figure 7.8. Assigning element attributes to element 1 and creating element 1.

Elem Attributes > change the material number to 2 > OK > Auto Numbered > Thru nodes Pick the 2nd and 3rd node > **OK**

Figure 7.9. Assigning element attributes to element 2 and creating element 2.

Elem Attributes > change the material number to 3 > OK > Auto Numbered > Thru nodes Pick the 3rd and 4th node > **OK**

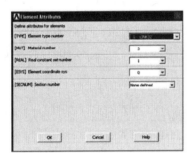

Figure 7.10. Assigning element attributes to element 3 and creating element 3.

Elem Attributed > change the element type to Link 34 > change the material number to 4 > change the Real constant set number to 2 > OK > Auto Numbered > Thru nodes Pick the 1st and 5th node > **OK**

Figure 7.11. Assigning element attributes to element 4 and creating element 4.

Apply the boundary conditions and temperature.
7. **Main Menu > Preprocessor > Loads > Define Loads > Apply > Thermal > Temperature > On Nodes** Pick the 4th node **> Apply > Click on TEMP and Enter Value = 20 > OK**

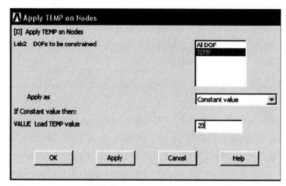

Figure 7.12. Applying temperature on node 4.

8. **Main Menu > Preprocessor > Loads > Define Loads > Apply > Thermal > Temperature > On Nodes** Pick the 5th node **> Apply > Click on TEMP and Enter Value = 800 > OK**

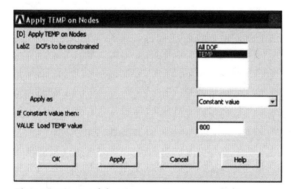

Figure 7.13. Applying temperature on node 5.

Solution

The interactive solution proceeds.
9. **Main Menu > Solution > Solve > Current LS > OK**
   The **/STATUS Command** window displays the problem parameters and **the Solve Current Load Step** window is shown. Check the solution options in the **/STATUS** window and if all is OK, select **File > Close**.
   In the **Solve Current Load Step** window, select **OK**, and when the solution is complete, **close the 'Solution is Done!'** window.

POST-PROCESSING

We can now plot the results of this analysis and also list the computed values.

10. **Main Menu > General Posrproc > Plot Results > Contour Plot > Nodal Solu > DOF Solution > Temperature > OK**

    This result is shown in Figure 7.1(c).

11. **Main Menu > General Postproc > List Results > Nodal Solu > Select Temperature > OK**

Example 7.2

Heat is generated in a large plate with $k = 0.75$ W/m°C at the rate 6000 W/m³. The plate is 40 cm thick. The outside surfaces of the plate are exposed to fluid at 35°C with a convective heat transfer coefficient of 15 W/m²°C. Determine temperature distribution in wall. The 2 element model to be used for solution.

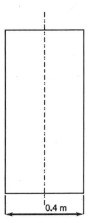

Figure 7.14. Example 7.2

Solution

(I) Analytical method

Governing equation is,

$$\frac{d^2T}{dx^2} = -\frac{q}{k}$$

$$\frac{dT}{dx} = -\frac{q \times x}{k} + c_1$$

At $x = 0$, $\quad \dfrac{dT}{dx} = 0 \Rightarrow c_1 = 0$

Thermal Analysis

$$\frac{dT}{dx} = \frac{-600 \times x}{0.75} + 0 = -8000x \tag{7.4}$$

$$T = -\frac{q}{k}\frac{x^2}{2} + c_2 = -4000x^2 + c_2. \tag{7.5}$$

Boundary conditions are,

At $x = L$,

$$-k\frac{dT}{dx} = h(T_3 - T_\infty)$$

$$-k(8000 \times x) = h(T_3 - T_\infty)$$

$$-0.75(8000 \times 0.2) = 15(T_3 - 35) \Rightarrow T_3 = 115°C.$$

We know at $x = 0.2$, $T_3 = 115°C$.
Substituting this in equation (7.5),

$$115 = -4000(0.2)^2 + c_2 \Rightarrow c_2 = 275.$$

Substituting c_2 in equation (7.5),

$$T = -4000 \times x^2 + 275$$

$$T_2 = T|_{x=0.1} = -4000(0.1)^2 + 275 = 235°C$$

$$T_1 = T|_{x=0} = 275°C.$$

(II) FEM by hand calculations [refer to Figure 7.14(a)]

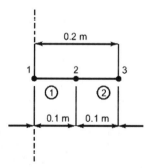

Figure 7.14(a). Symmetric finite element model for Example 7.2.

Given: $t = 40$ cm $= 0.4$ m, $T_\infty = 35°C$, $h = 15$ W/m²°C, $k = 0.75$ W/m°C

Governing equation is,

$$\frac{k}{L}\begin{bmatrix} 1 & -1 \\ -1 & 1 \end{bmatrix}\begin{Bmatrix} T_1 \\ T_2 \end{Bmatrix} = q\begin{Bmatrix} \frac{L}{2} \\ \frac{L}{2} \end{Bmatrix} + \begin{Bmatrix} -Q_1 \\ +Q_2 \end{Bmatrix}.$$

For element 1,

$$\frac{k_1}{L_1}\begin{bmatrix} 1 & -1 \\ -1 & 1 \end{bmatrix}\begin{Bmatrix} T_1 \\ T_2 \end{Bmatrix} = q\begin{Bmatrix} \frac{L}{2} \\ \frac{L}{2} \end{Bmatrix} + \begin{Bmatrix} -Q_1 \\ +Q_2 \end{Bmatrix}$$

$$\frac{0.75}{0.10}\begin{bmatrix} 1 & -1 \\ -1 & 1 \end{bmatrix}\begin{Bmatrix} T_1 \\ T_2 \end{Bmatrix} = 6000\begin{Bmatrix} 0.05 \\ 0.05 \end{Bmatrix} + \begin{Bmatrix} -Q_1 \\ +Q_2 \end{Bmatrix}$$

$$7.5\begin{bmatrix} \overset{1}{1} & \overset{2}{-1} \\ -1 & 1 \end{bmatrix}\begin{matrix}1\\2\end{matrix}\begin{Bmatrix} T_1 \\ T_2 \end{Bmatrix} = 6000\begin{Bmatrix} 0.05 \\ 0.05 \end{Bmatrix} + \begin{Bmatrix} -Q_1 \\ +Q_2 \end{Bmatrix}.$$

For element 2,

$$\frac{k_2}{L_2}\begin{bmatrix} 1 & -1 \\ -1 & 1 \end{bmatrix}\begin{Bmatrix} T_2 \\ T_3 \end{Bmatrix} = q\begin{Bmatrix} \frac{L}{2} \\ \frac{L}{2} \end{Bmatrix} + \begin{Bmatrix} -Q_2 \\ +Q_3 \end{Bmatrix}$$

$$\frac{0.75}{0.10}\begin{bmatrix} 1 & -1 \\ -1 & 1 \end{bmatrix}\begin{Bmatrix} T_2 \\ T_3 \end{Bmatrix} = 6000\begin{Bmatrix} 0.05 \\ 0.05 \end{Bmatrix} + \begin{Bmatrix} -Q_2 \\ +Q_3 \end{Bmatrix}$$

$$7.5\begin{bmatrix} \overset{2}{1} & \overset{3}{-1} \\ -1 & 1 \end{bmatrix}\begin{matrix}2\\3\end{matrix}\begin{Bmatrix} T_2 \\ T_3 \end{Bmatrix} = 6000\begin{Bmatrix} 0.05 \\ 0.05 \end{Bmatrix} + \begin{Bmatrix} -Q_2 \\ +Q_3 \end{Bmatrix}.$$

Assembling $\Rightarrow 7.5\begin{bmatrix} \overset{1}{1} & \overset{2}{-1} & \overset{3}{0} \\ -1 & 1+1 & -1 \\ 0 & -1 & 1 \end{bmatrix}\begin{matrix}1\\2\\3\end{matrix}\begin{Bmatrix} T_1 \\ T_2 \\ T_3 \end{Bmatrix} = 6000\begin{Bmatrix} 0.05 \\ 0.05+0.05 \\ 0.05 \end{Bmatrix} + \begin{Bmatrix} -Q_1 \\ 0 \\ +Q_3 \end{Bmatrix}.$

Boundary conditions are, $Q_1 = 0$ and $Q_3 = -h(T_3 - T_\infty) \Rightarrow Q_3 = -15(T_3 - 35) = -15T_3 + 525$

$$7.5\begin{bmatrix} 1 & -1 & 0 \\ -1 & 2 & -1 \\ 0 & -1 & 1 \end{bmatrix}\begin{Bmatrix} T_1 \\ T_2 \\ T_3 \end{Bmatrix} = 6000\begin{Bmatrix} 0.05 \\ 0.1 \\ 0.05 \end{Bmatrix} + \begin{Bmatrix} 0 \\ 0 \\ -15T_3 + 525 \end{Bmatrix}$$

$$\begin{bmatrix} 7.5 & -7.5 & 0 \\ -7.5 & 15 & -7.5 \\ 0 & -7.5 & 7.5 \end{bmatrix} \begin{Bmatrix} T_1 \\ T_2 \\ T_3 \end{Bmatrix} = \begin{Bmatrix} 300+0 \\ 600+0 \\ 300-15T_3+525 \end{Bmatrix}.$$

Now,

$$\begin{bmatrix} 7.5 & -7.5 & 0 \\ -7.5 & 15 & -7.5 \\ 0 & -7.5 & 7.5+15 \end{bmatrix} \begin{Bmatrix} T_1 \\ T_2 \\ T_3 \end{Bmatrix} = \begin{Bmatrix} 300 \\ 600 \\ 825 \end{Bmatrix}.$$

By solving the above matrix and simultaneous equations, we have temperature distribution as,

$$\begin{Bmatrix} T_1 \\ T_2 \\ T_3 \end{Bmatrix} = \begin{Bmatrix} 275 \\ 235 \\ 115 \end{Bmatrix}.$$

Therefore,

$$T_1 = 275°C$$
$$T_2 = 235°C$$
$$T_3 = 115°C.$$

(III) Software results

Due to symmetry of the geometry, only half of the finite element model is created for software analysis.

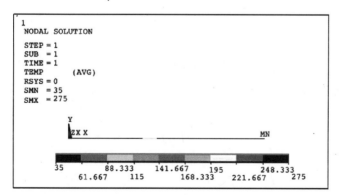

Figure 7.14(b). Temperature distribution in a large plate (refer to Appendix C for color figures).

Temperature values

NODE	TEMP
1	275.00
2	235.00
3	115.00
4	35.000

Answers for Example 7.2

Parameter	Analytical method	FEM-hand calculation	Software results
Temperature			
at node 1	275°C	275°C	275°C
at node 2	235°C	235°C	235°C
at node 3	115°C	115°C	115°C

Procedure for solving the problem using ANSYS® 11.0 academic teaching software.
For Example 7.2

PREPROCESSING

1. Main Menu > Preferences, then select Thermal > OK

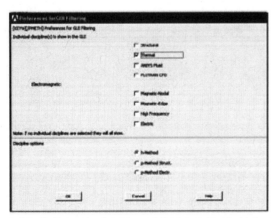

Figure 7.15. Selecting the preferences.

2. Main Menu > Preprocessor > Element Type > Add/Edit/Delete > Add > Click on Link > then on 2D conduction 32 > OK > Add > Click on Link > then on 3D convection 34 > OK > Close

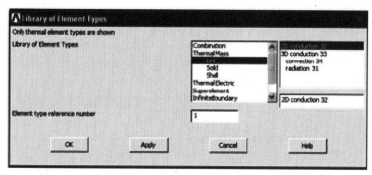

Figure 7.16. Selecting the element for conduction.

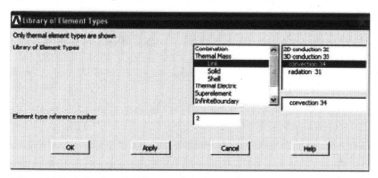

Figure 7.17. Selecting the element for convection.

3. Main Menu > Preprocessor > Real Constants > Add/Edit/Delete > Add > Click on Link 32 > OK

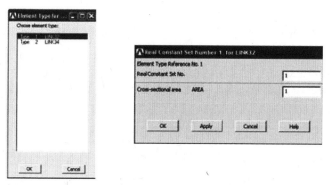

Figure 7.18. Enter the cross-sectional area for Link 32.

Enter cross-sectional area AREA > **Enter 1** > OK

Add > Click on Link 34 > OK
Enter cross-sectional area AREA > **Enter 1 > OK**

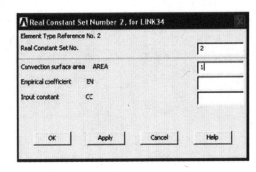

Figure 7.19. Enter the cross-sectional area for Link 34.

Enter the material properties.
4. **Main Menu > Preprocessor > Material Props > Material Models**
   Material Model Number 1,
   click Thermal > Conductivity > Isotropic
   Enter **KXX = 0.75 > OK**
   Then in the material model window, click on **Material menu > New Model > OK**
   Material Model Number 2,
   Click **Thermal > Convection or Film Coef.**
   Enter **HF = 15 > OK**
   (**Close** the Define Material Model Behavior window.)
   Create the nodes and elements. Due to geometric symmetry, only half of the model is created.
5. **Main Menu > Preprocessor > Modeling > Create > Nodes > In Active CS** Enter the coordinate of node 1 > **Apply** Enter the coordinates of node 2 > **Apply** Enter the coordinates of node 3 > **Apply** Enter the coordinates of node 4 > **OK**

Node Locations		
Node number	X coordinates	Y coordinates
1	0	0
2	0.1	0
3	0.0	0
4	0.3	0

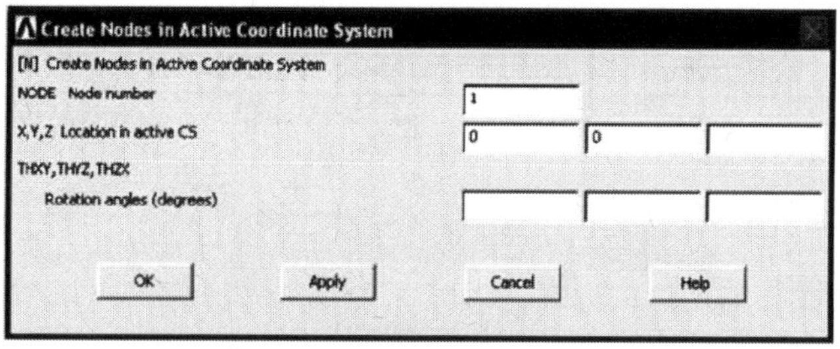

Thermal Analysis

Figure 7.20. Enter the node coordinates.

6. **Main Menu > Preprocessor > Modeling > Create > Elements > Elem Attributes > OK > Auto Numbered > Thru nodes** Pick the 1st and 2nd node > **Apply** > then Pick the 2nd and 3rd node **OK**

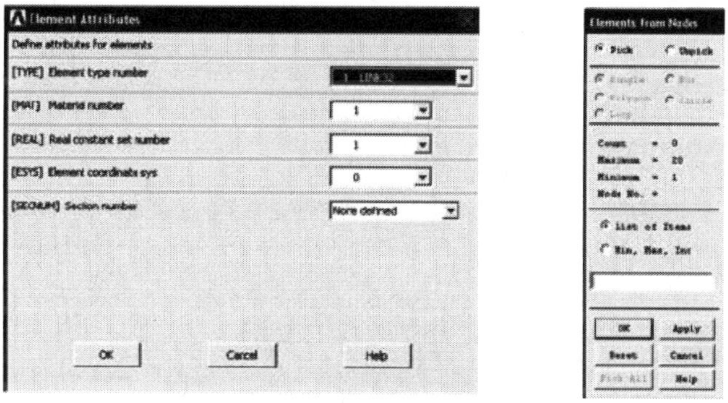

Figure 7.21. Assigning element attributes to elements 1 and 2 and creating elements 1 and 2.

Elem Attributes > change the element type to Link 34 > change the material number to 2 > change the Real constant set number to 2 > OK > Auto Numbered > Thru nodes Pick the 3rd and 4th node > **OK**

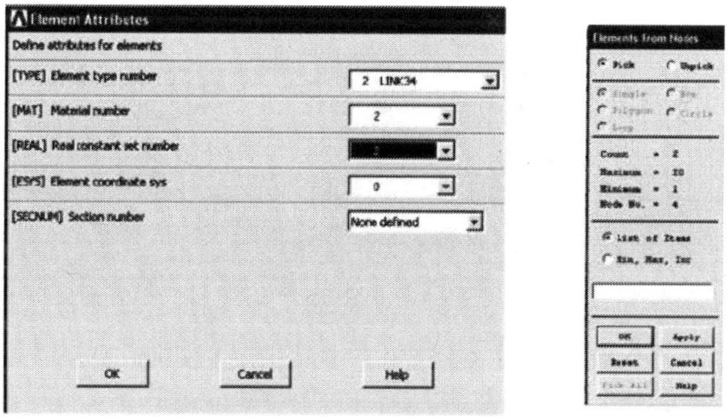

Figure 7.22. Assigning elements attributes to element 3 and creating element 3.

Apply the boundary conditions and temperature.

7. **Main Menu > Preprocessor > Loads > Define Loads > Apply > Thermal > Temperature > On Nodes** Pick the 4th node **> Apply > Click on TEMP and Enter Value = 35 > OK**

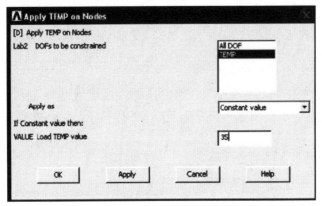

Figure 7.23. Applying temperature on node 4.

8. **Main Menu > Preprocessor > Loads > Define Loads > Apply > Thermal > Heat Generat > On Nodes** Pick the 1st, 2nd, and 3rd nodes **> Apply > Enter HGEN Value = 6000 > OK**

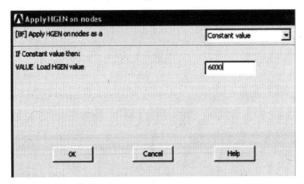

Figure 7.24. Assigning heat generation on nodes 1, 2, and 3.

Solution

The interactive solution proceeds.

9. **Main Menu > Solution > Solve > Current LS > OK**

    The **/STATUS Command** window displays the problem parameters and **the Solve Current Load Step** window is shown. Check the solution options in the **/STATUS** window and if all is OK, select **File > Close**.

    In the **Solve Current Load Step** window, select **OK**, and when the solution is complete, **close the 'Solution is Done!'** window.

POST-PROCESSING

We can now plot the results of this analysis and also list the computed values.

10. **Main Menu > General Posrproc > Plot Results > Contour Plot > Nodal Solu > DOF Solution > Temperature > OK**

    This result is shown in Figure 7.14(b).

11. **Main Menu > General Postproc > List Results > Nodal Solu > Select Temperature > OK**

Example 7.3

Compute the temperature distribution in a long steel cylinder with an inner radius of 125 mm and an outer radius of 250 mm. The interior of the cylinder is kept at 300°K and heat is lost on the exterior by convection to a fluid whose temperature is 280°K. The convection heat transfer coefficient h is 0.994 W/m²°K and the thermal conductivity for steel k is 0.031 W/m°K.

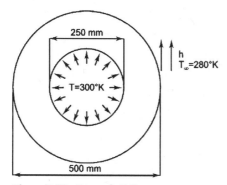

Figure 7.25. Example 7.3

Solution

(I) Analytical method

Here the problem is solved considering heat flow in radial direction.

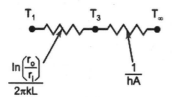

Figure 7.25(a). Analytical method for Example 7.3.

$$r_0 = 250 \text{ mm} = 0.25 \text{ m and } r_i = 125 \text{ mm} = 0.125 \text{ m}$$

Assume unit length of the cylinder

$$Q = \frac{(T_1 - T_\infty)}{\left(\dfrac{\ln\left(\dfrac{r_0}{r_1}\right)}{2\pi kL} + \dfrac{1}{hA}\right)} = \frac{(300 - 280)}{\left(\dfrac{\ln\left(\dfrac{250}{125}\right)}{2\pi \times 0.031 \times 1} + \dfrac{1}{0.994(2\pi \times 0.25)}\right)} = 4.762934 \text{ W}.$$

Now,

$$Q = \frac{(T_1 - T_3)}{\left(\dfrac{\ln\left(\dfrac{r_0}{r_1}\right)}{2\pi kL}\right)} \Rightarrow \frac{(300 - T_3)}{\left(\dfrac{\ln\left(\dfrac{250}{125}\right)}{2\pi \times 0.031 \times 1}\right)} = 4.762934 \Rightarrow T_3 = 280.51°C.$$

Let

$$T = T_2 \text{ at } r = 187.5 \text{ mm, then}$$

$$\frac{(300 - T_2)}{\left(\dfrac{\ln\left(\dfrac{250}{187.5}\right)}{2\pi \times 0.031 \times 1}\right)} = 4.762934 \Rightarrow T_2 = 283.1°C.$$

(II) FEM by hand calculations

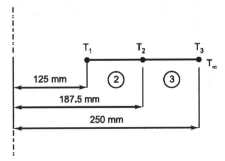

Figure 7.25(b). Finite element model for Example 7.3.

$$r_1 = 125 \text{ mm}, \ r_2 = 187.5 \text{ mm, and } r_3 = 250 \text{ mm}$$

Element matrices are,

$$k_{1C} = \frac{2\pi kL}{\ln\left(\frac{r_2}{r_1}\right)}\begin{bmatrix} 1 & -1 \\ -1 & 1 \end{bmatrix} = \frac{2\pi \times 0.031 \times 1}{\ln\left(\frac{187.5}{125}\right)}\begin{bmatrix} 1 & -1 \\ -1 & 1 \end{bmatrix} = 0.48\begin{bmatrix} 1 & -1 \\ -1 & 1 \end{bmatrix}\begin{matrix} T_1 \\ T_2 \end{matrix}$$

$$k_{2C} = \frac{2\pi kL}{\ln\left(\frac{r_3}{r_2}\right)}\begin{bmatrix} 1 & -1 \\ -1 & 1 \end{bmatrix} = \frac{2\pi \times 0.031 \times 1}{\ln\left(\frac{250}{187.5}\right)}\begin{bmatrix} 1 & -1 \\ -1 & 1 \end{bmatrix} = 0.68\begin{bmatrix} 1 & -1 \\ -1 & 1 \end{bmatrix}\begin{matrix} T_2 \\ T_3 \end{matrix}.$$

Global conduction matrix is,

$$[K_C] = \begin{bmatrix} 0.48 & -0.48 & 0 \\ -0.48 & 0.48+0.68 & -0.68 \\ 0 & -0.68 & 0.68 \end{bmatrix} = \begin{bmatrix} 0.48 & -0.48 & 0 \\ -0.48 & 1.16 & -0.68 \\ 0 & -0.68 & 0.68 \end{bmatrix}.$$

Global equation is,

$$\begin{bmatrix} 0.48 & -0.48 & 0 \\ -0.48 & 1.16 & -0.68 \\ 0 & -0.68 & 0.68 \end{bmatrix}\begin{Bmatrix} T_1 \\ T_2 \\ T_3 \end{Bmatrix} = \begin{Bmatrix} Q_1 \\ Q_2 \\ Q_3 \end{Bmatrix}.$$

Applying boundary conditions, $T_1 = 300°K$ and $Q_3 = -hA_0(T_3 - T_\infty)$

$$A_0 = 2\pi r_3 = 2\pi \times 0.25 = 1.57 \text{ m}^2.$$

Therefore, $Q_3 = -hA_0(T_3 - T_\infty) = -0.994 \times 1.57(T_3 - 280) = -(1.56T_3 - 437)$

$$\begin{bmatrix} 0.48 & -0.48 & 0 \\ -0.48 & 1.16 & -0.68 \\ 0 & -0.68 & 0.68 \end{bmatrix}\begin{Bmatrix} T_1 \\ T_2 \\ T_3 \end{Bmatrix} = \begin{Bmatrix} 0 \\ 0 \\ -(1.56T_3 - 437) \end{Bmatrix}$$

$$\begin{bmatrix} 0.48 & -0.48 & 0 \\ -0.48 & 1.16 & -0.68 \\ 0 & -0.68 & 0.68+1.56 \end{bmatrix}\begin{Bmatrix} T_1 \\ T_2 \\ T_3 \end{Bmatrix} = \begin{Bmatrix} 0 \\ 0 \\ 437 \end{Bmatrix}$$

$$\begin{bmatrix} 1.16 & -0.68 \\ -0.68 & 2.24 \end{bmatrix}\begin{Bmatrix} T_2 \\ T_3 \end{Bmatrix} + \begin{Bmatrix} 0 \\ 437 \end{Bmatrix} - \begin{Bmatrix} -0.48 \times 300 \\ 0 \end{Bmatrix} = \begin{Bmatrix} 144 \\ 437 \end{Bmatrix}.$$

Solving the above equation, we get $T_3 = 283.16°K$ and $T_2 = 290.13°K$.

(III) Software results

Due to the symmetry of the cylinder geometry, only a quarter of the geometry is drawn for finite element analysis.

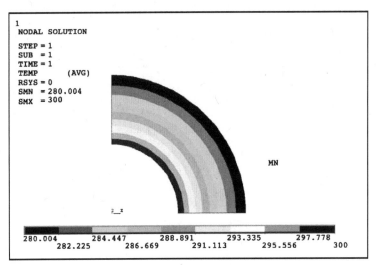

Figure 7.25(c). Temperature distribution in a long cylinder (refer to Appendix C for color figures).

The temperature in the interior is 300°K and on the outside wall, it is found to be 280.004°K.

Answers of Example 7.3

Parameter	Analytical method	FEM-hand calculation	Software results
Temperature on the interior surface	300°K	300°K	300°K
Temperature at radius 187.5 mm	238.1°K	290.13°K	288.891°K
Temperature on the outside wall	280.51°K	283.16°K	280.004°K

Procedure for solving the problem using ANSYS® 11.0 academic teaching software.
For Example 7.3

PREPROCESSING

1. **Main Menu > Preferences, then select Thermal > OK**

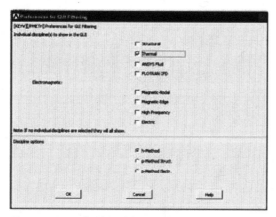

Figure 7.26. Selecting the preferences.

2. **Main Menu > Preprocessor > Element Type > Add/Edit/Delete > Add > Click on Solid > then on Quad 8 node 77 > OK > Close**

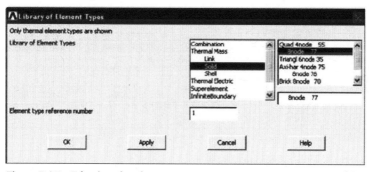

Figure 7.27. Selecting the element.

3. PLANE 77 does not require any real constant
   Enter the material properties.
4. **Main Menu > Preprocessor > Material Props > Material Models**
   Material Model Number 1,
   Click **Thermal > Conductivity > Isotropic**
   Enter **KXX = 0.031 > OK**
   (**Close** the Define Material Model Behavior window.)
   Recognize symmetry of the problem, and a quadrant of a section through the cylinder is created.

5. **Main Menu > Preprocessor > Modeling > Create > Areas > Circles > Partial Annulus**

   Enter the data as shown below.

   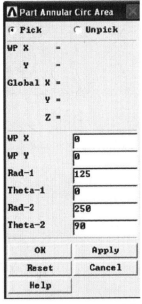

   Figure 7.28. Create partial annular area.

   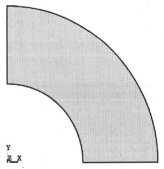

   Figure 7.29. Quadrant of a cylinder.

6. **Main Menu > Preprocessor > Meshing > Mesh Tool**

   The **Mesh Tool** dialog box appears. In that dialog box, click on the **Smart Size** and move the slider available below the **Smart Size to 2** (i.e., towards **Fine** side). Then close the Mesh Tool box.

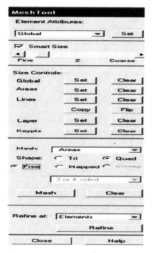

Figure 7.30. Mesh tool box.

7. **Main Menu > Preprocessor > Meshing > Mesh > Areas > Free.** Pick the quadrant > **OK**

Figure 7.31. Quad element mesh.

8. **Main Menu > Preprocessor > Loads > Define loads > Apply > Thermal > Temperatures > On Lines**

   Select the line on the interior and set the temperature to 300.

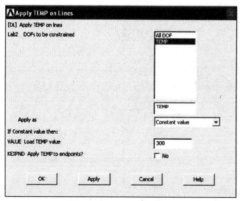

Figure 7.32. Setting the temperature on the interior of the cylinder.

9. **Main Menu > Preprocessor > Loads > Apply > Convection > On Lines**
   Select the lines defining the outer surface and set the convection coefficient to 0.994 and the fluid temp to 280.

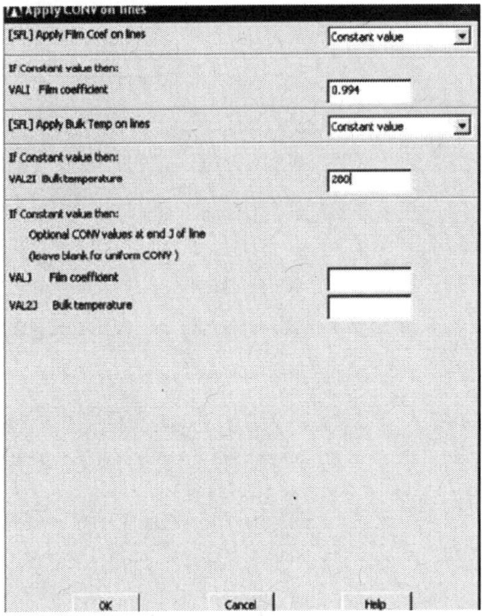

Figure 7.33. Setting the convection coefficient on outer surface.

10. **Main Menu > Preprocessor > Loads > Apply > Heat Flux > On Lines**
    To account for symmetry, select the vertical and horizontal lines of symmetry and set the heat flux to zero.

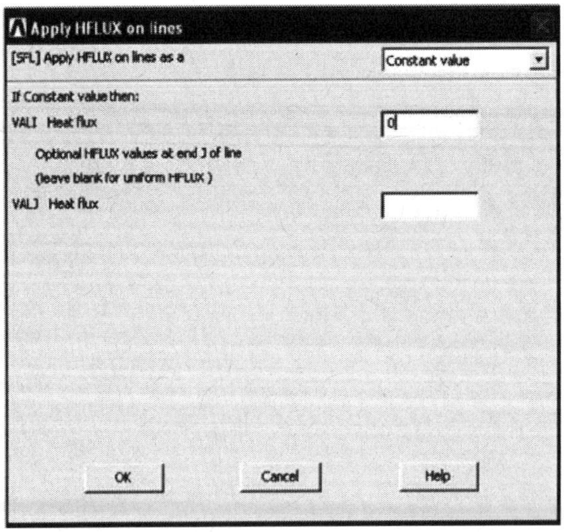

Figure 7.34. Setting the heat flux.

THERMAL ANALYSIS

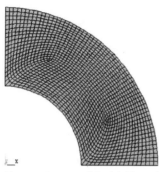

Figure 7.35. Model with boundary conditions.

Solution

The interactive solution proceeds.
11. **Main Menu > Solution > Solve > Current LS > OK**
    The **/STATUS Command** window displays the problem parameters and **the Solve Current Load Step** window is shown. Check the solution options in the **/STATUS** window and if all is OK, select **File > Close**.
    In the **Solve Current Load Step** window, select **OK**, and when the solution is complete, **close the 'Solution is Done!' window**.

POST-PROCESSING

We can now plot the results of this analysis and also list the computed values.
12. **Main Menu > General Posrproc > Plot Results > Contour Plot > Nodal Solu > DOF Solution > Temperature > OK**
    This result is shown in Figure 7.25(b).
13. **Main Menu > General Postproc > List Results > Nodal Solu > Select Temperature > OK**

7.4 TWO–DIMENSIONAL PROBLEM WITH CONDUCTION AND WITH CONVECTION BOUNDARY CONDITIONS

Example 7.4

A body having rectangular cross-section is subjected to boundary conditions as shown in Figure 7.36. The thermal conductivity of the body is 1.5 W/m°. On one side of the body, it is insulated and on the other side, convection takes place with

$h = 50$ W/m²°C and $T_\infty = 35$°C. The top and bottom sides are maintained at a uniform temperature of 180°C. Determine the temperature distribution in the body.

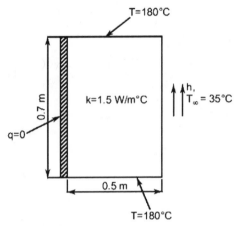

Figure 7.36. Example 7.4

Solution

(I) Software results

The temperature at the top and bottom edges is found to be 180°C and at the right edge the temperature is found to be 46.802°C.

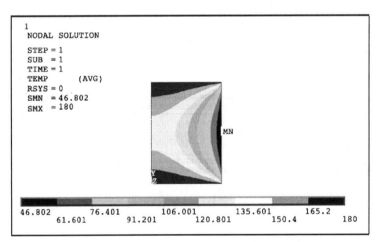

Figure 7.36(a). Temperature distribution in a body of rectangular cross-section (refer to Appendix C for color figures).

PROBLEMS

1. Define conduction and convection.
2. Write the formulas for the rate of heat flow in *x*-direction by conduction and the rate of heat flow by convection.
3. Determine the temperature distribution for the two-dimensional body shown in Figure 7.37, subjected to boundary conditions as shown in the figure. The top and bottom edges are insulated. The left side of the body is maintained at a temperature of 45°C. On the right side, the convection process takes place with heat transfer coefficient h = 100 W/m²°C and T_∞ = 20°C. The thermal conductivity of the body is k = 45 W/m°C.

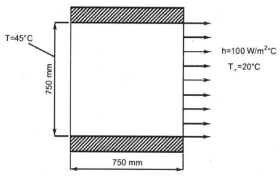

Figure 7.37. Problem 3

4. Determine the temperature distribution for the two dimensional body shown in Figure 7.38. The temperature of 200°C is maintained at the top and bottom edges. The left and right edges are insulated. Heat is generated at the rate of q = 2000 W/m³ in a body as shown in the figure. Let k = 35 W/m°C.

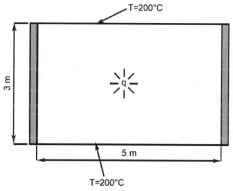

Figure 7.38. Problem 4

5. Determine the temperature distribution for the two-dimensional body shown in Figure 7.39, subjected to boundary conditions as shown in the figure. The top and bottom edges are insulated. The left side of the body is maintained at a temperature of 50°C. On the right side, the convection process takes place with heat transfer coefficient h = 150 W/m²°C and T_∞ = 25°C. The thermal conductivity of the body is k = 50 W/m°C.

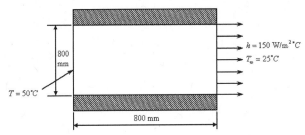

Figure 7.39. Problem 5

6. Determine the temperature distribution for the two dimensional body shown in Figure 7.40. The temperature of 200°C is maintained at the top and bottom edges. The left and right edges are insulated. Heat is generated at the rate of q = 2100 W/m³ in a body as shown in figure. Let k = 45 W/m°C.

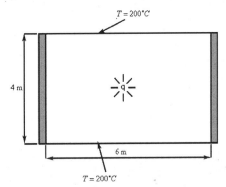

Figure 7.40. Problem 6

7. Determine the load matrix and the global load matrix for Figure 7.41. The top and bottom edges are insulated.

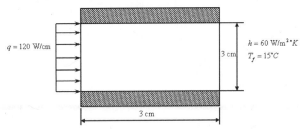

Figure 7.41. Problem 7

8. Consider the rectangular plate shown in Figure 7.42. The outer temperature is $T_0 = 30°C$. Convection heat transfer takes place on the inner surface of the wall with $T_\infty = 80°C$ and $h = 50$ W/m²°K. Determine the temperature distribution in the wall. Take the thermal conductivity value $k = 160$ W/m°K.

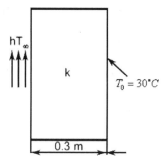

Figure 7.42. Problem 8

9. Consider a composite wall consisting of 2 materials shown in Figure 7.43. The outer temperature is $T_0 = 30°C$. Convection heat transfer takes place on the inner surface of the wall with $T_\infty = 80°C$ and $h = 50$ W/m²°K. Determine the temperature distribution in the wall. Take the thermal conductivity value $k_1 = 40$ W/m°C and $k_2 = 60$ W/m°C.

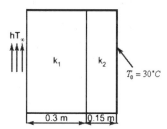

Figure 7.43. Problem 9

REFERENCES

1. L. Segrlind, "Applied Finite Element Analysis, Second Edition," Jon Wiley and Sons, 1984.
2. S. Moaveni, "Finite Element Analysis: Theory and Application with ANSYS, Third Edition," Prentice Hall, 2008.
3. D. L. Logan, "A First Course in the Finite Element Method, Fifth Edition," Cengage Learning, 2012.

4. F. P. Incropera and D. P. DeWitt, "Fundamentals of Heat and Mass Transfer, Fourth Edition," Wiley, 1996.
5. F. P. Incropera, D. P. DeWitt, T. L. Bergman, and A. S. Lavine, "Introduction to Heat Transfer, Fifth Edition," Wiley, 2007.
6. S. S. Rao, "The Finite Element Method in Engineering, Fifth Edition," Butterworth-Heinemann, 2011.

Chapter 8

Fluid Flow Analysis

8.1 INTRODUCTION

A substance (liquid or gas) that will deform continuously by applied surface (shearing) stresses is called a *fluid*. The magnitude of shear stress depends on the magnitude of angular deformation. Indeed, different fluids have different relations between stress and the rate of deformation. Also, fluids are classified as *compressible* (usually gas) and *incompressible* (usually liquid).

The terms of velocities and accelerations of fluid particles at different times and different points throughout the fluid filled space are used to describe the flow field. The fluid is called ideal when the fluid has zero viscosity and is incompressible. A fluid is said to be *incompressible* if the volume change is zero (i.e., ρ = constant).

$$\nabla \cdot \mathbf{v} = 0,$$

where,
$\mathbf{v}$ is velocity vector.

Depending on the importance of the viscosity of the fluid in the analysis, a flow can be termed as *inviscid* or *viscous*. An inviscid flow is a frictionless flow characterized by zero viscosity, that is, there is no real fluid. In other words, a fluid is called *inviscid* if the viscosity is zero (i.e., $\mu = 0$).

A viscous flow is a flow in which the fluid is assumed to have nonzero viscosity. An *irrotational flow* is a flow in which the particles of the fluid are not rotating, the rotation is zero. In other words, an *irrotational flow* is a flow with negligible angular velocity, if

$$\nabla \times \mathbf{v} = 0.$$

On the other hand, a potential *flow* is an irrotational flow of an ideal fluid (i.e., ρ = constant and $\mu = 0$).

A line that connects a series of points in space at a given instant where all particles falling on the line at that instant have velocities whose vectors are tangent to the line is called a *streamline*.

The flow is *steady* which means that the flow pattern or streamlines do not change over time and the streamlines represent the trajectory of the fluid's particles. But, when the flow is *ideal* that means that the fluid has zero velocity.

This chapter covers the finite element solution of ideal or potential flow (inviscid, incompressible flow) problems. Typical examples of potential flow are flow over a cylinder, flow around an airfoil, and flow out of an orifice.

The two-dimensional potential flow (irrotational flow) problems can be formulated in terms of a velocity potential function (φ) or a stream function (Ψ). The selection between velocity and stream function formulations in the finite element analysis depends on the ease of applying boundary conditions. If the geometry is simple, any one function can be used.

Fluid elements (e.g., FLUID141) are used in the steady state or transient analysis of fluid systems. Pressure, velocity, and temperature distributions can be obtained using these elements.

Two-dimensional fluid elements are defined using 3 (triangular element) or 4 (quadrilateral element) nodes added by isotropic properties. Inputs to these elements are nodal coordinates, real constants, material properties, surface and body loads, etc. Outputs of interest are nodal values of pressure and velocity.

8.2 PROCEDURE OF FINITE ELEMENT ANALYSIS (RELATED TO FLUID FLOW PROBLEMS)

Step 1. Select element type—the basic 3 node triangular element can be used.

Step 2. Choose a potential function.

Step 3. Define the gradient/potential and velocity/gradient relationships.

Step 4. Derive the element stiffness matrix and equations.

Step 5. Assemble the element equations to obtain the global equations and introduce boundary conditions.

Step 6. Solve for the nodal potentials.

Step 7. Solve for the element velocities and volumetric rates.

The finite element solution using software of potential flow problems is illustrated below. Only potential function formulation is considered. Two cases are considered in this chapter.

8.3 POTENTIAL FLOW OVER A CYLINDER

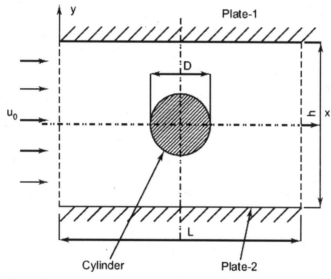

Figure 8.1. Potential flow over a cylinder.

The previous figure depicts the steady-state irrotational flow of an ideal fluid over a cylinder, confined between 2 parallel plates. We assume that, at the inlet, velocity is uniform, say u_0. Here, we have to determine the flow velocities near the cylinder.

Flow past a fixed circular cylinder can be obtained by combining uniform flow with a doublet.

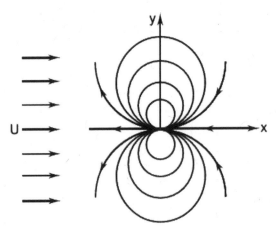

Figure 8.2. Superposition of a uniform flow and a doublet.

The superimposed stream function and velocity potential are given by,

$$\Psi = \Psi_{\text{uniform flow}} + \Psi_{\text{doublet}} = U \times r \times \sin\theta - K \times \frac{\sin\theta}{r} \quad (8.1)$$

and

$$\Phi = \Phi_{\text{uniform flow}} + \Phi_{\text{doublet}} = U \times r \times \cos\theta - K \times \frac{\cos\theta}{r}, \text{ respectively,} \quad (8.2)$$

where, U is velocity.

Because the streamline that passes through the stagnation point has a value of zero, the stream function on the surface of the cylinder of radius 'a' is then given by,

$$\Psi = U \times a \times \sin\theta - K \times \frac{\sin\theta}{a} = 0 \quad (8.3)$$

which gives the strength of the doublet as,

$$K = U \times a^2. \quad (8.4)$$

The stream function and velocity potential for flow past a fixed circular cylinder becomes

$$\Psi = U \times r \left(1 - \left(\frac{a}{r}\right)^2\right) \sin\theta \quad (8.5)$$

and

$$\Phi = U \times r\left(1-\left(\frac{a}{r}\right)^2\right)\cos\theta, \text{ respectively.} \qquad (8.6)$$

The plot of the streamlines is shown in Figure 8.3.

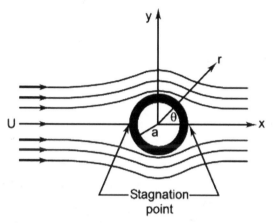

Figure 8.3. Streamlines for flow past a fixed cylinder.

The velocity components can be determined by,

$$v_r = \frac{1}{r}\frac{\partial \Psi}{\partial \theta} = U\left(1-\left(\frac{a}{r}\right)^2\right)\cos\theta \qquad (8.7)$$

$$v_\theta = \frac{\partial \Psi}{\partial r} = -U\left(1-\left(\frac{a}{r}\right)^2\right)\sin\theta. \qquad (8.8)$$

Along the cylinder ($r = a$), the velocity components reduce to $v_r = 0$ and $v_\theta = -2U\sin\theta$.

The radial velocity component is always zero along the cylinder while the tangential velocity component varies from 0 at the stagnation point ($\theta = \pi$) to a maximum velocity of $2U$ at the top and bottom of the cylinder ($\theta = \frac{\pi}{2}$ or $\theta = -\frac{\pi}{2}$).

8.4 POTENTIAL FLOW AROUND AN AIRFOIL

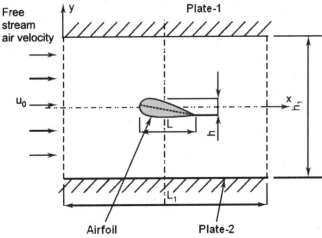

Figure 8.4. Potential flow around an airfoil.

The x- and y-components of fluid's velocity respectively can be expressed in a stream function $\Psi(x, y)$ as

$$v_x = \frac{\partial \Psi}{\partial y} \text{ and } v_y = -\frac{\partial \Psi}{\partial x}. \tag{8.9}$$

The x- and y-components of fluid's velocity with irrotational flows respectively can be expressed in a potential function $\phi(x, y)$ as

$$v_x = \frac{\partial \phi}{\partial x} \text{ and } v_y = \frac{\partial \phi}{\partial y}. \tag{8.10}$$

Example 8.1

Flow over a circular cylinder between 2 parallel plates is shown in Figure 8.5. Assume unit thickness. Find the velocity distribution for the flow over a circular cylinder. Consider the flow of a liquid over a circular cylinder. Take liquid as water. Water density = 1000 kg/m³ and viscosity = 0.001 N-s/m².

FLUID FLOW ANALYSIS

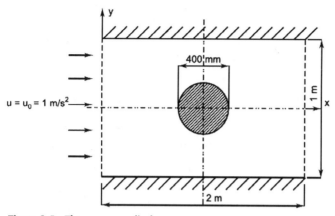

Figure 8.5. Flow over a cylinder.

Solution

(I) Software results

Procedure for solving the problem using ANSYS® 11.0 academic teaching software.

PREPROCESSING

1. Main Menu > Preferences, then select FLOTRAN CFD > OK

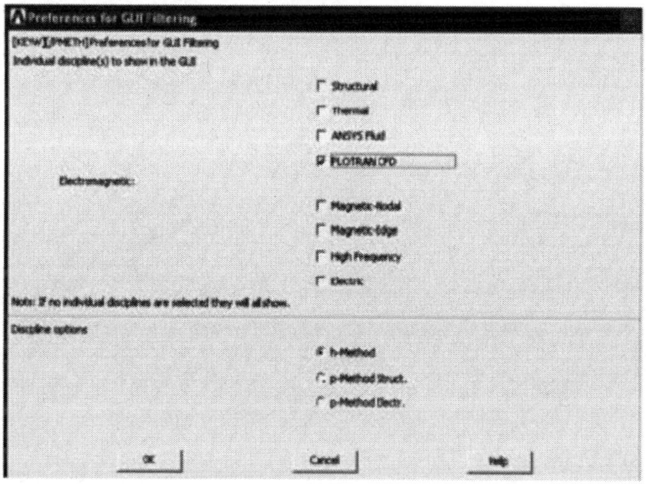

Figure 8.6. Selecting the preferences.

2. Main Menu > Preprocessor > Element Type > Add/Edit/Delete > Add > FLOTRAN CFD > 2D FLOTRAN 141 > OK

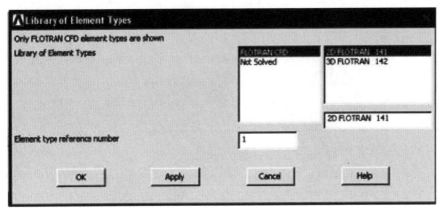

Figure 8.7. Element selection.

3. Main Menu > Preprocessor > Modeling > Create > Areas > Rectangle > By 2 Corners
   Enter (lower left corner) WP X = 0.0, WP Y = 0.0 and Width = 2, Height = 1 > OK
4. Main Menu > Preprocessor > Modeling > Create > Areas > Circle > Solid Circle. Enter WP X = 1, WP Y = 0.5 and Radius = 200e-3 > OK

Figure 8.8. Create areas.

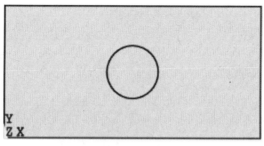

Figure 8.9. Rectangle and circle.

Now **subtract** the circle from the rectangle. (Read the messages in the window at the bottom of the screen as necessary.)

5. **Main Menu > Preprocessor > Modeling > Operate > Booleans > Subtract > Areas >**
Pick the rectangle > **OK,** then pick the circle > **OK**

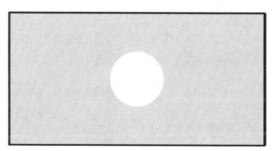

Figure 8.10. Geometry for the flow over a cylinder.

Create a mesh of quadrilateral elements over the area.

6. **Main Menu > Preprocessor > Meshing > Mesh Tool**
The **Mesh Tool** dialog box appears. Close the **Mesh Tool** box.

7. **Main Menu > Preprocessor > Meshing > Mesh > Areas > Free** Pick the area > **OK**

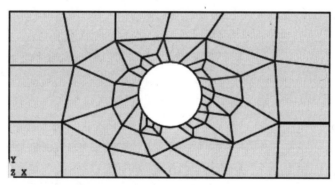

Figure 8.11. Quadrilateral element mesh.

Apply the velocity boundary conditions and pressure.

8. **Main Menu > Preprocessor > Loads > Define Loads > Apply > Fluid/CFD > Velocity
   > On Lines** Pick the left edge of the plate > **OK > Enter VX = 1 > OK**
   (VX = 1 means an initial velocity of 1 m/s^2)

9. **Main Menu > Preprocessor > Loads > Define Loads > Apply > Fluid/CFD > Velocity
   > On Lines** Pick the edges around the cylinder > **OK > Enter VX = 0 and VY = 0 > OK**

10. **Main Menu > Preprocessor > Loads > Define Loads > Apply > Fluid/CFD > Pressure
    DOF > On Lines** Pick the top, bottom and right edges of the plate > **OK > OK**
    Once all the boundary conditions are applied, the cylinder with plate will look like Figure 8.12.

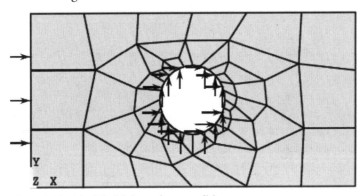

Figure 8.12. Model with boundary conditions.

The model-building step is now complete, and we can proceed to the solution. First, save the model.

Solution

The interactive solution proceeds.

11. **Main Menu > Solution > FLOTRAN Set Up > Fluid Properties** > A dialog in that select against density as liquid and against viscosity as liquid > **OK**
    Then another dialog box appears and, in that, enter the value of density = 1000 value = 0.001 > **OK**

12. **Main Menu > Solution > FLOTRAN Set Up > Execution Ctrl** > a dialog in that Enter in the first row "Global iterations EXEC" = 200

13. **Main Menu > Solution > Run FLOTRAN**
    When the solution is complete, close the **'Solution is Done!'** window.

POST-PROCESSING

We can now plot the results of this analysis and also list the computed values.

14. Main Menu > General Postproc > Read Results > Last Set

15. General Postproc > Plot Results > Contour Plot > Nodal Solu
Select **DOF Solution** and **Fluid Velocity** and click **OK**
This is what the solution should look like:

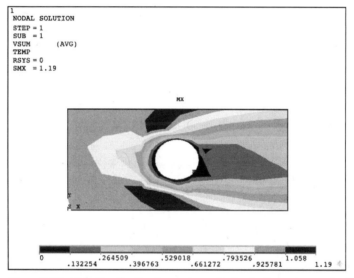

Figure 8.13. Velocity distribution over a cylinder (refer to Appendix C for color figures).

16. Next, go to **Main Menu > General Postproc > Plot Results > Vector Plot > Predefined.** One window will appear then click **OK**

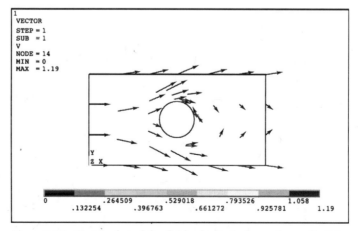

Figure 8.14. Vector plot of the fluid velocity (refer to Appendix C for color figures).

17. **General Postproc > Plot Results > Contour Plot > Nodal Solu**
    Select **DOF Solution** and **Pressure** and Click **OK**

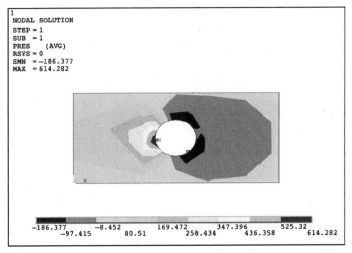

Figure 8.15. Pressure distribution over a cylinder (refer to Appendix C for color figures).

PROBLEMS

1. Define a fluid, inviscid flow, viscous flow, and irrotational flow.
2. What are the 2 fluid classifications?
3. Define streamline in a graphic of fluid motion?
4. What we mean when we say the flow is steady and ideal?
5. Define irrotational *flow and potential flow*?
6. Compute and plot velocity distribution over the airfoil as shown in Figure 8.16. Assume unit thickness. Take density of air = 1.23 kg/m^3 and viscosity = 1.79×10^{-5} N-s/m^2.

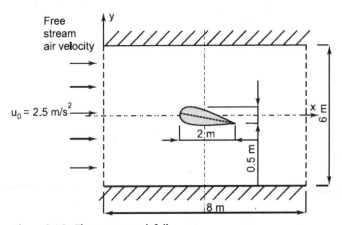

Figure 8.16. Flow over an airfoil.

7. Flow over a circular cylinder between 2 parallel plates is as shown in Figure 8.17. Assume unit thickness. Find the velocity distribution for the flow over a circular cylinder. Consider the flow of a liquid over a circular cylinder. Take liquid as water. Water density = 1000 kg/m^3 and viscosity = 0.001 N-s/m^2, $u = u_0 = 2$ m/s^2, $h = 2$ m, and $L = 4$m.

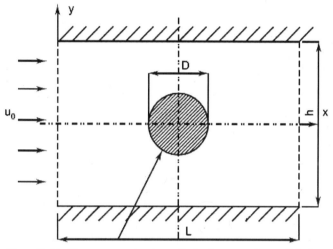

Figure 8.17. Flow over a circular cylinder between 2 parallel plates.

8. Compute and plot velocity distribution over the airfoil as shown in Figure 8.18. Assume unit thickness. Take density of air = 1.23 kg/m^3 and viscosity = 1.79 × 10^{-5}N-s/m^2, $L = 4$ m, $L_1 = 20$ m, $h_1 = 18$ m, and $u = u_0 = 3$ m/s^2.

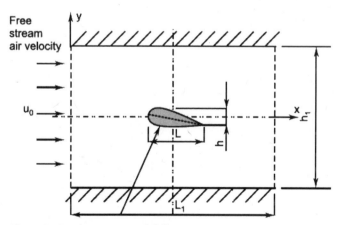

Figure 8.18. Flow over an airfoil.

9. Flow over an elliptical cylinder between 2 parallel plates is shown in Figure 8.19. Assume unit thickness. Find the velocity distribution for the flow over a circular cylinder. Consider the flow of a liquid over a circular cylinder. Take liquid as water. Water density = 1000 kg/m^3 and viscosity = 0.001 N-s/m^2, $u = u_0 = 1$ m/s^2, $D = 2$ m, $b = 1$ m, $h = 4$ m, and $L = 8$ m.

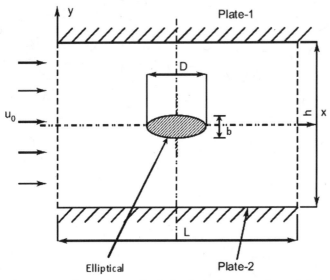

Figure 8.19. Flow over an elliptical cylinder between 2 parallel plates.

REFERENCES

1. J. G. Knudsen and D. L. Katz, "Fluid Dynamics and Heat Transfer," McGraw-Hill, 1958.
2. J. W. Daily and D. R. F. Harleman, "Fluid Dynamics," Addison-Wesley, Reading, 1966.
3. S. S. Rao, "The Finite Element Method in Engineering, Fifth Edition," Butterworth-Heinemann, 2011.
4. T. J. Chung, "Finite Element Analysis in Fluid Dynamics," McGraw-Hill, 1978.
5. D. L. Logan, "A First Course in the Finite Element Method, Fifth Edition," Cengage Learning, 2012.
6. J. N. Reddy, "An Introduction to the Finite Element Method, Third Edition," McGraw-Hill, 2004.

Chapter 9
Dynamic Analysis

9.1 INTRODUCTION

A dynamic system is a system that has mass and components, or parts, that are capable of relative motion. Structural dynamics encompass modal analysis, harmonic response analysis, and transient response analysis. Modal analysis consists of finding natural frequencies and corresponding modal shapes of structures. Finding amplitude of vibration when the loads vary sinusoidal with time is known as *harmonic response analysis*. Finding the structural response to arbitrary time dependent loading is referred to as transient response analysis.

In this chapter, one-dimensional problems relating to these topics are covered. In vibration analysis, mass matrix and damping matrix will also be discussed in addition to stiffness matrix.

Governing equation of undamped free vibration assumes the form,

$$([k] - \omega^2 [m])\{q\} = 0. \tag{9.1}$$

The nontrivial solution of equation (9.1) is the determinate,

$$\left|([k] - \omega^2 [m])\right| = 0 \tag{9.2}$$

where
ω = radian (or natural) frequency.

The solution of equation (9.2) gives natural frequencies (ω). Substituting the value of ω back into the governing equation gives modal shapes (or amplitudes of the displacements) defined by $\{q\}$.

The governing equation for the complete structure in global coordinate is

$$([K] - \omega^2 [M])\{q\} = 0.$$

Mass matrices for bar element and beam elements are given by,

$$[m]_{Bar} = \rho A L \begin{bmatrix} \frac{1}{3} & \frac{1}{6} \\ \frac{1}{6} & \frac{1}{3} \end{bmatrix} \qquad (9.3)$$

$$[m]_{Beam} = \frac{\rho A L}{420} \begin{bmatrix} 156 & 22L & 54 & -13L \\ 22L & 4L^2 & 13L & -3L^2 \\ 54 & 13L & 156 & -22L \\ -13L & -3L^2 & -22L & 4L^2 \end{bmatrix}, \qquad (9.4)$$

where
- ρ = density of the element material
- A = cross-sectional area
- L = length.

9.2 PROCEDURE OF FINITE ELEMENT ANALYSIS (RELATED TO DYNAMIC PROBLEMS)

Step 1. Select element type.
Step 2. Select a displacement function.
Step 3. Define the strain/displacement and stress/strain relationships.
Step 4. Derive the element stiffness and mass matrices and equations.
Step 5. Assemble the element equations to obtain the global equations and introduce boundary conditions.
Step 6. Solve for the natural requencies and mode shapes.

9.3 FIXED-FIXED BEAM FOR NATURAL FREQUENCY DETERMINATION

Example 9.1

Determine the first 2 natural frequencies for the fixed-fixed beam shown in Figure 9.1. The beam is made of steel with modulus of elasticity E = 209 GPa, Poisson's ratio = 0.3, length L = 0.75 m, cross-section area A = 625 mm², mass density ρ = 7800 kg/m³, moment of inertia I = 34700 mm⁴.

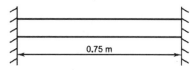

Figure 9.1. Fixed-fixed beam for Example 9.1.

Solution

(I) Analytical method

$$\therefore \omega_1 = \frac{22.4}{L^2} \sqrt{\frac{EI}{\rho A}} \tag{9.5}$$

$$\omega_1 = \frac{22.4}{(0.75)^2} \sqrt{\frac{209 \times 10^9 \times 34700 \times 10^{-12}}{7800 \times 625 \times 10^{-6}}} = 1535.95 \text{ rad/s.}$$

Frequency,

$$f_1 = \frac{\omega_1}{2\pi} \tag{9.6}$$

$$f_1 = \frac{1535.95}{2\pi} = 244.45 \text{ Hz}$$

$$\therefore \omega_2 = \frac{61.7}{L^2} \sqrt{\frac{EI}{\rho A}} \tag{9.7}$$

$$\omega_2 = \frac{61.7}{(0.75)^2} \sqrt{\frac{209 \times 10^9 \times 34700 \times 10^{-12}}{7800 \times 625 \times 10^{-6}}} = 1535.95 \text{ rad/s.}$$

Frequency,

$$f_2 = \frac{\omega_2}{2\pi} \tag{9.8}$$

$$f_2 = \frac{4230.71}{2\pi} = 673.34 \text{ Hz.}$$

(II) FEM by hand calculations

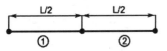

Figure 9.1(a). Finite element model.

Mass matrices are,

$$[M_1] = [M_2] = \frac{\rho \times A \times \frac{L}{2}}{420} \begin{bmatrix} 156 & 22\left(\frac{L}{2}\right) & 54 & -13\left(\frac{L}{2}\right) \\ 22\left(\frac{L}{2}\right) & 4\left(\frac{L}{2}\right)^2 & 13\left(\frac{L}{2}\right) & -3\left(\frac{L}{2}\right)^2 \\ 54 & 13\left(\frac{L}{2}\right) & 156 & -22\left(\frac{L}{2}\right) \\ -13\left(\frac{L}{2}\right) & -3\left(\frac{L}{2}\right)^2 & -22\left(\frac{L}{2}\right) & 4\left(\frac{L}{2}\right)^2 \end{bmatrix}$$

$$[M_1] = [M_2] = \frac{\rho \times A \times L}{840} \begin{bmatrix} 156 & 11L & 54 & -6.5L \\ 11L & L^2 & 6.5L & -3\left(\frac{L^2}{4}\right) \\ 54 & 13\left(\frac{L}{2}\right) & 156 & -11L \\ -6.5L & -3\left(\frac{L^4}{4}\right) & -11L & L^2 \end{bmatrix}.$$

Stiffness matrices are,

$$[k_1] = [k_2] = \frac{EI}{\left(\frac{L}{2}\right)^3} \begin{bmatrix} 12 & 6\left(\frac{L}{2}\right) & -12 & 6\left(\frac{L}{2}\right) \\ & 4\left(\frac{L}{2}\right)^2 & -6\left(\frac{L}{2}\right) & 2\left(\frac{L}{2}\right)^2 \\ & & 12 & -6\left(\frac{L}{2}\right) \\ \text{Symmetric} & & & 4\left(\frac{L}{2}\right)^2 \end{bmatrix}$$

$$= \frac{8EI}{L^3} \begin{bmatrix} 12 & 3L & -12 & 3L \\ & L^2 & -3L & \frac{L^2}{2} \\ & & 12 & -3L \\ \text{Symmetric} & & & L^2 \end{bmatrix}.$$

Dynamic Analysis

Global mass matrix,

$$[M] = \frac{\rho AL}{840}\begin{bmatrix} 156+156 & -11L+11L \\ -11L+11L & L^2+L^2 \end{bmatrix} = \frac{\rho AL}{840}\begin{bmatrix} 312 & 0 \\ 0 & 2L^2 \end{bmatrix}.$$

Global stiffness matrix,

$$[K] = \frac{8EI}{L^3}\begin{bmatrix} 12+12 & 3L+-3L \\ 3L+-3L & L^2+L^2 \end{bmatrix} = \frac{8EI}{L^3}\begin{bmatrix} 24 & 0 \\ 0 & 2L^2 \end{bmatrix}.$$

Governing equation is,

$$\left([K] - \omega^2[M]\right)\{q\} = 0,$$

$$\left(\frac{8EI}{L^3}\begin{bmatrix} 24 & 0 \\ 0 & 2L^2 \end{bmatrix} - \omega^2 \frac{\rho AL}{840}\begin{bmatrix} 312 & 0 \\ 0 & 2L^2 \end{bmatrix}\right)\{q\} = 0$$

$$\frac{L^3}{8EI} \times \left(\left(\frac{8EI}{L^3}\begin{bmatrix} 24 & 0 \\ 0 & 2L^2 \end{bmatrix} - \omega^2 \frac{\rho AL}{840}\begin{bmatrix} 312 & 0 \\ 0 & 2L^2 \end{bmatrix}\right)\{q\} = 0\right)$$

$$\left(\begin{bmatrix} 24 & 0 \\ 0 & 2L^2 \end{bmatrix} - \frac{\omega^2 \rho AL L^3}{840 \times 8EI}\begin{bmatrix} 312 & 0 \\ 0 & 2L^2 \end{bmatrix}\right)\{q\} = 0$$

$$\left(\begin{bmatrix} 24-312a & 0 \\ 0 & 2L^2 - 2L^2 a \end{bmatrix}\right)\{q\} = 0$$

where $a = \dfrac{\omega^2 \rho AL^4}{6720 EI}$.

For nontrivial solution,

$$\left|\left([K] - \omega^2[M]\right)\right| = 0$$

$$\left\|\begin{bmatrix} 24-312a & 0 \\ 0 & 2L^2 - 2L^2 a \end{bmatrix}\right\| = 0.$$

Solving, we get

$$a = 1 \quad \text{or} \quad a = 0.076923$$

$$1 = \frac{\omega^2 \rho AL^4}{6720 EI} \quad \text{or} \quad 0.076923 = \frac{\omega^2 \rho AL^4}{6720 EI}$$

$$\therefore \omega_1 = \frac{22.74}{L^2}\sqrt{\frac{EI}{\rho A}} \quad \therefore \omega_2 = \frac{81.98}{L^2}\sqrt{\frac{EI}{\rho A}}.$$

Given:

$$E = 209 \text{ GPa}$$
$$A = 625 \times 10^{-6} \text{ m}^2$$
$$I = 34700 \times 10^{-12} \text{ m}^4$$
$$\rho = 7800 \text{ kg/m}^3$$
$$L = 0.75 \text{ m}.$$

Substituting, we get

$$\therefore \omega_1 = \frac{22.74}{(0.75)^2} \sqrt{\frac{209 \times 10^9 \times 34700 \times 10^{-12}}{7800 \times 625 \times 10^{-6}}} = 1559.26 \text{ rad/s}.$$

Frequency,

$$f_1 = \frac{\omega_1}{2\pi} = \frac{1559.26}{2\pi} = 248.164 \text{ Hz}$$

$$\therefore \omega_2 = \frac{81.98}{(0.75)^2} \sqrt{\frac{209 \times 10^9 \times 34700 \times 10^{-12}}{7800 \times 625 \times 10^{-6}}} = 5621.29 \text{ rad/s}.$$

Frequency,

$$f_2 = \frac{\omega_2}{2\pi} = \frac{5621.29}{2\pi} = 894166 \text{ Hz}.$$

(III) Software results

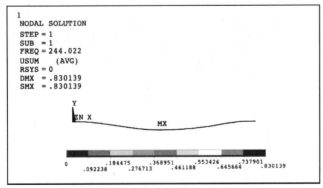

Figure 9.1(b). Deflection pattern for a fixed-fixed beam for mode 1 (refer to Appendix C for color figures).

Frequency values (in Hz)

SET	TIME/FREQ	LOAD STEP	SUB STEP	CUMULATIVE
1	244.02	1	1	1
2	671.69	1	2	2

The following are the mode shapes:

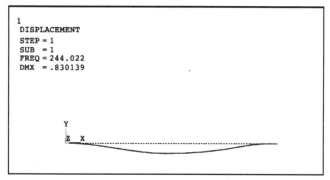

Figure 9.1(c). Mode 1 for fixed-fixed beam (refer to Appendix C for color figures).

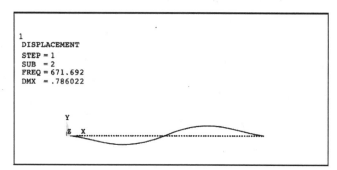

Figure 9.1(d). Mode 2 for fixed-fixed beam (refer to Appendix C for color figures).

Answers for Example 9.1

Parameter	Analytical method	FEM–hand calculation (with 2 elements)	Software results (with 10 elements)
Natural frequency			
f_1	244.45 Hz	248.16 Hz	244.02 Hz
f_2	673.34 Hz	894.66 Hz	671.69 Hz

Procedure for solving the problem using ANSYS® 11.0 academic teaching software For Example 9.1

PREPROCESSING

1. **Main Menu > Preprocessor > Element Type > Add/Edit/Delete > Add > Beam > 2D elastic 3 > OK > Close**

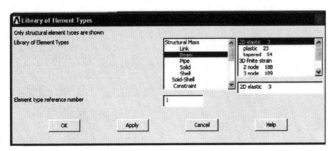

Figure 9.2. Element selection.

2. **Main Menu > Preprocessor > Real Constants > Add/Edit/Delete > Add > OK**

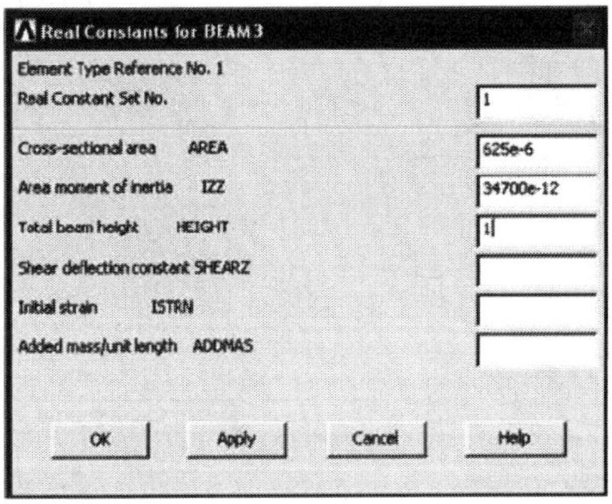

Figure 9.3. Enter the area and moment of inertia.

Cross-sectional area AREA > **Enter 625e-6**
Area moment of inertia IZZ > **Enter 34700e-12**
Total beam height HEIGHT > **Enter 1 > OK > Close**
Enter the material properties.

3. **Main Menu > Preprocessor > Material Props > Material Models**
Material Model Number 1, **click Structural > Linear > Elastic > Isotropic**
Enter **EX = 209E9 and PRXY = 0.3 > OK**
click **Structural > Linear > Density**
Enter DENS = 7800 > OK
(**Close** the Define Material Model Behavior window.)
Create the keypoints and lines as shown in the figure.

4. **Main Menu > Preprocessor > Modeling > Create > Keypoints > In Active CS,** Enter the coordinates of keypoint 1 > **Apply** Enter the coordinates of keypoint 2 > **OK**

Keypoint locations		
Keypoint number	X coordinate	Y coordinate
1	0	0
2	0.75	0

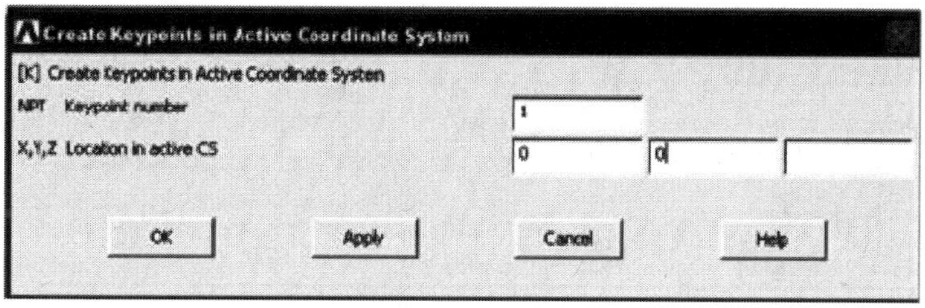

Figure 9.4. Enter the keypoint coordinates.

5. **Main Menu > Preprocessor > Modeling > Create > Lines > Lines > Straight Line,** Pick the 1st and 2nd keypoint > **OK**

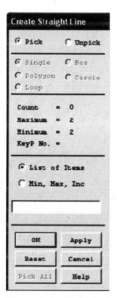

Figure 9.5. Pick the keypoints to create lines.

6. **Main Menu > Preprocessor > Meshing > Size Cntrls > Manual Size > Lines > All Lines >** Enter **NDIV No. of element divisions = 10**

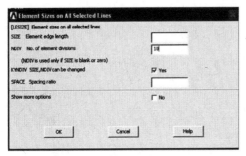

Figure 9.6. Specify element length.

7. **Main Menu > Preprocessor > Meshing > Mesh > Lines > Click Pick All**

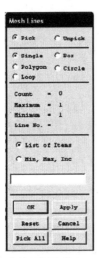

Figure 9.7. Create elements by meshing.

8. **Main Menu > Solution > Analysis Type > New Analysis > Select Modal > OK**

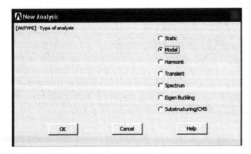

Figure 9.8. Define the type of analysis.

9. **Main Menu > Preprocessor > Loads > Define Loads > Apply > Structural > Displacement > On Nodes** Pick the left most node and right most node > **Apply > Select All DOF > OK**

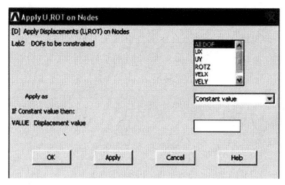

Figure 9.9. Apply the displacement constraint.

10. **Main Menu > Solution > Analysis Type > Analysis Options > Select PCG Lanczos option**
    Enter No. of modes to extract = 2
    NMODE No. of modes to expand = 2 > OK
    After OK one more window will appear, for that also click OK

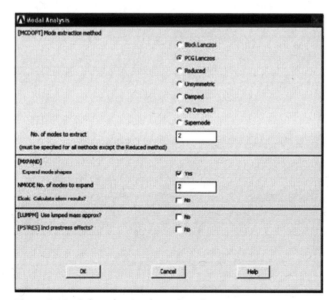

Figure 9.10. Select the number of modes to extract.

Solution

The interactive solution proceeds.
11. **Main Menu > Solution > Solve > Current LS > OK**
    The **/STATUS Command** window displays the problem parameters and **the Solve Current Load Step** window is shown. Check the solution options in the **/STATUS** window and if all is OK, select **File > Close**.
    In the **Solve Current Load Step** window, select **OK**, and when the solution is complete, **close** the **'Solution is Done!'** window.

POST-PROCESSING

12. **Main Menu > General Postproc > Results Summary**
    This result is shown as frequency values in Hz.
13. **Main Menu > General Postproc > Read Results > First Set**
14. **Main Menu > General Postproc > Plot Results > Deformed Shape > Click Def + undeformed > OK**
    This result is the first mode shown in Figure 9.1(c).
15. **Main Menu > General Postproc > Read Results > Next Set**
16. **Main Menu > General Postproc > Plot Results > Deformed Shape > Click Def + undeformed > OK**
    This result is the second mode shown in Figure 9.1(d).

9.4 TRANSVERSE VIBRATIONS OF A CANTILEVER BEAM

Example 9.2

Determine the first 4 natural frequencies for the cantilever beam shown in Figure 9.11. The beam is made of steel with modulus of elasticity, E = 207 GPa, Poisson's ratio = 0.3, length L = 0.75 m, cross-section area A = 625 mm^2, mass density ρ = 7800 kg/m^3, moment of inertia I = 34700 mm^4.

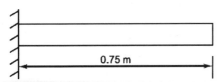

Figure 9.11. Cantilever beam for Example 9.2.

Dynamic Analysis

Solution

(1) Analytical solution

$$\omega_1 = \frac{3.52}{L^2}\sqrt{\frac{EI}{\rho A}}$$

$$\omega_1 = \frac{3.52}{(0.75)^2}\sqrt{\frac{207\times 10^9 \times 34700 \times 10^{-12}}{7800\times 625 \times 10^{-6}}} = 240 \text{ rad/s.}$$

Frequency,

$$f_1 = \frac{\omega_1}{2\pi}$$

$$f_1 = \frac{240}{2\pi} = 38.197 \text{ Hz}$$

$$\omega_2 = \frac{22}{L^2}\sqrt{\frac{EI}{\rho A}}$$

$$\omega_2 = \frac{22}{(0.75)^2}\sqrt{\frac{207\times 10^9 \times 34700 \times 10^{-12}}{7800\times 625 \times 10^{-6}}} = 1501 \text{ rad/s.}$$

Frequency,

$$f_2 = \frac{\omega_2}{2\pi}$$

$$f_2 = \frac{1501}{2\pi} = 238.89 \text{ Hz}$$

$$\omega_3 = \frac{61.7}{L^2}\sqrt{\frac{EI}{\rho A}}$$

$$\omega_2 = \frac{61.7}{(0.75)^2}\sqrt{\frac{207\times 10^9 \times 34700 \times 10^{-12}}{7800\times 625 \times 10^{-6}}} = 4210 \text{ rad/s.}$$

Frequency,

$$f_3 = \frac{\omega_3}{2\pi}$$

$$f_3 = \frac{4210}{2\pi} = 670.04 \text{ Hz}$$

$$\omega_4 = \frac{121}{L^2}\sqrt{\frac{EI}{\rho A}}$$

$$\omega_2 = \frac{121}{(0.75)^2}\sqrt{\frac{207\times 10^9 \times 34700\times 10^{-12}}{7800\times 625\times 10^{-6}}} = 8257 \text{ rad/s}.$$

Frequency,

$$f_4 = \frac{\omega_4}{2\pi}$$

$$f_4 = \frac{8257}{2\pi} = 1314.14 \text{ Hz}.$$

(II) FEM by hand calculations

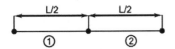

Figure 9.11(a). Finite element model.

Stiffness matrices are,

$$[k_1]=[k_2]=\frac{EI}{\left(\frac{L}{2}\right)^3}\begin{bmatrix} 12 & 6\left(\frac{L}{2}\right) & -12 & 6\left(\frac{L}{2}\right) \\ 6\left(\frac{L}{2}\right) & 4\left(\frac{L}{2}\right)^2 & -6\left(\frac{L}{2}\right) & 2\left(\frac{L}{2}\right)^2 \\ -12 & -6\left(\frac{L}{2}\right) & 12 & -6\left(\frac{L}{2}\right) \\ 6\left(\frac{L}{2}\right) & 2\left(\frac{L}{2}\right)^2 & -6\left(\frac{L}{2}\right) & 4\left(\frac{L}{2}\right)^2 \end{bmatrix}.$$

Global stiffness matrix,

$$[K]=\frac{8EI}{L^3}\begin{bmatrix} 24 & 0 & -12 & 3L \\ 0 & 2L^2 & -3L & \frac{L^2}{2} \\ -12 & -3L & 12 & -3L \\ 3L & \frac{L^2}{2} & -3L & L^2 \end{bmatrix} = 136209.07 \begin{bmatrix} 24 & 0 & -12 & 2.25 \\ 0 & 1.125 & -2.25 & 0.28125 \\ -12 & -2.25 & 12 & -2.25 \\ 2.25 & 0.28125 & -2.25 & 0.5625 \end{bmatrix}.$$

Mass matrices are,

$$[M_1] = [M_2] = \frac{\rho \times A \times \frac{L}{2}}{420} \begin{bmatrix} 156 & 22\left(\frac{L}{2}\right) & 54 & -13\left(\frac{L}{2}\right) \\ 22\left(\frac{L}{2}\right) & 4\left(\frac{L}{2}\right)^2 & 13\left(\frac{L}{2}\right) & -3\left(\frac{L}{2}\right)^2 \\ 54 & 13\left(\frac{L}{2}\right) & 156 & -22\left(\frac{L}{2}\right) \\ -13\left(\frac{L}{2}\right) & -3\left(\frac{L}{2}\right)^2 & -22\left(\frac{L}{2}\right) & 4\left(\frac{L}{2}\right)^2 \end{bmatrix}$$

$$[M_1] = [M_2] = \frac{\rho \times A \times L}{840} \begin{bmatrix} 156 & 11L & 54 & -6.5L \\ 11L & L^2 & 6.5L & -3\left(\frac{L^2}{4}\right) \\ 54 & 13\left(\frac{L}{2}\right) & 156 & -11L \\ -6.5L & -3\left(\frac{L^4}{4}\right) & -11L & L^2 \end{bmatrix}.$$

Global mass matrix is,

$$[M] = \frac{\rho \times A \times L}{840} \begin{bmatrix} 312 & 0 & 54 & -6.5L \\ 0 & L^2 & 6.5L & -3\left(\frac{L^2}{4}\right) \\ 54 & 13\left(\frac{L}{2}\right) & 156 & -11L \\ -6.5L & -3\left(\frac{L^4}{4}\right) & -11L & L^2 \end{bmatrix}$$

$$[M] = 4.352676 \times 10^{-3} \begin{bmatrix} 312 & 0 & 54 & -4.875 \\ 0 & 1.125 & 4.875 & -0.421875 \\ 54 & 4.875 & 156 & -8.25 \\ -4.875 & -0.421875 & -8.25 & 0.5625 \end{bmatrix}.$$

Governing equation is,

$$([K] - \omega^2 [M])\{q\} = \begin{Bmatrix} \omega_2 \\ \theta_2 \\ \omega_3 \\ \theta_3 \end{Bmatrix} = 0.$$

For a nontrivial solution

$$\text{Det}\left([K]-\omega^2[M]\right)=0 \Rightarrow \left|([K]-\omega^2[M])\right|=0.$$

Substituting and solving, we get

$$\omega^2 = \begin{Bmatrix} \omega_1^2 \\ \omega_2^2 \\ \omega_3^2 \\ \omega_4^2 \end{Bmatrix} = 10^8 \begin{Bmatrix} 0.0006 \\ 0.0230 \\ 0.2630 \\ 2.2159 \end{Bmatrix}$$

$$\omega_1 = 245 \text{ rad/s} \Rightarrow f_1 = \frac{\omega_1}{2\pi} = \frac{245}{2\pi} = 38.993 \text{ Hz}$$

$$\omega_2 = 1517 \text{ rad/s} \Rightarrow f_2 = \frac{\omega_2}{2\pi} = \frac{1517}{2\pi} = 241.44 \text{ Hz}$$

$$\omega_3 = 5128 \text{ rad/s} \Rightarrow f_3 = \frac{\omega_3}{2\pi} = \frac{5128}{2\pi} = 816.147 \text{ Hz}$$

$$\omega_3 = 14885.9 \text{ rad/s} \Rightarrow f_4 = \frac{\omega_4}{2\pi} = \frac{14885.9}{2\pi} = 2369.16 \text{ Hz}.$$

(III) Software results

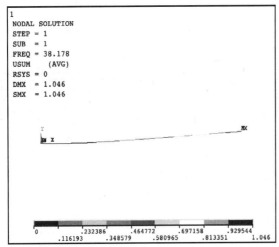

Figure 9.11(b). Deflection pattern for a fixed-fixed beam for mode 1 (Refer to Appendix C for color figures).

Frequency values (in Hz)

SET	TIME/FREQ	LOAD STEP	SUB STEP	CUMULATIVE
1	38.178	1	1	1
2	238.94	1	2	2
3	667.71	1	3	3
4	1305.2	1	4	4

The following are the mode shapes:

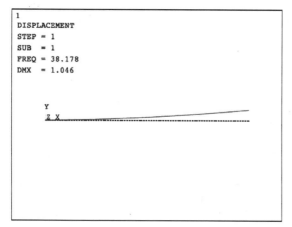

Figure 9.11(c). Mode 1 for cantilever beam (refer to Appendix C for color figures).

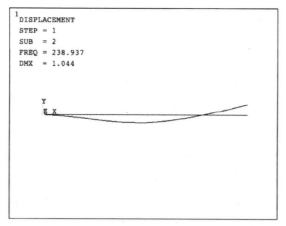

Figure 9.11(d). Mode 2 for cantilever beam (refer to Appendix C for color figures).

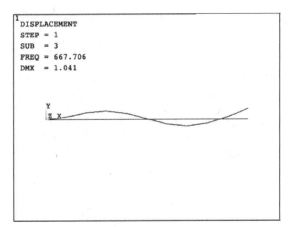

Figure 9.11(e). Mode 3 for cantilever beam (refer to Appendix C for color figures).

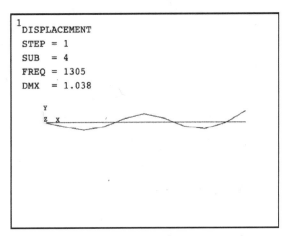

Figure 9.11(f). Mode 4 for cantilever beam (refer to Appendix C for color figures).

Answers of Example 9.2

Parameter	Analytical method	FEM–hand calculation (with 2 elements)	Software results (with 10 elements)
Natural frequency			
f_1	38.197 Hz	38.993 Hz	38.178 Hz
f_2	238.89 Hz	241.44 Hz	238.94 Hz
f_3	670.04 Hz	816.147 Hz	667.71 Hz
f_4	1314.14 Hz	2369.16 Hz	1305.2 Hz

9.5 FIXED-FIXED BEAM SUBJECTED TO FORCING FUNCTION

Example 9.3

For the fixed-fixed beam subjected to the time dependent forcing function shown in Figure 9.12, determine the displacement response for 0.2 seconds. Use time step integration of 0.01 sec. Let E = 46 GPa, Poisson's ratio = 0.35, length of beam L = 5 m, cross-section area A = 1 m², mass density, ρ = 1750 kg/m³, moment of inertia I = 4.2 × 10⁻⁵ m⁴.

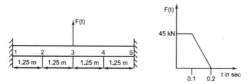

Figure 9.12. Fixed-fixed beam subjected to the time dependent forcing function for Example 9.3

Solution

(I) Software results

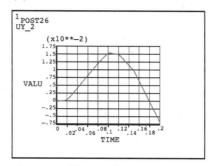

Figure 9.12(a). Displacement response for 0.2 sec for node 2 (refer to Appendix C for color figures).

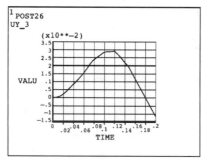

Figure 9.12(b). Displacement response for 0.2 sec for node 3 (refer to Appendix C for color figures).

Displacement values (in meters) for node 2

TIME	2 UY
	UY_2
0.0000	0.00000
0.10000E–01	0.421220E–05
0.20000E–01	0.284618E–03
0.50000E–01	0.602161E–02
0.80000E–01	0.121677E–01
0.10000	0.153042E–01
0.12000	0.148820E–01
0.15000	0.979873E–02
0.18000	0.868368E–04
0.20000	–0.649350E–02

Displacement values (in meters) for node 3

TIME	3 UY
	UY_3
0.0000	0.00000
0.10000E–01	0.505126E–03
0.20000E–01	0.218959E–02
0.50000E–01	0.113766E–01
0.80000E–01	0.241211E–01
0.10000	0.286233E–01
0.12000	0.292504E–01
0.15000	0.183799E–01
0.18000	–0.205644E–03
0.20000	–0.117477E–01

Procedure for solving the problem using ANSYS® 11.0 academic teaching software.
For Example 9.3

PREPROCESSING

1. Main Menu > Preferences > Select Structural > OK

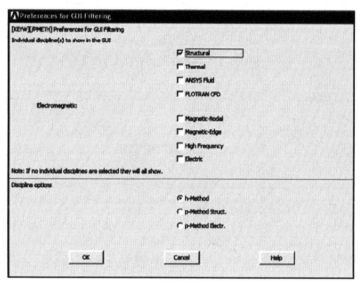

Figure 9.13. Selecting the preferences.

2. Main Menu > Preprocessor > Element Type > Add/Edit/Delete > Add > Beam > 2D elastic 3 > OK > Close

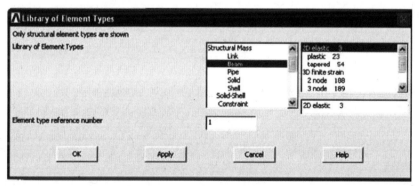

Figure 9.14. Element selection.

3. **Main Menu > Preprocessor > Real Constants > Add/Edit/Delete > Add > OK**

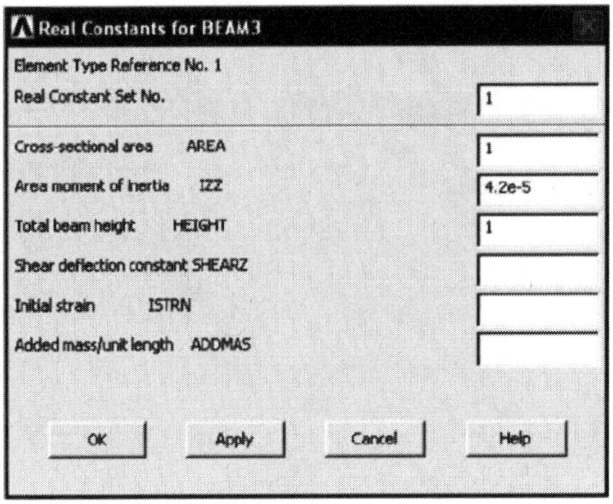

Figure 9.15. Enter the area and moment of inertia.

Cross-sectional area AREA > **Enter 1**
Area moment of inertia IZZ > **Enter 4.2e–5**
Total beam height HEIGHT > **Enter 1 > OK > Close**
Enter the material properties.

4. **Main Menu > Preprocessor > Material Props > Material Models**
Material Model Number 1, click **Structural > Linear > Elastic > Isotropic**
Enter **EX = 46E9 and PRXY = 0.35 > OK**
Click **Structural > Linear > Density**
Enter DENS = 1750 > OK
(**Close** the Define Material Model Behavior window.)
Create the nodes and elements as shown in the figure.

5. **Main Menu > Preprocessor > Modeling > Create > Nodes > In Active CS**
Enter the coordinates of node 1 > **Apply** > Enter the coordinates of node 2 > **Apply** > **Enter** the coordinates of node 3 > **Apply** > Enter the coordinates of node 4 > **Apply** Enter the coordinates of node 5 > **OK**

Node locations		
Node number	X coordinate	Y coordinate
1	0	0
2	1.25	0
3	2.5	0
4	3.75	0
5	5	0

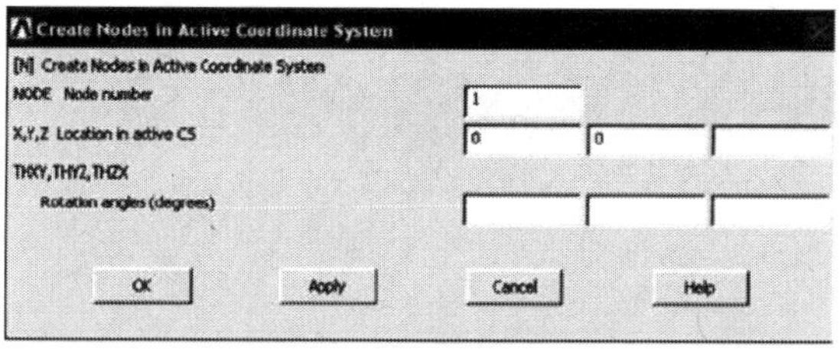

Figure 9.16. Enter the node coordinates.

6. **Main Menu > Preprocessor > Modeling > Create > Elements > Auto Numbered > Thru**
Nodes Pick the 1st and 2nd node > Apply > Pick the 2nd and 3rd node > Apply > Pick the 3rd and 4th node > Apply > Pick the 4th and 5th node > **OK**

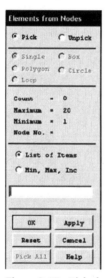

Figure 9.17. Pick the nodes to create elements.

7. **Main Menu > Solution > Analysis Type > New Analysis > Select Transient > OK**

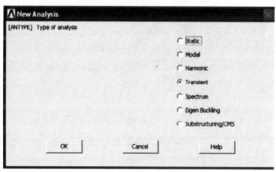

Figure 9.18. Define the type of analysis.

then select > **Reduced** > **OK**

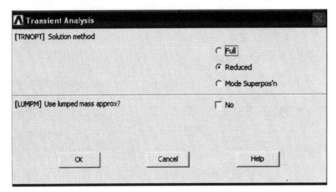

Figure 9.19. Define the type of transient analysis.

8. **Main Menu** > **Preprocessor** > **Loads** > **Define Loads** > **Apply** > **Structural** > **Displacement** > **On Nodes** > Pick the left most node and right most node > **Apply** > **Select All DOF** > **OK**

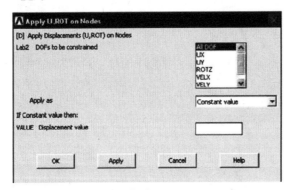

Figure 9.20. Apply the displacement constraint.

9. **Main Menu** > **Solution** > **Master DOFs** > **User Selected** > **Define** > Pick 2nd, 3rd, and 4th node > **Apply** > Select UY from Lab 1 1st degree of freedom > **OK**

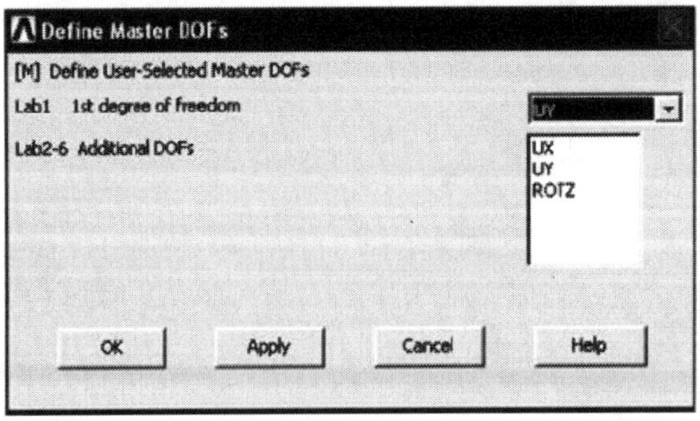

Figure 9.21. Defining master DOF.

Dynamic Analysis

10. **Main Menu > Solution > Load Step Opts > Time/Frequenc > Time-Time Step**
    Enter [TIME] Time at end of load step – 0
    Enter [DELTIM] Time step size – 0.01 < **OK**

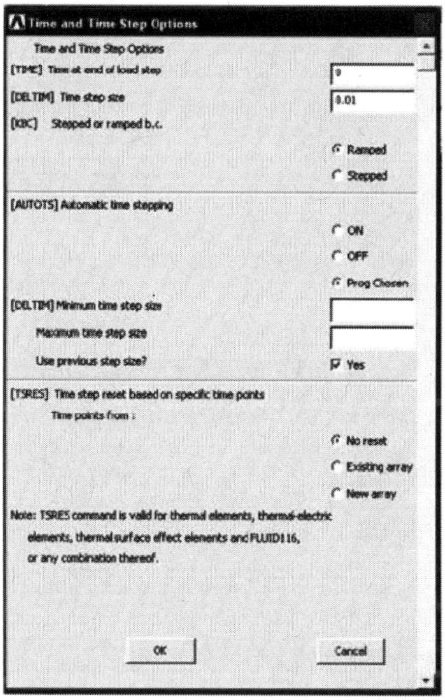

Figure 9.22. Defining time step size.

11. **Main Menu > Solution > Load Step Opts > Write LS File**
    Enter **LSNUM** Load step fine number n = 1 > **OK**

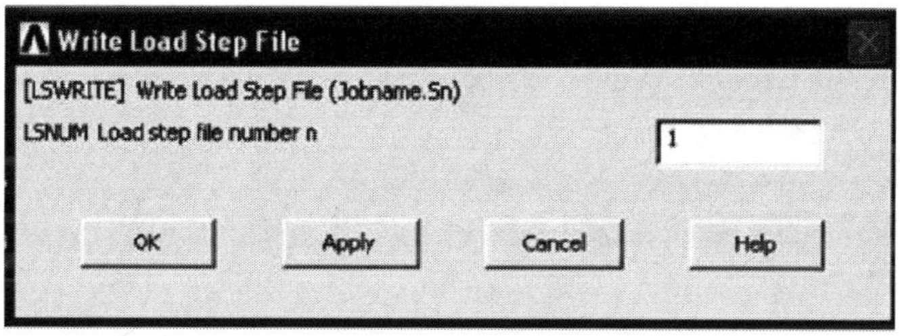

Figure 9.23. Creating LS file.

12. **Main Menu > Solution > Define Loads > Apply > Structural > Force/Moment > On Nodes** > Pick the middle or 3rd node **Apply > Enter FY = 45e3 > OK**

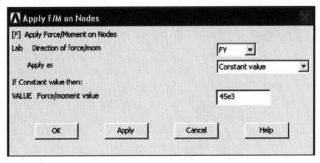

Figure 9.24. Applying force on node.

13. **Main Menu > Solution > Load Step Opts > Time/Frequenc > Time-Time Step**
    Enter [TIME] Time at end of load step – 0.01 > **OK**

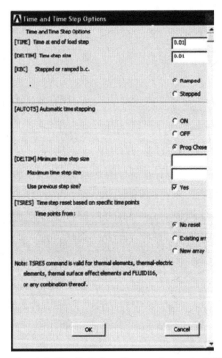

Figure 9.25. Defining time at the end of 1st load step.

14. **Main Menu > Solution > Load Step Opts > Write LS File**
    Enter LSNUM Load step fine number n = 2 > **OK**

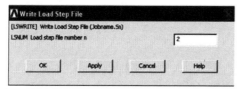

Figure 9.26. Creating LS file for 1st load step.

Similarly repeat Steps 13 and 14 for Time at end of load step of 0.02, 0.05, 0.08, and 0.1 and each time create a LS file with next numbers (n), i.e., 3, 4, 5, and 6.

15. **Main Menu > Solution > Define Loads > Delete > Structural > Force/Moment > On Nodes >** Pick the middle or 3rd node **Apply > OK**
16. **Main Menu > Solution > Define Loads > Apply > Structural > Force/Moment > On Nodes >** Pick the middle or 3rd node **Apply > Enter FY = 36e3 > OK**

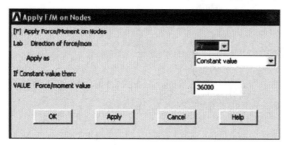

Figure 9.27. Applying force on node.

17. **Main Menu > Solution > Load Step Opts > Time/Frequenc > Time-Time Step**
    Enter [TIME] Time at end of load step– 0.12 **> OK**

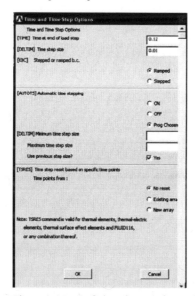

Figure 9.28. Defining time at the end of 6th load step.

18. **Main Menu > Solution > Load Step Opts > Write LS File**
    Enter **LSNUM** Load step fine number $n = 7$ > **OK**

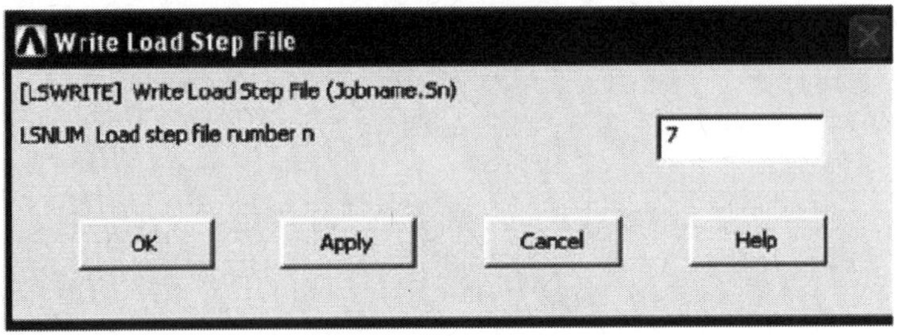

Figure 9.29. Creating LS file for 6th load step.

Repeat Step 15 and delete the force.
Then apply the force of 22.5 kN (i.e., 22.5e3) and define the Time at end of load step of 0.15 and create a LS file with number $(n) = 8$.
Again, repeat Step 15 and delete the force.
Then apply the force of 9 kN (i.e., 9e3) and define the Time at end of load step of 0.18 and create a LS file with number $(n) = 9$.
Again, repeat Step 15 and delete the force.
Define the Time at end of load step of 0.2 and create a LS file with number $(n) = 10$.

19. **Main Menu > Solution > Solve > From LS Files**
    Enter LSMIN Starting LS file number = 1
    Enter LSMAX Ending LS file number = 10
    LSINC File number increment = 1

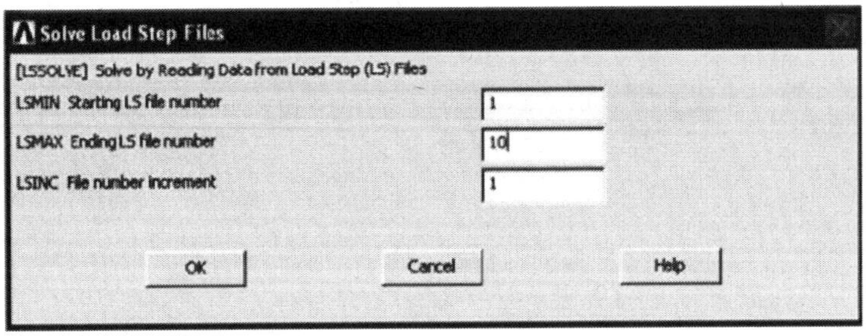

Figure 9.30. Solving from LS files.

20. **Main Menu > Time Hist Post pro**
    The following dialog box will appear

Dynamic Analysis

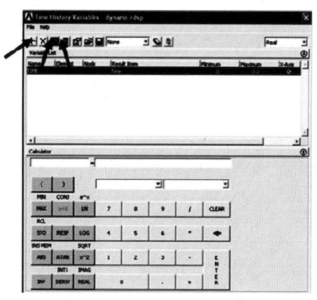

Figure 9.31. Time hist dialog box.

In that dialog box click on the first icon, i.e., on Add data, one more dialog box will appear as shown below. **Then click on DOF Solution > y-Component of displacement > OK.**

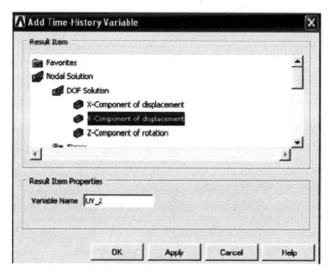

Figure 9.32. Selecting the displacement in *y*-direction.

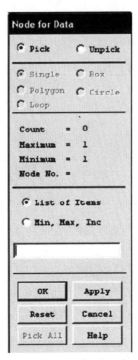

Figure 9.33. Selecting the node.

It asks for the node to pick, so pick the node 3 or middle node < OK.

Then, in the Time hist dialog box, click on 4th icon, i.e., List Data (refer to Figure 9.31).

This result is shown as displacement values for node 3 in the software results of the problem. Then, in the Time hist dialog box, click on 3rd icon, i.e., Graph Data (refer to Figure 9.31).

The result is shown in Figure 9.12(b) for node 3 in the software results of the problem.

Maximum displacement values (in meters)

Name	Element	Node	Result Item	Minimum	Maximum	X-Axis
TIME			Time	0	0.2	●
UY_2		2	Y-Component of displacement	-0.0064935	0.0153042	
UY_3		3	Y-Component of displacement	-0.0117477	0.0292504	●

Figure 9.34. Values of displacement.

9.6 AXIAL VIBRATIONS OF A BAR

Example 9.4

For the bar shown in Figure 9.35, determine the first 2 natural frequencies. Let E = 207 GPa, Poisson's ratio = 0.3, length L = 2.5 m, cross-section area A = 1 m², mass density ρ = 7800 kg/m³.

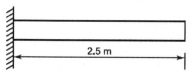

Figure 9.35. The Bar for Example 9.4.

Solution

(I) Analytical method

$$\therefore \omega_1 = \frac{1.57}{L}\sqrt{\frac{E}{\rho}}$$

$$\omega_1 = \frac{1.57}{2.5}\sqrt{\frac{207 \times 10^9}{7800}} = 3235.17 \text{ rad/s.}$$

Frequency,

$$f_1 = \frac{\omega_1}{2\pi}$$

$$f_1 = \frac{3235.17}{2\pi} = 514.89 \text{ Hz}$$

$$\therefore \omega_2 = \frac{4.71}{L}\sqrt{\frac{E}{\rho}}$$

$$\omega_2 = \frac{4.71}{2.5}\sqrt{\frac{207 \times 10^9}{7800}} = 9705.52 \text{ rad/s.}$$

Frequency,

$$f_2 = \frac{\omega_2}{2\pi}$$

$$f_2 = \frac{9705.52}{2\pi} = 1544.68 \text{ Hz.}$$

(II) FEM by hand calculations

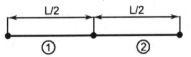

Figure 9.35(a). Finite element model.

Mass matrices are,

$$[M_1]=[M_2]=\frac{\rho A\left(\dfrac{L}{2}\right)}{6}\begin{bmatrix}2 & 1 \\ 1 & 2\end{bmatrix}=\frac{\rho AL}{12}\begin{bmatrix}2 & 1 \\ 1 & 2\end{bmatrix}.$$

Stiffness matrices are,

$$[k_1]=[k_2]=\frac{EA}{\left(\dfrac{L}{2}\right)}\begin{bmatrix}1 & -1 \\ -1 & 1\end{bmatrix}=\frac{2EA}{L}\begin{bmatrix}1 & -1 \\ -1 & 1\end{bmatrix}.$$

Global mass matrix is,

$$[M]=\frac{\rho AL}{12}\begin{bmatrix}2 & 1 & 0 \\ 1 & 4 & 1 \\ 0 & -1 & 2\end{bmatrix}.$$

Global stiffness matrix is,

$$[K]=\frac{2EA}{L}\begin{bmatrix}1 & -1 & 0 \\ -1 & 2 & -1 \\ 0 & -1 & 1\end{bmatrix}.$$

Governing equation,

$$\left([K]-\omega^2[M]\right)\begin{Bmatrix}u_1 \\ u_2 \\ u_3\end{Bmatrix}=[0]$$

$$\left(\frac{2EA}{L}\begin{bmatrix}1 & -1 & 0 \\ -1 & 2 & -1 \\ 0 & -1 & 1\end{bmatrix}-\omega^2\times\frac{\rho AL}{12}\begin{bmatrix}2 & 1 & 0 \\ 1 & 4 & 1 \\ 0 & -1 & 2\end{bmatrix}\right)\begin{Bmatrix}u_1 \\ u_2 \\ u_3\end{Bmatrix}=0.$$

Boundary conditions are, $u_1=0$.

Dynamic Analysis

Applying boundary conditions and for a nontrivial solution,

$$\left\| \begin{bmatrix} 2 & -1 \\ -1 & 1 \end{bmatrix} - \frac{\omega^2 \rho L^2}{24E} \begin{bmatrix} 4 & -1 \\ -1 & 2 \end{bmatrix} \right\| = 0,$$

i.e.,

$$\left\| \begin{bmatrix} 2 & -1 \\ -1 & 1 \end{bmatrix} - \begin{bmatrix} 4a & a \\ a & 2a \end{bmatrix} \right\| = 0,$$

where

$$a = \frac{\omega^2 \rho L^2}{24E}.$$

By solving, $\left\| \begin{bmatrix} 2-4a & -1-a \\ -1-a & 1-2a \end{bmatrix} \right\| = 0$, we get

$$a = 0.1081941 = \frac{\omega^2 \rho L^2}{24E} \text{ or } a = 1.3203772 = \frac{\omega^2 \rho L^2}{24E}$$

$$\therefore \omega_1 = \frac{1.61}{L}\sqrt{\frac{E}{\rho}}$$

$$\omega_1 = \frac{1.61}{2.5}\sqrt{\frac{207 \times 10^9}{7800}} = 3317.6 \text{ rad/s}.$$

Frequency,

$$f_1 = \frac{\omega_1}{2\pi}$$

$$f_1 = \frac{3317.6}{2\pi} = 528.01 \text{ Hz}$$

$$\therefore \omega_2 = \frac{5.63}{L}\sqrt{\frac{E}{\rho}}$$

$$\omega_2 = \frac{5.63}{2.5}\sqrt{\frac{207 \times 10^9}{7800}} = 11601.3 \text{ rad/s}.$$

Frequency,

$$f_2 = \frac{\omega_2}{2\pi}$$

$$f_2 = \frac{11601.3}{2\pi} = 1846.4 \text{ Hz}.$$

(III) Software results

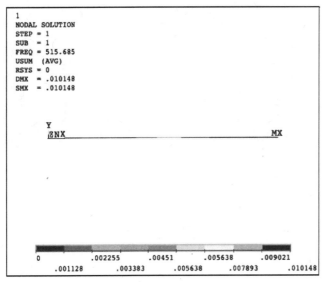

Figure 9.35(b). Deflection pattern for a bar (refer to Appendix C for color figures).

Frequency values (in Hz)

SET	TIME/FREQ	LOAD STEP	SUBSTEP	CUMULATIVE
1	515.68	1	1	1
2	1559.8	1	2	2

Answers for Example 9.4

Parameter	Analytical method	FEM–hand calculation (with 2 elements)	Software results (with 10 elements)
Natural frequency			
f_1	514.89 Hz	528.01 Hz	515.68 Hz
f_2	1544.68 Hz	1846.4 Hz	1559.8 Hz

Procedure for solving the problem using ANSYS® 11.0 academic teaching software.
For Problem 9.4

PREPROCESSING

1. **Main Menu > Preprocessor > Element Type > Add/Edit/Delete > Add > Link > 2D spar 1 > OK > Close**

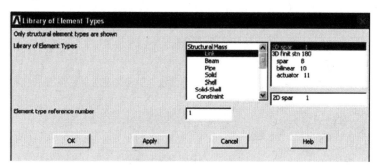

Figure 9.36. Element selection.

2. **Main Menu > Preprocessor > Real Constants > Add/Edit/Delete > Add > OK**

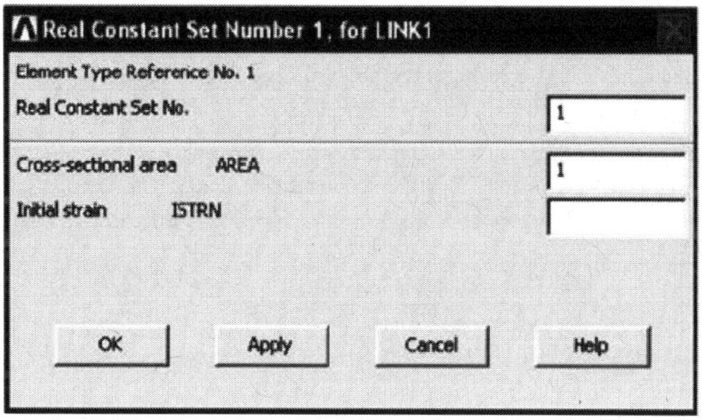

Figure 9.37. Enter the cross-sectional area.

Cross-sectional area AREA > **Enter 1 > OK > Close**
Enter the material properties.

3. **Main Menu > Preprocessor > Material Props > Material Models**
Material Model Number 1, click **Structural > Linear > Elastic > Isotropic**
Enter **EX = 207E9 and PRXY = 0.3 > OK**
Click **Structural > Linear > Density**
Enter DENS = 7800
(**Close** the Define Material Model Behavior window.)
Create the keypoints and lines as shown in the figure.

340   FINITE ELEMENT ANALYSIS

4. **Main Menu > Preprocessor > Modeling > Create > Keypoints > In Active CS** Enter the coordinates of keypoint 1 > **Apply** Enter the coordinates of keypoint 2 > **OK**

Keypoint locations		
Keypoint number	X coordinate	Y coordinate
1	0	0
2	2.5	0

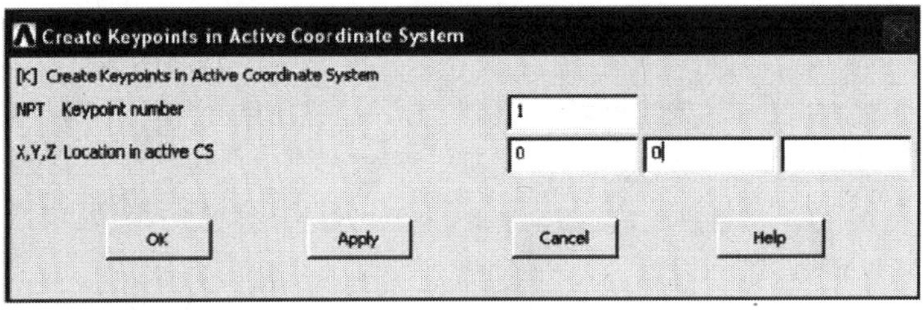

Figure 9.38. Enter the keypoint coordinates.

5. **Main Menu > Preprocessor > Modeling > Create > Lines > Lines > Straight Line** Pick the 1st and 2nd keypoint > **OK**

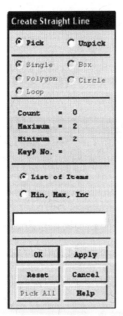

Figure 9.39. Pick the keypoints to create lines.

Dynamic Analysis

6. Main Menu > Preprocessor > Meshing > Size Cntrls > Manual Size > Lines > All Lines > Enter NDIV No. of element divisions = 10

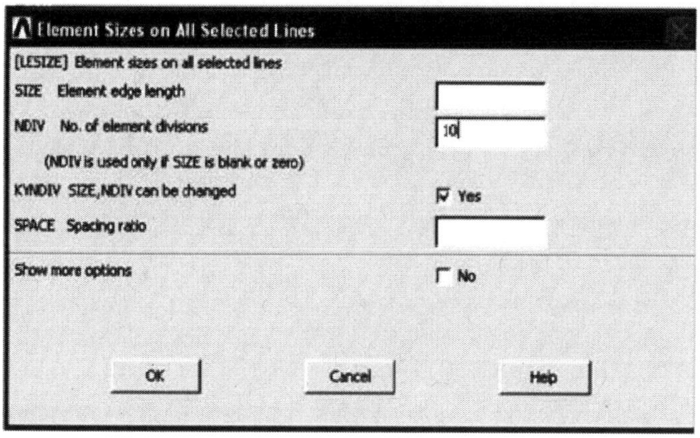

Figure 9.40. Specify element length.

7. Main Menu > Preprocessor > Meshing > Mesh > Lines > Click Pick All

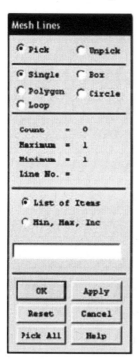

Figure 9.41. Create elements by meshing.

8. **Main Menu > Solution > Analysis Type > New Analysis > Select Modal > OK**

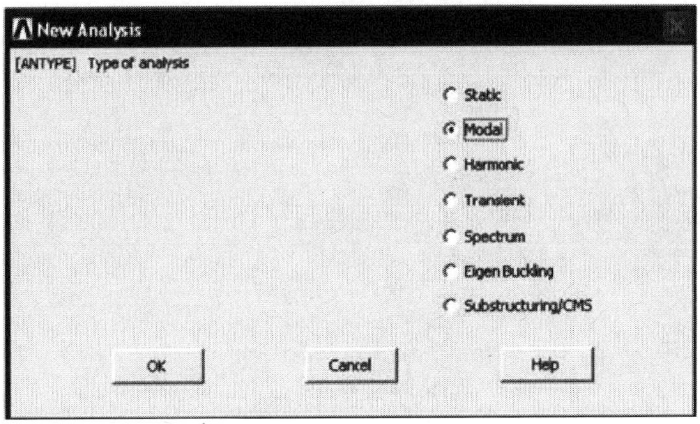

Figure 9.42. Define the type of analysis.

9. **Main Menu > Preprocessor > Loads > Define Loads > Apply > Structural > Displacement > On Nodes** Pick the left most node > **Apply > Select All DOF > OK**

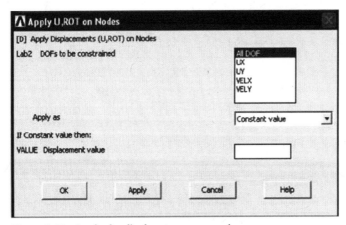

Figure 9.43. Apply the displacement constraint.

10. **Main Menu > Solution > Analysis Type > Analysis Options > Select Reduced option**
    Enter No. of modes to extract = 2
    NMODE No. of modes to expand = 2 > OK

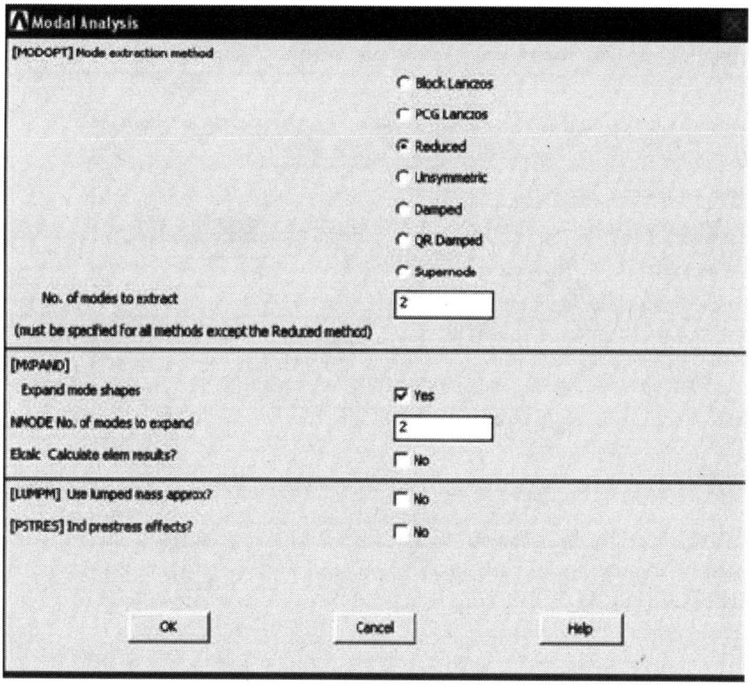

Dynamic Analysis

Figure 9.44. Select the number of modes to extract.

Enter FREQE Frequency range 0 2500 > OK

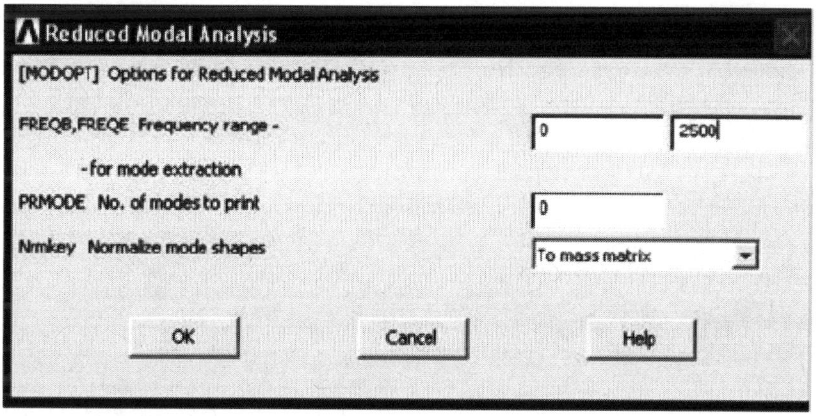

Figure 9.45. Enter the frequency range.

11. **Main Menu > Solution > Master DOFs > User Selected > Define > Pick all nodes except left most node > OK > Select UX from Lab 1 1st degree of freedom > OK**

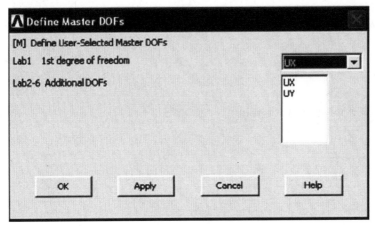

Figure 9.46. Defining the master degree of freedom.

Figure 9.47. Model with master DOF applied.

Solution

The interactive solution proceeds.

12. **Main Menu > Solution > Solve > Current LS > OK**

    The **/STATUS Command** window displays the problem parameters and the **Solve Current Load Step** window is shown. Check the solution options in the /STATUS window and if all is OK, select **File > Close**.

    In the **Solve Current Load Step** window, select **OK**, and when the solution is complete, **close** the **'Solution is Done!'** window.

POST-PROCESSING

13. **Main Menu > General Postproc > Results Summary**
    This result is shown as frequency values in Hz.
14. **Main Menu > General Postproc > Read Results > First Set**
15. **Main Menu > General Postproc > Plot Results > Contour Plot > Nodal Solu > DOF Solution, click on Displacement vector sum > OK**
    This result is shown in Figure 9.35(b).

9.7 BAR SUBJECTED TO FORCING FUNCTION

Example 9.5

The bar shown in Figure 9.48 is subjected to time dependent forcing function as shown, determine the nodal displacements for 5 time steps using 2 finite elements. Let E = 207 GPa, Poisson's ratio = 0.3, length of beam L = 5 m, cross-section area A = 625 × 10–6 m², mass density ρ = 7800 kg/m³. Use time step of integration 0.00025 seconds.

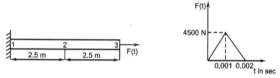

Figure 9.48. The bar for Example 9.5.

Solution

(I) Software results

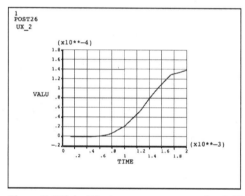

Figure 9.48(a). Displacement response for 0.00025 sec for node 2 (refer to Appendix C for color figures).

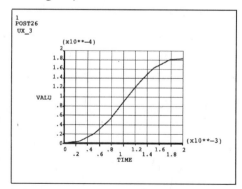

Figure 9.48(b). Displacement response for 0.00025 sec for node 3 (refer to Appendix C for color figures).

Displacement values (in meters) for node 2

TIME	2 UX
	UX_2
0.0000	0.00000
0.25000E–03	–0.467370E–06
0.50000E–03	–0.821457E–06
0.75000E–03	0.396081E–05
0.10000E–02	0.210563E–04
0.12500E–02	0.535055E–04
0.15000E–02	0.950064E–04
0.17500E–02	0.128841E–03
0.20000E–02	0.138387E–03

Displacement values (in meters) for node 3

TIME	3 UX
	UX_3
0.0000	0.00000
0.25000E–03	0.375512E–05
0.50000E–03	0.191517E–04
0.75000E–03	0.488709E–04
0.10000E–02	0.889759E–04
0.12500E–02	0.130597E–03
0.15000E–02	0.161991E–03
0.17500E–02	0.179673E–03
0.20000E–02	0.184097E–03

Maximum displacement values (in meters)

Name	Element	Node	Result Item	Minimum	Maximum	X-Axis
TIME			Time	0	0.002	⊙
UX_2		2	X-Component of displacement	-8.21457e-007	0.000138387	○
UX_3		3	X-Component of displacement	0	0.000184097	●

Figure 9.48(c). Values of displacement.

PROBLEMS

1. What is the governing equation of undamped free vibration and its nontrivial solution?
2. What are the mass matrices for bar element and beam elements?
3. Determine the first 5 natural frequencies for the fixed-fixed beam shown in Figure 9.49. The beam is made of steel with E = 200 GPa, Poisson's ration = 0.3, length = 2 m, cross-section area = 60 cm^2, mass density ρ = 7800 kg/m^3, moment if inertia I = 200 mm^4.

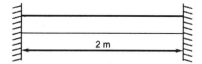

Figure 9.49. Fixed-fixed beam for problem 3.

4. For the bar shown in Figure 9.50, determine nodal displacements for the 5 time finite elements. Let E = 70 GPa, ρ = 2700 kg/m^3, A = 645 mm^2, and L = 2.5 m.

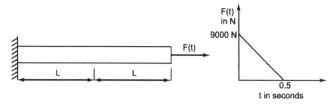

Figure 9.50. The bar for problem 4.

5. The beam shown in Figure 9.51 is subjected to the forcing functions shown, determine the maximum deflections. Let E = 207 GPa, ρ = 7800 kg/m^3, A = 0.0194 m^2, I = 8.2 × 10^{-5} m^4, L = 6 m. Take time step of 0.05 seconds.

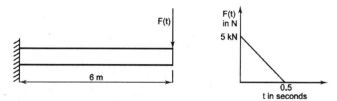

Figure 9.51. The beam for problem 5.

6. Determine the natural frequencies of vibrations for the cantilever beam shown in Figure 9.52.

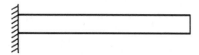

Figure 9.52. Cantilever beam for problem 6.

$$[K] = \frac{EI}{L^3}\begin{bmatrix} 16 & -6L \\ -6L & 4L^2 \end{bmatrix}, \quad [M] = \frac{\rho AL}{420}\begin{bmatrix} 156 & -22L \\ -22L & 4L^2 \end{bmatrix}$$

7. For the bar shown in Figure 9.53, determine nodal displacements for the 5 time finite elements. Let E = 210 GPa, ρ = 2800 kg/m³, A = 825 mm² and L = 3 m.

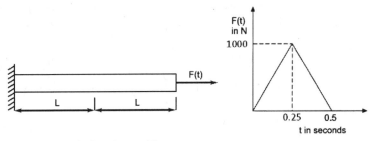

Figure 9.53. The bar for problem 7.

8. For the beam shown in Figure 9.54, determine the mode shapes. Let E = 310 ×10⁶ psi, ρ = 0.283 lbf/in³, A = 1 in², v = 0.3, and L = 30 in.

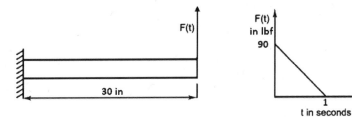

Figure 9.54. The beam for problem 8.

9. For the bar shown in Figure 9.55, subjected to the forcing functions shown, determine the nodal displacement, velocities, acceleration, and the maximum deflections for 5 time steps using 2 finite elements. Let E = 2 × 10⁶ psi, ρ = 2 lb-s² in⁴, A = 2 in², I = 322.83 in⁴, L = 10 in.

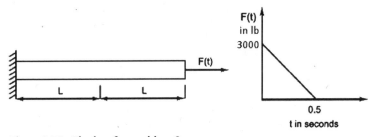

Figure 9.55. The bar for problem 9.

REFERENCES

1. D. L. Logan, "A First Course in the Finite Element Method, Fifth Edition," Cengage Learning, 2012.
2. S. Moaveni, "Finite Element Analysis: Theory and Application with ANSYS, Third Edition," Prentice Hall, 2008.
3. C. T. F. Ross, "Finite Element Method in Structural Mechanics," Ellis Horwood Limited Publishers, 1985.
4. S. S. Rao, "The Finite Element Method in Engineering, Fifth Edition," Butterworth-Heinemann, 2011.
5. G. L. Narasaiah, "Finite Element Analysis," CRC Press, 2009.
6. W. T. Thompson and M. D. Dahleh, "Theory of Vibrations with Applications, Fifth Edition," Prentice–Hall, 1998.

Chapter 10 ENGINEERING ELECTROMAGNETICS ANALYSIS

10.1 INTRODUCTION TO ELECTROMAGNETICS

Electromagnetics (EM) govern many applications in engineering such as the transmission lines system. Therefore, it is essential to understand the fundamental concepts of EM in order to properly design and model electrical systems and devices using the finite element method (FEM). Furthermore, EM becomes more useful in designing engineering systems with recent technologies, especially due to the increasing speeds of digital devices and the increased use of modern electronics circuits such as printed-circuit-board and communications systems such as cellular phones. The most important equations in EM theory are Maxwell's equations, which are known as the foundation of EM theory.

10.2 MAXWELL'S EQUATIONS AND CONTINUITY EQUATION

In electromagnetic analysis on a macroscopic level, it is based on solving the *Maxwell's equations* issue on certain boundary conditions. Also, there is another fundamental equation that can specify the conservation (indestructibility) of electric charge know as the *equation of continuity*. Maxwell's equations and continuity equation can be written in both differential and integral forms. We choose to start here with the differential form because it leads to differential equations that the finite element method (FEM) can handle.

10.2.1 Maxwell's Equations and Continuity Equation in Differential Form

Now, we can present the 4 Maxwell's equations in differential form in time-varying EM fields as:

$$\nabla \times \mathbf{H} = \mathbf{J}_e + \frac{\partial \mathbf{D}}{\partial t} \quad \text{(Ampere's law)} \tag{10.1}$$

$$\nabla \times \mathbf{E} = -\frac{\partial \mathbf{B}}{\partial t} - \mathbf{J}_m \quad \text{(Faraday's law of induction)} \tag{10.2}$$

$$\nabla \cdot \mathbf{D} = \rho_v \quad \text{(Gauss's law-for electric field)} \tag{10.3}$$

$$\nabla \cdot \mathbf{B} = 0 \quad \text{(Gauss's law-for magnetic field)} \tag{10.4}$$

where

$\mathbf{E}$ = Electric field intensity, (in volt/meter) $-V/m^2$
$\mathbf{D}$ = Electric flux density (or electric displacement), (in coulomb/meter^2) $-C/m^2$
$\mathbf{H}$ = Magnetic field intensity, (in ampere/meter) $-A/m^2$
$\mathbf{B}$ = Magnetic flux density, (in tesla or weber/meter^2)$-T$ or Wb/m^2
$\mathbf{J}_e$ = Electric Current density or charge flux (surface), (in ampere/meter^2) $-A/m^2$
$\mathbf{J}_m$ = The magnetic conductive current density, (in volt/meter^2) $-V/m^2$,
  where $\mathbf{J}_m = \sigma_m \mathbf{H}$
σ_m = The magnetic conductive resistivity (in ohm/meter) $-\Omega/m$
ρ_v = Electric charge density (volume), (in coulomb/meter^2) $-C/m^2$.

Now, the *equation of continuity* can be written in differential form as

$$\nabla \cdot \mathbf{J}_e = -\frac{\partial \rho_v}{\partial t} \quad \text{(Continuity equation)}. \tag{10.5}$$

There are 3 *independent* equations from the above 5 equations. They are either equations 1, 2, and 3, or equations 1, 2, and 5. The other two equations 4 and 5, or equations 3 and 4 can be derived from the independent equations, and therefore are called *dependent equations*. Additionally, equation 5 can be derived the divergence of equation 1 and using equation 3.

10.2.2 Maxwell's Equations and Continuity Equation in Integral Form

Furthermore, let us now look to the 4 Maxwell's equations and the continuity equation in integral form in time-varying EM fields. The integrals are taken over in an open surface S or its boundary contour L as shown in Figure 10.1, where I is the electric current that flows through the path L.

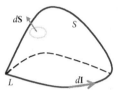

Figure 10.1. The surface S and contour L for the integral form of Maxwell's equations.

$$\oint_L \mathbf{H} \cdot d\mathbf{l} = \int_S \left(\mathbf{J}_e + \frac{\partial \mathbf{D}}{\partial t} \right) \cdot d\mathbf{S} \quad \text{(Ampere's law)} \quad (10.6)$$

$$\oint_L \mathbf{E} \cdot d\mathbf{l} = -\int_S \left(\frac{\partial \mathbf{B}}{\partial t} + \mathbf{J}_m \right) \cdot d\mathbf{S} \quad \text{(Faraday's law of induction)} \quad (10.7)$$

$$\oint_S \mathbf{D} \cdot d\mathbf{S} = \oint_v \rho_v dv \quad \text{(Gauss's law-for electric field)} \quad (10.8)$$

$$\oint_S \mathbf{B} \cdot d\mathbf{S} = 0 \quad \text{(Gauss's law-for magnetic field)} \quad (10.9)$$

$$-\int_S \mathbf{J}_e \cdot d\mathbf{S} = \frac{\partial}{\partial t} \int_v \rho_v dv \quad \text{(Continuity equation)} \quad (10.10)$$

where the surface S encloses the volume v, while the contour L encloses the surface S. $\mathbf{l}$ is the line vector over the contour L and $\mathbf{S}$ is the surface vector. Note that, the direction of $d\mathbf{l}$ must be consistent with the direction of the $d\mathbf{S}$ in agreement with the right-hand rule.

10.2.3 Divergence and Stokes Theorems

Indeed, equations 6 through 10, the integral forms can be derived from the differential forms or vice versa. This can be done by using either divergence (Gauss's) theorem or Stokes's theorem,

$$\oint_S \mathbf{F} \cdot d\mathbf{S} = \int_v \nabla \cdot \mathbf{F} dv \quad \text{(Divergence theorem)} \quad (10.11)$$

$$\oint_L \mathbf{F} \cdot d\mathbf{l} = \int_v \nabla \times \mathbf{F} \cdot d\mathbf{S} \quad \text{(Stokes's theorem)}, \quad (10.12)$$

where **F** is any arbitrary vector field.

10.2.4 Maxwell's Equations and Continuity Equation in Quasi-Statics Case

So far, we did the Maxwell's equations in fully dynamic case. Now, we can express Maxwell's equations in quasi-statics case which the displacement current (**D**) is neglected. That is,

$$\nabla \times \mathbf{H} = \mathbf{J}_e. \qquad (10.13)$$

Whereas equations (10.2), (10.3), and (10.4) remain the same. Also, we can write the continuity equation (10.5) in quasi-statics case as

$$\nabla \cdot \mathbf{J}_e = 0. \qquad (10.14)$$

Indeed, the quasi-static approximation is mainly used for time-varying fields in various conducting media. This is due to that, for good conductors, the conduction current greatly exceeds the displacement current, **D**, for the frequencies.

10.2.5 Maxwell's Equations and Continuity Equation in Statics Case

In the statics field case, the displacement current term $\left(\frac{\partial \mathbf{D}}{\partial t}\right)$ and the time-varying magnetic flux density term $\left(\frac{\partial \mathbf{B}}{\partial t}\right)$ are neglected (the field quantities do not vary with time). Therefore, the Maxwell's equations in static form are expressed as

$$\nabla \times \mathbf{E} = 0 \qquad (10.15)$$

whereas equations (10.3), (10.4), and (10.13) still hold. Also, the continuity equation (10.14) remains the same.

To emphasize, there is no interaction between the electric and the magnetic fields. Thus, the static case can be divided into 2 separate cases, *electrostatic case* and *magnetostatic case*.

In electrostatic case, it can be described by equations (10.3) and (10.15), while, for magnetostatic case, it can be described by equations (10.4) and (10.13).

10.2.6 Maxwell's Equations and Continuity Equation in Source-Free Regions of Space Case

The sources of the electromagnetic fields can be the volume charge density (ρ_v) and the electric current density ($\mathbf{J}_e$). In fact, these densities are localized in space. Also, these sources can make the generated electric and magnetic fields to radiate away from them and they can make the generated electric and magnetic fields to propagate to larger distances to the receiving destination. Therefore, Maxwell's equations can be written in *source-free* regions of space (away from the source) as:

$$\nabla \times \mathbf{H} = \frac{\partial \mathbf{D}}{\partial t} \tag{10.16}$$

$$\nabla \times \mathbf{E} = -\frac{\partial \mathbf{B}}{\partial t} \tag{10.17}$$

$$\nabla \cdot \mathbf{D} = 0 \tag{10.18}$$

whereas equation (10.4) remains the same. With this in mind, the continuity equation (10.14) also remains the same.

10.2.7 Maxwell's Equations and Continuity Equation in Time-Harmonic Fields Case

So far, we considered the arbitrary time variation of electromagnetic fields. Here, we consider only the steady-state (equilibrium) solution of electromagnetic fields when produced by sinusoidal currents. The time-harmonic (sinusoidal steady-state) field for Maxwell's equations exists when the field quantities in the equations are harmonically oscillating functions with a single sinusoidal frequency ω. The time-harmonic fields case is the most regularly used in electrical engineering. Now, an arbitrary time-dependent field $\mathbf{F}(x, y, z, t)$ or $\mathbf{F}(\mathbf{r}, t)$ can be written as

$$\mathbf{F}(\mathbf{r}, t) = \mathrm{Re}\left(\mathbf{F}_s(\mathbf{r})e^{jwt}\right) \tag{10.19}$$

where e^{jwt} is the time convention, ω is the angular frequency (rad/s) of the sinusoidal excitation, $\mathbf{F}_s(\mathbf{r}) = \mathbf{F}_s(x, y, z)$ is the phasor form of $\mathbf{F}(\mathbf{r}, t)$ and it is in general complex, and Re ( ) indicates taking the real part of quantity in the parenthesis. Furthermore, the electromagnetic field quantities can be expressed in phasor notation as

$$\begin{bmatrix} \mathbf{H}(\mathbf{r}, t) \\ \mathbf{E}(\mathbf{r}, t) \\ \mathbf{D}(\mathbf{r}, t) \\ \mathbf{B}(\mathbf{r}, t) \end{bmatrix} = \begin{bmatrix} \mathbf{H}(\mathbf{r}) \\ \mathbf{E}(\mathbf{r}) \\ \mathbf{D}(\mathbf{r}) \\ \mathbf{B}(\mathbf{r}) \end{bmatrix} e^{jwt}. \tag{10.20}$$

For example, the fields can be expresses in time dependent e^{jwt}, as in equation (10.20), $\mathbf{H}(\mathbf{r}, t) = \mathbf{H}(\mathbf{r})e^{jwt}$ and $\mathbf{E}(\mathbf{r}, t) = \mathbf{E}(\mathbf{r})e^{jwt}$, etc.

As a result, using the phasor representation can allow us to replace the time derivations $\frac{\partial}{\partial t}$ by $j\omega$, because

$$\frac{\partial e^{j\omega t}}{\partial t} = j\omega e^{j\omega t}. \tag{10.21}$$

Therefore, the Maxwell's equations can be expressed in time-harmonic as

$$\nabla \times \mathbf{H}_s = \mathbf{J}_{es} + j\omega \mathbf{D}_s \tag{10.22}$$

$$\nabla \times \mathbf{E}_s = -\frac{\partial \mathbf{B}_s}{\partial t} - \mathbf{J}_{ms} \tag{10.23}$$

$$\nabla \cdot \mathbf{D}_s = \rho_{vs} \tag{10.24}$$

$$\nabla \cdot \mathbf{B}_s = 0. \tag{10.25}$$

Now, the continuity equation can be presented as

$$\nabla \cdot \mathbf{J}_{es} = -j\omega \rho_{vs}. \tag{10.26}$$

On the other hand, a nonsinusoidal field can be presented as

$$\mathbf{F}(\mathbf{r},t) = \text{Re}\left(\int_{-\infty}^{\infty} \mathbf{F}_s(\mathbf{r},t) e^{j\omega t} d\omega\right). \tag{10.27}$$

Therefore, the solutions to Maxwell's equations for a nonsinusoidal field can be found by assuming that all the Fourier components $\mathbf{F}_s(\mathbf{r}, \omega)$ over ω.

10.3 LORENTZ FORCE LAW AND CONTINUITY EQUATION

The Lorentz Force $\mathbf{F}$ is the force on a charge q with a vector velocity $\mathbf{u}$ in the present electric filed $\mathbf{E}$ and magnetic field $\mathbf{B}$ and can be obtained as

$$\mathbf{F} = q(\mathbf{E} + \mathbf{u} \times \mathbf{B}). \tag{10.28}$$

In addition, the volume charge ρ_v and the current distribution $\mathbf{J}$ can be subjected to the forces in the presence of fields. Thus, Lorentz Force $\mathbf{F}$ per unit volume acting on the volume charge and the current distribution can be expressed as

$$\mathbf{F} = \rho_v \mathbf{E} + \mathbf{J} \times \mathbf{B}. \tag{10.29}$$

However, if the current distribution **J** occurs from the motion of the charges q within the volume charge ρ_v, then current distribution **J** can be formed as $\mathbf{J} = \rho_v \mathbf{v}$. This can make the Lorentz Force **F** as

$$\mathbf{F} = \rho_v (\mathbf{E} + \mathbf{v} \times \mathbf{B}). \tag{10.30}$$

Moreover, the Lorentz Force law is essential to understand the interaction between EM fields and matter. Indeed, the law is used in the design of many electrical devices.

Furthermore, the continuity equation which expresses the conservation of electric charge can be written as

$$\nabla \cdot \mathbf{J} = -\frac{\partial \rho_v}{\partial t}. \tag{10.31}$$

Equation (10.31) is implicit in Maxell's equations.

10.4 CONSTITUTIVE RELATIONS

In addition to the Maxwell's equations and the continuity equation, there are constitutive relations which describe the macroscopic properties of the medium in which the fields exist. In other words, constitutive relations describe the relationship between the EM fields through the properties of the medium. Indeed, Maxwell's equations and constitutive relations are used to obtain the solutions of EM fields that exist in any microwave structures. The constitutive relations can be presentenced in vacuum (free space) as

$$\mathbf{D} = \varepsilon_0 \mathbf{E} \tag{10.32}$$

$$\mathbf{B} = \mu_0 \mathbf{H} \tag{10.33}$$

$$\mathbf{J}_e = \sigma_e \mathbf{E} \tag{10.34}$$

$$\mathbf{J}_m = \sigma_m \mathbf{M} \tag{10.35}$$

where

ε_0 = the permittivity of vacuum
μ_0 = the permeability of vacuum
σ_e = the electrical conductivity
M = magnetization field.

The numerical values of ε_0 and μ_0 are written as

$$\varepsilon_0 = 8.854 \times 10^{-12}\, Farad/m \cong \frac{1}{36\pi} \times 10^{-9}\, F/m,\ \mu_0 = 12.6 \times 10^{-7}\, Henry/m \quad (10.36)$$

$$= 4\pi \times 10^{-7}\, H/m.$$

We can use these 2 quantities to define the *speed of light* (c_0) and the *characteristic impedances in vacuum* (η_0) as:

$$c_0 = \frac{1}{\sqrt{\varepsilon_0 \mu_0}} = 3 \times 10^8\, m/sec.,\ \eta_0 = \sqrt{\frac{\mu_0}{\varepsilon_0}} = 377\,\Omega. \quad (10.37)$$

To emphasis, the constitutive relations are needed to solve for EM fields quantities using Maxwell's equations.

For simple homogenous isotropic dielectric and for magnetic material (linear and isotropic media), the constitutive relations are given as

$$\mathbf{D} = \varepsilon \mathbf{E} \quad (10.38)$$

$$\mathbf{B} = \mu \mathbf{E} \quad (10.39)$$

where as equations (10.34) and (10.35) remain the same.

Where, ε is the permittivity of the material, and μ is the permeability of the material.

For inhomogeneous media, the constitutive relations are functions of the position.

The permittivity of the material ε and the permeability of the material μ can be presented as

$$\begin{aligned} \varepsilon &= \varepsilon_0(1+\chi_e) \\ \mu &= \mu_0(1+\chi_m) \end{aligned} \quad (10.40)$$

where χ_e is the electric susceptibility of the material which is the measure of the electric polarization property of material (dimensionless scalar), and χ_m is the magnetic susceptibility of the material which is the measure of the magnetic polarization property of material (dimensionless scalar).

Moreover, the speed of light in the material c and the characteristic impedance of the material η is expressed as

$$c = \frac{1}{\sqrt{\varepsilon \mu}},\ \eta = \sqrt{\frac{\mu}{\varepsilon}}. \quad (10.41)$$

The relative permittivity ε_r of a material, the relative permeability μ_r of a material, and the refractive index n of a material are formed as

$$\varepsilon_r = \frac{\varepsilon}{\varepsilon_0} = 1 + \chi_e, \mu_r = \frac{\mu}{\mu_0} = 1 + \chi_m, n = \sqrt{\varepsilon_r \mu_r} \to n^2 = \varepsilon_r \mu_r. \quad (10.42)$$

By using equation (10.41) and (10.42), we get

$$c = \frac{c_0}{n} \text{ and } \eta = \frac{\eta_0 n}{\varepsilon_r}. \quad (10.43)$$

It is good to know that for nonmagnetic material $\mu_r = 1$ or $\mu_r = \mu_0$, and $\eta = \frac{\eta_0}{n}$.
Now, the constitutive relations for time-harmonic fields in a simple media are:

$$\mathbf{D} = \varepsilon_0 \varepsilon_r(\omega)\mathbf{E} = \varepsilon(\omega)\mathbf{E} \quad (10.44)$$

$$\mathbf{B} = \mu_0 \mu_r(\omega)\mathbf{H} = \mu(\omega)\mathbf{H} \quad (10.45)$$

$$\mathbf{J}_e = \sigma_e(\omega)\mathbf{E}. \quad (10.46)$$

Furthermore, both the electric polarization $\mathbf{P}$ (Coulomb/m^2) which describes how the material is polarized when an electric field $\mathbf{E}$ is present and the magnetization $\mathbf{M}$ (Ampere/m) which describes how the material is magnetized when a magnetic field $\mathbf{H}$ can be included in the constitutive relations in any material as

$$\mathbf{D} = \varepsilon_0 \mathbf{E} + \mathbf{P} \quad (10.47)$$

$$\mathbf{B} = \mu_0 (\mathbf{H} + \mathbf{M}) \quad (10.48)$$

$$\mathbf{J}_e = \sigma_e \mathbf{E} \quad (10.49)$$

$$\mathbf{J}_m = \sigma_m \mathbf{M} \quad (10.50)$$

where $\mathbf{P} = \varepsilon_0 \chi_e \mathbf{E}$ and $\mathbf{M} = \chi_m \mathbf{H}$.

Next, for nonlinear material, the constitutive relationships can be presented as

$$\mathbf{D} = \varepsilon_0 \varepsilon_r \mathbf{E} + \mathbf{D}_{re} \quad (10.51)$$

$$\mathbf{B} = \mu_0 \mu_r \mathbf{H} + \mathbf{B}_{re} \quad (10.52)$$

$$\mathbf{J}_e = \sigma_e \mathbf{E} + \mathbf{J}_{ex} \quad (10.53)$$

where $\mathbf{D}_{re}$ is the remanent displacement that is the displacement when the electric field is not present, $\mathbf{B}_{re}$ is the remanent magnetic flux density that is the magnetic flux density when the magnetic field is not present, and $\mathbf{J}_{ex}$ is an externally generated current.

It is beneficial to know that the Maxwell's equations can be expressed in an approach that ensures the contribution of the medium in terms of the fields **E** and **B** as

$$\nabla \times \mathbf{B} = \varepsilon_0 \mu_0 \frac{\partial \mathbf{E}}{\partial t} + \mu_0 \left(\mathbf{J} + \frac{\partial \mathbf{P}}{\partial t} + \nabla \times \mathbf{M} \right) \tag{10.54}$$

$$\nabla \times \mathbf{E} = -\frac{\partial \mathbf{B}}{\partial t} \tag{10.55}$$

$$\nabla \cdot \mathbf{E} = \frac{1}{\varepsilon_0}(\rho_v - \nabla \cdot \mathbf{P}) \tag{10.56}$$

$$\nabla \cdot \mathbf{B} = 0. \tag{10.57}$$

Example 10.1

Given $\mathbf{H} = He^{j(\omega t + \beta z)}\mathbf{a}_x$ in free space, calculate **E**.

Solution

We know $\mathbf{D} = \varepsilon \mathbf{E}$ and $\nabla \times \mathbf{H} = \frac{\partial \mathbf{D}}{\partial t}$, therefore

$$\frac{\partial \mathbf{D}}{\partial t} = \frac{\partial}{\partial z} He^{j(\omega t + \beta z)} \mathbf{a}_y$$

$$\frac{\partial \mathbf{D}}{\partial t} = j\beta He^{j(\omega t + \beta z)} \mathbf{a}_y$$

$$\mathbf{D} = \frac{\beta H}{\omega} e^{j(\omega t + \beta z)} \mathbf{a}_y$$

$$\mathbf{E} = \frac{\beta H}{\varepsilon \omega} e^{j(\omega t + \beta z)} \mathbf{a}_y.$$

10.5 POTENTIAL EQUATIONS

Often under certain circumstances, it can be essential to formulate EM problems in terms of potential functions, that is, the scalar electric potential V_e and vector magnetic potential **A**. These potential functions are arbitrary and they are required to satisfy Maxwell's equations. They are described by

$$\mathbf{B} = \nabla \times \mathbf{A} \tag{10.58}$$

$$\mathbf{E} = -\nabla V_e - \frac{\partial \mathbf{A}}{\partial t}. \tag{10.59}$$

In fact, equation (10.55) is a direct consequence of the magnetic Gauss' law and equation (10.55) is a result from Faraday's law. In the magnetostatic case (there are no currents present), Ampere's law reduces to

$$\nabla \times \mathbf{H} = 0. \tag{10.60}$$

Indeed, when equation (10.57) holds, we can present the scalar magnetic potential V_m by

$$\mathbf{H} = -\nabla V_m. \tag{10.61}$$

It is clear that, equations (10.58) and (10.59) satisfy the Maxwell's equations (1.2) and (1.4). Now, to relate the potential functions to other two Maxwell's equations (1.1) and (1.3), by assuming the Lorentz condition hold, that is,

$$\nabla \cdot \mathbf{A} = -\varepsilon\mu \frac{\partial V_e}{\partial t}. \tag{10.62}$$

These equations can be written in the case of linear and homogenous medium as

$$\nabla^2 \cdot V_e - \varepsilon\mu \frac{\partial^2 V_e}{\partial t^2} = -\frac{\rho_v}{\varepsilon} \tag{10.63}$$

$$\nabla^2 \cdot \mathbf{A} - \varepsilon\mu \frac{\partial^2 \mathbf{A}}{\partial t^2} = -\mu \mathbf{J}. \tag{10.64}$$

Equations (10.63) and (10.64) as wave equations and the integral solutions to these equations are known as the *retarded* potential solutions, i.e.,

$$V_e = \int \frac{[\rho_v] dv}{4\pi\varepsilon R} \tag{10.65}$$

$$\mathbf{A} = \int \frac{\mu [\mathbf{J}] dv}{4\pi R} \tag{10.66}$$

where R is the distance from the source point to the field point at which the potential is required, and the square brackets [ ] denote that ρ_v and $\mathbf{J}$ are specified at a time $R\sqrt{\varepsilon\mu}$ earlier than for which V_e or $\mathbf{A}$ is being formed.

10.6 BOUNDARY CONDITIONS

The material medium in which an electromagnetic field exists is usually characterized by its constitutive parameters σ, ε, and μ. If σ, ε, and μ are independent of **E** and **H**, the medium is *linear*. Also, if σ, ε, and μ are dependent of **E** and **H**, the medium is *nonlinear*. Now, if σ, ε, and μ are functions of space variables, the medium is *inhomogeneous*. But, if σ, ε, and μ are not functions of space variables, the medium is *homogeneous*. Additionally, if σ, ε, and μ are independent of direction (scalars), the medium is *isotropic*. If σ, ε, and μ are dependent of direction (vectors), the medium is *anisotropic*. Indeed, most of substrates used in electronic circuits are homogenous, isotropic, and linear.

The boundary conditions at the interface separating 2 different media 1 and 2, with parameters $(\varepsilon_1, \mu_1, \sigma_1)$ and $(\varepsilon_2, \mu_2, \sigma_2)$, respectively, as shown in Figure 10.2.

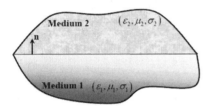

Figure 10.2. Interface between two media.

The boundary conditions for the EM fields across material boundaries are derived from the integral form of Maxwell's equations. They are given by

$$\mathbf{n} \times (\mathbf{E}_1 - \mathbf{E}_2) = 0 \text{ or } E_{1t} - E_{2t} = 0 \tag{10.67}$$

$$\mathbf{n} \cdot (\mathbf{D}_1 - \mathbf{D}_2) = \rho_s \text{ or } D_{1n} - D_{2n} = \rho_s \tag{10.68}$$

$$\mathbf{n} \times (\mathbf{H}_1 - \mathbf{H}_2) = \mathbf{J}_s \text{ or } H_{1t} - H_{2t} = J_s \tag{10.69}$$

$$\mathbf{n} \cdot (\mathbf{B}_1 - \mathbf{B}_2) = 0 \text{ or } B_{1n} - B_{2n} = 0 \tag{10.70}$$

where **n** is a unit normal vector directed from medium 1 to medium 2, subscript t and n denote tangent and normal components of the fields, respectively, ρ_s and $\mathbf{J}_s$ are surface electric charge density (coulomb/m^2) and surface current density (ampere/m), respectively. Furthermore, equations (10.67) and (10.70) state that the tangential components of **E** and the normal components of **B** are continuous across the boundary. But, equation (10.68) states that the discontinuity in the normal component **D** is the same as the surface electric charge density ρ_s on the boundary. However, equation (1.69) states that the tangential component of **H** is discontinuous by the surface current density $\mathbf{J}_s$ on the boundary. In many interface problems, only 2 of Maxwell's equations are used, equations (10.68) and (10.70), when a medium is source free ($\mathbf{J} = 0, \rho_v = 0$), since the other 2 boundary conditions are implied. In such a case, the boundaries conditions may be written as

$$E_{1t} = -E_{2t} \tag{10.71}$$

$$D_{1n} = D_{2n} \tag{10.72}$$

$$H_{1t} = H_{2t} \tag{10.73}$$

$$B_{1n} = B_{2n}. \tag{10.74}$$

Moreover, Maxwell's equations under the source free condition are applicable to passive microwave structures such as transmissions lines.

However, when one of the media is a perfect conductor, boundary conditions are different. A perfect conductor has infinite electrical conductivity and thus no internal electric field (full of free charges). Or else, it would produce an infinite current density according to the third constitutive relations. When an EM field is applied to a perfect conductor medium, the free charges which are pushed to the applied EM field, move themselves in such a way that they produce an opposite EM field that completely cancels the applied EM field. Indeed, this causes the creation of the surface charges and currents on the boundary of the perfect conductor. At an interface between a dielectric and a perfect conductor, the boundary conditions for $\mathbf{E}$ and $\mathbf{D}$ fields are simplified. Now, assume that medium 1 is a perfect conductor, then $\mathbf{E}_1 = 0$ and $\mathbf{D}_1 = 0$. Also, if it is a time-varying case, then $\mathbf{H}_1 = 0$ and $\mathbf{B}_1 = 0$, and, in addition, as a correspondence of Maxwell's equations. Therefore, the boundary conditions for the fields in the dielectric medium for the time-varying at the surface are

$$-\mathbf{n} \times \mathbf{E}_2 = 0 \tag{10.75}$$

$$-\mathbf{n} \cdot \mathbf{D}_2 = \rho_s \tag{10.76}$$

$$-\mathbf{n} \times \mathbf{H}_2 = \mathbf{J}_s \tag{10.77}$$

$$-\mathbf{n} \cdot \mathbf{B}_2 = 0. \tag{10.78}$$

Furthermore, we can apply the integral form of the continuity equation (10.10) to the surface at the interface between lossy media (i.e., $\sigma_1 \neq 0, \sigma_2 \neq 0$) or lossy dielectric (i.e., $\sigma_1 \neq \sigma_2$ and $\varepsilon_1 \neq \varepsilon_2$), or perfect conductor (i.e., no fields inside the media). Therefore, the interface condition for current density $\mathbf{J}$ can be obtained as

$$\mathbf{n} \cdot (\mathbf{J}_1 - \mathbf{J}_2) = -\frac{\partial \rho_s}{\partial t} \text{ or } (J_{1n} - J_{2n}) = -\frac{\partial \rho_s}{\partial t}. \tag{10.79}$$

Equation (10.79) states that the normal component $\mathbf{J}$ is continuous, except where the time-varying surface electric charge density ρ_s on the boundary may exist.

10.7 LAWS FOR STATIC FIELDS IN UNBOUNDED REGIONS

Coulomb's law and Biot-Savart's law are the 2 fundamental laws governing the static fields in unbounded regions.

10.7.1 Coulomb's Law and Field Intensity

Coulomb's law is an experimental law that deals with the force a point charge exerts on another point charge. In other words, Coulomb's law states that the force F (in newtons) between two points charges Q_1 (in coulombs) and Q_2 is

$$F = \frac{Q_1 Q_2}{4\pi\varepsilon_0 R^2} \qquad (10.80)$$

where R (in meter) is the distance between the 2 charges. We can define the electrostatic field intensity $\mathbf{E}$ as the force F applied by 1 charge Q on a unit positive point charge as

$$\mathbf{E} = \frac{Q \mathbf{a}_R}{4\pi\varepsilon_0 R^2}. \qquad (10.81)$$

Knowing that, the point at which the charge Q is located is called the source point, and the point at which the electrostatic field intensity $\mathbf{E}$ is taken is called the field point. Thus, here $\mathbf{a}_R$ is the unite vector in the direction from the source point toward the field point, and R is the distance between the source point and the field point.

Now, it is possible to obtain a continuous charge along a line, on a surface, or in a volume, respectively as

$$\mathbf{E} = \int_L \frac{\rho_l \mathbf{a}_R}{4\pi\varepsilon_0 R^2} \, dl \qquad (10.82)$$

$$\mathbf{E} = \int_S \frac{\rho_s \mathbf{a}_R}{4\pi\varepsilon_0 R^2} \, dS \qquad (10.83)$$

$$\mathbf{E} = \int_v \frac{\rho_v \mathbf{a}_R}{4\pi\varepsilon_0 R^2} \, dv, \qquad (10.84)$$

where L is the line along which the charge is distributed, S is the surface which the charge is distributed, v is the volume enclosed by a surface S. ρ_l, ρ_s, and ρ_v, are the line, surface, and volume charge density, respectively.

10.7.2 Bio-Savart's Law and Field Intensity

The Bio-Savart's law is a magnetostatic law used to express the static magnetic field as a summation over elementary current sources. Now, we can obtain the Bio-Savart law for the line current, surface current, and volume current, respectively in terms of the distributed current sources as

$$\mathbf{H} = \int_L \frac{Id\mathbf{l} \times \mathbf{a}_R}{4\pi R^2} \tag{10.85}$$

$$\mathbf{H} = \int_S \frac{\mathbf{J}_s dS \times \mathbf{a}_R}{4\pi R^2} \tag{10.86}$$

$$\mathbf{H} = \int_v \frac{\mathbf{J}_v dv \times \mathbf{a}_R}{4\pi R^2}, \mathbf{J}_s ds \tag{10.87}$$

where $\mathbf{I}$ is the line current density, $\mathbf{J}_s$ is the surface charge density, $\mathbf{J}_v$ is the volume charge density, and $\mathbf{a}_R$ is a unit vector pointing from the differential elements of current to the point of interest. Indeed, the source elements are related as

$$Id\mathbf{l} \equiv \mathbf{J}_s ds \equiv \mathbf{J}_v dv. \tag{10.88}$$

10.8 ELECTROMAGNETIC ENERGY AND POWER FLOW

The electric energy W_e is defined as

$$W_e = \int_v \left(\int_0^D \mathbf{E} \cdot d\mathbf{D} \right) dv = \int_v \left(\int_0^T \mathbf{E} \cdot \frac{\partial \mathbf{D}}{\partial t} dt \right) dv \tag{10.89}$$

where D is the magnitude of electric displacement, and T is the period.

The electrostatic energy present in an assembly of charges can be written as

$$W_e = \frac{1}{2} \sum_{k=1}^n Q_k V_k \tag{10.90}$$

where V is the potential, and Q is the point charge. Now, instead of point charges, the region has a continuous charge distribution, the summation equation (10.90) becomes integrations for line charge, surface charge, and volume charge, respectively as

$$W_e = \frac{1}{2}\int_L \rho_l V dl \qquad (10.91)$$

$$W_e = \frac{1}{2}\int_S \rho_s V dS \qquad (10.91)$$

$$W_e = \frac{1}{2}\int_v \rho_v V dv. \qquad (10.93)$$

In the meantime, $\rho_v = \nabla \cdot \mathbf{D}$, $\mathbf{E} = -\nabla V$, and $\mathbf{D} = \varepsilon_0 \mathbf{E}$, and by using the identity for vector and scalar rules and applying divergence theorem, and knowing that in a simple medium, whose constitutive parameters (μ, ε, and σ) do not change with time, we have

$$\mathbf{E} \cdot \frac{\partial \mathbf{D}}{\partial t} = \mathbf{E} \cdot \frac{\partial(\varepsilon \mathbf{E})}{\partial t} = \frac{1}{2}\frac{\partial(\varepsilon \mathbf{E} \cdot \mathbf{E})}{\partial t} = \frac{\partial}{\partial t}\left(\frac{1}{2}\varepsilon E^2\right). \qquad (10.94)$$

We can obtain electrostatic energy as

$$W_e = \frac{1}{2}\int_v \mathbf{D} \cdot \mathbf{E}\, dv = \frac{1}{2}\int_v \varepsilon_0 E^2\, dv. \qquad (10.95)$$

Also, the electrostatic energy density w_e (in J/m²) can be obtained as

$$w_e = \frac{1}{2}\mathbf{D} \cdot \mathbf{E} = \frac{1}{2}\varepsilon_0 E^2 = \frac{D^2}{2\varepsilon_0}. \qquad (10.96)$$

When a wave propagates in a medium, it carries the electric field and power. However, the time derivatives of equation (10.89) is the electric power which is written as

$$P_e = \int_v \mathbf{E} \cdot \frac{\partial \mathbf{D}}{\partial t} dv. \qquad (10.97)$$

Furthermore, the magnetic energy can be defined as

$$W_m = \int_v \left(\int_0^B \mathbf{H} \cdot d\mathbf{B}\right) dv = \int_v \left(\int_0^T \mathbf{H} \cdot \frac{\partial \mathbf{B}}{\partial t} dt\right) dv, \qquad (10.98)$$

where B is the magnitude of magnetic flux density, and T is the period.

The magnetostatic energy present in an assembly of currents k can be written as

$$W_m = \frac{1}{2} \sum_{k=1}^{n} I_k \Phi_k \qquad (10.99)$$

where I_k is the kth current, and Φ_k is k^{th} magnetic flux.

Note, knowing that in a simple medium, whose constitutive parameters (μ, ε, and σ) do not change with time, we have

$$\mathbf{H} \cdot \frac{\partial \mathbf{B}}{\partial t} = \mathbf{H} \cdot \frac{\partial (\mu \mathbf{H})}{\partial t} = \frac{1}{2} \frac{\partial (\mu \mathbf{H} \cdot \mathbf{H})}{\partial t} = \frac{\partial}{\partial t}\left(\frac{1}{2}\mu H^2\right). \qquad (10.100)$$

We can obtain magnetostatic energy as

$$W_e = \frac{1}{2}\int_v \mathbf{B} \cdot \mathbf{H}\, dv = \frac{1}{2}\int_v \mu H^2\, dv. \qquad (10.101)$$

Also, the magnetostatic energy density w_m (in J/m²) can be obtained as

$$w_m = \frac{1}{2}\mathbf{B} \cdot \mathbf{H} = \frac{1}{2}\mu H^2 = \frac{B^2}{2\mu}. \qquad (10.102)$$

When a wave propagates in a medium, it carries the magnetic field and power. However, the time derivatives of equation (10.98) is the magnetic power that is written as

$$P_m = \int_v \mathbf{H} \cdot \frac{\partial \mathbf{B}}{\partial t}\, dv. \qquad (10.103)$$

The instantaneous power density vector associated with the electromagnetic field at a given point is known as the *Poynting vector* $\mathbf{P}_{ov}$ (in W/m²), which is written as

$$\mathbf{P}_{ov} = \mathbf{E} \times \mathbf{H}. \qquad (10.104)$$

For more practical value than $\mathbf{P}_{ov}$, we determine the time-average instantaneous Poynting vector (or power average density) (in W/m²) over the period $T = \frac{2\pi}{\omega}$ as

$$\mathbf{P}_{ov\text{-}ave}(z) = \frac{1}{T}\int_0^T \mathbf{P}_{ov}(z,t)\, dt. \qquad (10.105)$$

In addition, for time-harmonic fields, we can defined a phasor *Poynting vector* as

$$\mathbf{P}_{ovs} = \mathbf{E}_s \times \mathbf{H}_s^* \qquad (10.106)$$

Where $\mathbf{H}_s^*$ is the complex conjugate of $\mathbf{H}_s$. Now, for a phasor *Poynting vector*, we can define the time-average power which is equivalent of equation (10.106) as

$$\mathbf{P}_{ov\text{-}ave}(z) = \frac{1}{2}\operatorname{Re}(\mathbf{E}_s \times \mathbf{H}_s^*) \qquad (10.107)$$

where Re( ) stands for the real part of a complex quantity. Furthermore, the total time-average power crossing a given surface S is given by

$$\mathbf{P}_{tave} = \frac{1}{2}\operatorname{Re}\int_S (\mathbf{E} \times \mathbf{H}) \cdot d\mathbf{S} = \int_S \mathbf{P}_{ov-ave} \cdot d\mathbf{S}. \qquad (10.108)$$

The electric and magnetic powers quantities are related through *Poynting's theorem* as

$$-\int_v \left(\mathbf{E} \cdot \frac{\partial \mathbf{D}}{\partial t} + \mathbf{H} \cdot \frac{\partial \mathbf{B}}{\partial t} \right) dv = \int_v \mathbf{J} \cdot \mathbf{E}\, dv + \oint_S (\mathbf{E} \times \mathbf{H}) \cdot d\mathbf{S} \qquad (10.109)$$

where

$\int_v \mathbf{J} \cdot \mathbf{E}\, dv$ is called *resistive losses* which result in heat dissipation in the material.

$\oint_S (\mathbf{E} \times \mathbf{H}) \cdot d\mathbf{S}$ is called the *radiative losses*.

However, the *Poynting's theorem* as presented in equation (10.109) can be written as

$$\oint_S (\mathbf{E} \times \mathbf{H}) \cdot d\mathbf{S} = -\frac{\partial}{\partial t}\int_v \left(\frac{1}{2}\varepsilon E^2 + \frac{1}{2}\mu H^2 \right) dv - \int_v \sigma E^2\, dv \qquad (10.110)$$

where

$\oint_S (\mathbf{E} \times \mathbf{H}) \cdot d\mathbf{S}$ is the total power leaving the volume.

$-\frac{\partial}{\partial t}\int_v \left(\frac{1}{2}\varepsilon E^2 + \frac{1}{2}\mu H^2 \right) dv$ is the rate of decrease in energy stored in electric and magnetic fields.

$-\int_v \sigma E^2\, dv$ is the decrease in ohmic power density (dissipated).

Indeed, under the material is linear and isotropic, it holds that

$$\mathbf{E} \cdot \frac{\partial \mathbf{D}}{\partial t} = \mathbf{E} \cdot \frac{\partial (\varepsilon \mathbf{E})}{\partial t} = \frac{1}{2} \frac{\partial (\varepsilon \mathbf{E} \cdot \mathbf{E})}{\partial t} \qquad (10.111)$$

$$\mathbf{H} \cdot \frac{\partial \mathbf{B}}{\partial t} = \frac{1}{\mu} \mathbf{B} \cdot \frac{\partial \mathbf{B}}{\partial t} = \frac{\partial}{\partial t}\left(\frac{1}{2\mu} \mathbf{B} \cdot \mathbf{B}\right). \qquad (10.112)$$

Therefore, based on equations (10.111) and (10.112), the equation (10.109) can be written as

$$-\frac{\partial}{\partial t}\int_v \left(\frac{1}{2}\varepsilon \mathbf{E} \cdot \mathbf{E} + \frac{1}{2\mu} \mathbf{B} \cdot \mathbf{B}\right) dv = \int_v \mathbf{J} \cdot \mathbf{E} dv + \oint_S (\mathbf{E} \times \mathbf{H}) \cdot d\mathbf{S}. \qquad (10.113)$$

Now, by integrating the left-hand side of equation (10.113) is the total electromagnetic energy density w_t

$$w_t = w_e + w_m = \frac{1}{2}\left(\varepsilon \mathbf{E} \cdot \mathbf{E} + \frac{1}{\mu} \mathbf{B} \cdot \mathbf{B}\right). \qquad (10.114)$$

10.9 LOSS IN MEDIUM

The electronic circuits have dielectrics that are always not perfect. Thus, there is always loss in any practical nonmagnetic dielectrics that is known as *dielectric loss*. This dielectric loss is due to a nonzero conductivity of the medium. Now, we can write the time harmonic Maxwell's equation (10.22), making use the time-harmonic constitutive relations (10.44) and (10.46), as

$$\nabla \times \mathbf{H}_s = j\omega\varepsilon\left(1 - j\frac{\sigma}{\omega\varepsilon}\right)\mathbf{E} \qquad (10.115)$$

or

$$\nabla \times \mathbf{H}_s = j\omega\varepsilon(1 - j\tan\delta)\mathbf{E} \qquad (10.116)$$

where

$$\tan\delta = \frac{\sigma}{\omega\varepsilon}. \qquad (10.117)$$

Equation (10.117) is called the *loss tangent* of the medium, which is usually used to characterize the medium's loss. In addition, now we can define a complex dielectric constant of a lossy medium $\hat{\varepsilon}$ as

$$\hat{\varepsilon} = \varepsilon' - j\varepsilon'' \tag{10.118}$$

where the real part ε' of the complex dielectric constant is the dielectric property that contributes to the stored electric energy in the medium and it is defined as

$$\varepsilon' = \varepsilon = \varepsilon_0 \varepsilon_r \tag{10.119}$$

and the imaginary part ε'' contains the finite conductivity and results in loss in the medium which is defined as

$$\varepsilon' = \frac{\sigma}{\omega} = \varepsilon \tan \delta. \tag{10.120}$$

For example, the loss tangent for GaAs material is 0.006 at frequency 10Ghz, relative dielectric constant equal to 12.9, and temperature 25°C. Also, the loss tangent for silicon material is 0.004 at frequency 10Ghz, relative dielectric constant equal to 11.9, and temperature 25°C.

10.10 SKIN DEPTH

The measure of the depth to which the electromagnetic wave can penetrate the medium is known as *skin depth* (or *depth of penetration*). Skin depth is one of the most important parameters of a medium, because it presents the distance from the medium surface over which the magnitude of the fields of a wave traveling in the medium are reduced to e^{-1} (or 0.368) of those at the medium's surface. The skin depth δ of a good conductor is approximately written as

$$\delta = \sqrt{\frac{2}{\omega \mu \sigma}} = \frac{1}{\sqrt{\pi f \mu \sigma}} \tag{10.121}$$

where $\omega = 2\pi f$.

It is essential to know that the skin depth of good conductors is very small, especially at high frequencies. Thus, it results a low conduction loss.

Example 10.2

Calculate the skin depth, δ, for aluminum in 1.6×10^6 Hz field ($\sigma = 38.2 \times 10^6$ S/m and $\mu = 1$).

Solution

$$\delta = \sqrt{\frac{2}{\omega\mu\sigma}} = \frac{1}{\sqrt{\pi f \mu \sigma}} = \frac{1}{\sqrt{\pi \times 1.6 \times 10^6 \times 1 \times 38.2 \times 10^6}} = 64.4 \ \mu m.$$

10.11 POISSON'S AND LAPLACE'S EQUATIONS

Poisson's and Laplace's equations are derived from Gauss's law (for a linear, isotropic material medium)

$$\nabla \cdot \mathbf{D} = \nabla \cdot \varepsilon \mathbf{E} = \rho_v \qquad (10.122)$$

and

$$\mathbf{E} = -\nabla V. \qquad (10.123)$$

By substituting equation (10.123) into equation (10.122), we get

$$\nabla \cdot (-\varepsilon \nabla V) = \rho_v \qquad (10.124)$$

for an inhomogeneous medium. Equation (10.124) can be obtained for a homogeneous medium as

$$\nabla^2 V = -\frac{\rho_v}{\varepsilon}. \qquad (10.125)$$

Equation (10.125) is known as *Poisson's equation*.

Now, *Laplace's equation* is a special case of Poisson's equation when $\rho_v = 0$ (i.e., for a charge free region), and it can be described as

$$\nabla^2 V = 0. \qquad (10.126)$$

Laplace's equation is used to determine the static or quasi-static characteristic impedance and effective relative dielectric constant of a transmission line.

10.12 WAVE EQUATIONS

We used so far Maxwell's equations and constitutive relations directly to determine the EM fields. However, it can very convenient to obtain the EM fields by solving wave equations.

When the electromagnetic wave is in a simple (linear, isotropic, and homogenous) nonconducting medium (ε, μ, and $\sigma = 0$), the homogenous vector wave equations can be presented as

$$\nabla^2 \mathbf{E} - \frac{1}{c^2} \frac{\partial^2 \mathbf{E}}{\partial t^2} = 0 \qquad (10.127)$$

and

$$\nabla^2 \mathbf{H} - \frac{1}{c^2} \frac{\partial^2 \mathbf{H}}{\partial t^2} = 0. \qquad (10.128)$$

On the other hand, the relation between scalar potential V and vector potential $\mathbf{A}$ is called the Lorentz condition (or Lorentz gauge) for potentials that is expressed as

$$\nabla \cdot \mathbf{A} + \mu \varepsilon \frac{\partial V}{\partial t} = 0. \qquad (10.129)$$

The nonhomogenous wave equation for vector potential $\mathbf{A}$ is given by

$$\nabla^2 \mathbf{A} - \mu \varepsilon \frac{\partial^2 \mathbf{A}}{\partial t^2} = -\mu \mathbf{J}. \qquad (10.130)$$

But, the nonhomogenous wave equation for scalar potential V is given by

$$\nabla^2 V - \mu \varepsilon \frac{\partial^2 V}{\partial t^2} = -\frac{\rho}{\varepsilon}. \qquad (10.131)$$

The time-harmonic wave equations for vector potential $\mathbf{A}$ and scalar potential V equations can be obtained, respectively as

$$\nabla^2 \mathbf{A} + k^2 \mathbf{A} = -\mu \mathbf{J} \qquad (10.132)$$

and

$$\nabla^2 V + k^2 V = -\frac{\rho}{\varepsilon} \qquad (10.133)$$

where

$$k = \omega \sqrt{\mu \varepsilon} = \frac{\omega}{c}. \qquad (10.134)$$

Equation (10.134) is called *wave number*, and equations (10.132) and (10.133) are known as *nonhomogenous Helmholtz's equations*.

However, when the EM wave in a simple, nonconducting source free medium (characterize by $\rho = 0$, $\mathbf{J} = 0$, $\sigma = 0$) and the time-harmonic wave equations can be obtained as

$$\nabla^2 \mathbf{E} + k^2 \mathbf{E} = 0 \qquad (10.135)$$

and

$$\nabla^2 \mathbf{H} + k^2 \mathbf{H} = 0. \qquad (10.136)$$

Equations (10.135) and (10.136) are known as the homogenous vector *Helmholtz's equations*.

10.13 ELECTROMAGNETIC ANALYSIS

Due to the cost effectiveness of experiments and testing, the development of transmission lines in integrated circuit systems is time consuming. Today, researchers, designers, and engineers used several numerical and analytical methods to study and investigate the parameters variations and properties of designing high-speed integrated circuits (microwave circuits) and electromagnetic (EM) problems. The most common analytical methods used for exact solutions in electromagnetic are conformal mapping, integral solutions, separation of variables, and series expansion. Also, the most popular numerical methods used for approximate solutions are methods called moment methods, methods of line, finite difference methods, and finite element methods (FEM).

Finite element method has a great success in electromagnetic analysis compared to other methods. In contrast to other numerical methods, it is very useful for solving problems in complex geometries and inhomogeneous media. In this chapter, we show an overview of the finite element method. FEM requires that any problem involved in the geometrical region to be subdivided into finite number of smaller regions or elements. An approximate solution for the partial differential equation can be developed for each of these elements. In addition, the total solution is generated by assembling together the individual solutions taking care in order to ensure continuity at the interelement boundaries. Basically, there are four steps used in FEM: *first*, creating and discretizing the solution region (domain) into a finite number of subregions or elements; that is, divide the problem into nodes and elements and assume a shape function to represent the physical behavior of an element; *second*, developing equations for an element; *third*, assembling all the elements to represent in solution region, constructing the global coefficient matrix and applying boundary conditions and initial conditions; *fourth*, solving the system of equations to obtain the important information of the problem.

10.13.1 One-Dimensional Elements

10.13.1.1 The Approach to FEM Standard Steps Procedure

The *first* step is the discretization step, that is, the solution domain is divided into finite elements. Figure 10.3 provides an example of elements employed in one dimension. It shows the points of intersection of the lines that make up the sides of the elements called *nodes* and the sides themselves are known as *nodal lines*.

Figure 10.3. Example of elements in one-dimensional (1D).

The *second* step is the development of equations to approximate the solution for each element. It can be done by choosing an approximate function with unknown coefficients that will be used to approximate the solution. We use a first-order polynomial (straight line) as a linear variation of potential between the nodes over element m, i.e.,

$$V^{(m)}(x) = a + bx, \qquad (10.137)$$

where $V(x)$ is the dependent variable (potential function); a and b are constants; x is the independent variable.

We can find the two constants a and b by using the two nodes to satisfy the equation at the location of the two nodes as:

$$V_1^{(m)} = a + bx_1 \qquad (10.138)$$

and

$$V_2^{(m)} = a + bx_2 \qquad (10.139)$$

where $V_1^{(m)} = V^{(m)}(x_1)$ and $V_2^{(m)} = V^{(m)}(x_2)$. By using Cramer's rule, we can solve equations (10.138) and (10.139), i.e.,

$$a = \frac{V_1^{(m)} x_2 - V_2(m) x_1}{x_2 - x_1} \qquad (10.140)$$

$$b = \frac{V_2^{(m)} - V_1^{(m)}}{x_2 - x_1}. \qquad (10.141)$$

Equations (10.140) and (10.141) can be substituted into equation (10.137) to give the approximate (or shape) function $V(x)$ in terms of the interpolation functions, H_1 and H_2 over element m, that is,

$$V^{(m)}(x) = a_1^{(m)}(x)V_1^{(m)} + a_2^{(m)}(x)V_2^{(m)} \tag{10.142}$$

where

$$a_1^{(m)}(x) = \frac{x_2 - x}{x_2 - x_1} \tag{10.143}$$

$$a_2^{(m)}(x) = \frac{x - x_1}{x_2 - x_1}. \tag{10.144}$$

Indeed, equation (10.142) is a first-order interpolating polynomial. In addition, it provides a means to calculate intermediate values between the given values V_1 and V_2 at the nodes.

The shape function, along with the corresponding interpolation functions, is presented in Figure 10.4. Moreover, the sum of the interpolation functions, a_1 and a_2, that is, $\sum_{i=1}^{2} a_i = 1$.

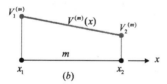

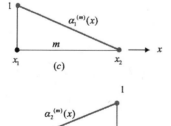

Figure 10.4. (a) a line element, (b) a linear approximation (or shape) function, (c) the corresponding interpolation function $a_1^{(m)}(x)$ for $V^{(m)}(x)$, and (d) the corresponding interpolation function $a_2^{(m)}(x)$ for $V^{(m)}(x)$.

Furthermore, it follows that,

$$\frac{dV^{(m)}}{dx} = \frac{da_1^{(m)}}{dx}V_1^{(m)} + \frac{da_2^{(m)}}{dx}V_2^{(m)} \qquad (10.145)$$

$$\frac{da_1^{(m)}}{dx} = \frac{-1}{x_2 - x_1} \qquad (10.146)$$

and

$$\frac{da_2^{(m)}}{dx} = \frac{1}{x_2 - x_1}. \qquad (10.147)$$

Thus,

$$\frac{dV^{(m)}}{dx} = \frac{\left(-V_1^{(m)} + V_2^{(m)}\right)}{x_2 - x_1}. \qquad (10.148)$$

Now, the integral of $V^{(m)}$ is:

$$\int_{x_1}^{x_2} V^{(m)} dx = \int_{x_1}^{x_2} \left(a_1^{(m)} V_1^{(m)} + a_2^{(m)} V_2^{(m)}\right) dx = \frac{\left(V_1^{(m)} + V_2^{(m)}\right)(x_2 - x_1)}{2}. \qquad (10.149)$$

Now, we evaluate the coefficients so that the function approximates the solution in a best approach. The most common methods used for this propose are the variational approaches, the weighted residuals, and the direct approaches. These methods can specify the relationships between the unknowns in equation (10.142) that satisfy the partial differential equation in an optimal approach. The resulting element equations can be expressed in a set of linear equations in matrix form, i.e.,

$$\left[C^{(m)}\right]\left\{V^{(m)}\right\}_c = \left\{\Psi_c^{(m)}\right\} \qquad (10.150)$$

where
$[C^{(m)}]$ is element property (stiffness) matrix; $\{V_c^{(m)}\}$ is a column vector of unknowns at the nodes over element m; and $\{\psi_c^{(m)}\}$ is a column vector reflecting the effect of any external influences applied at the node over element m.

Third, we assemble all the elements to represent in the solution region. The solutions for closest elements are matched so that the unknown values at their common nodes are equivalent. Therefore, the total sum will be continuous. Then, the assembled system need to be modified for its boundary condition. The system can be expressed as:

$$[C_a^{(m)}]\{V_{ac}^{(m)}\} = \{\Psi_{ac}^{(m)}\} \quad (10.151)$$

where
$[C_a^{(m)}]$ is the assemblage element property (stiffness) matrix; $\{V_{ac}^{(m)}\}$ is the assemblage column vector of unknowns at the nodes over element m; and $\{\Psi_{ac}^{(m)}\}$ is the assemblage column vector reflecting the effect of any external influences applied at the node over element m.

Fourth, solving the system of equations (10.151) to obtain the important information of the problem, it can be obtained by LU decomposition technique.

10.13.1.2 Application to Poisson's Equation in One-Dimension

In this section, we solve the one-dimensional (1D) Poisson's equation for the potential distribution $V(x)$

$$\frac{d^2}{dx^2}V = -\frac{\rho_v}{\varepsilon} \quad (10.152)$$

with boundary conditions (BCs) $V(a) = v_1$, $V(b) = v_2$.

Using the same essential four steps as in the previous section with FEM, we focus here on the source term and only the major differences.

We will use the variational principle and the weighted residuals method to obtain the solution of one-dimensional (1D) Poisson's equation.

(1) Variational Approach

The deriving element governing equations step. We look for the potential distribution $V(x)$ that can minimize an energy function $F(V)$ as

$$F(V) = \int_a^b \left(\frac{1}{2}\left(\frac{dV}{dx}\right)^2 - \frac{\rho_m}{\varepsilon}V(x) \right) dx. \quad (10.153)$$

Two nodal values of $V(x)$ are required to define uniquely a line variation of $V^{(m)}(x)$ over an element (m). Hence, the linear variation of $V^{(m)}(x)$ can be presented as

$$V^{(m)}(x) = a_1(x)V_1 + a_2(x)V_2 \quad (10.154)$$

where the interpolation functions $a_1(x)$ and $a_2(x)$ are presented as

$$a_1(x) = \frac{x_2 - x}{x_2 - x_1}, \quad a_2(x) = \frac{x - x_1}{x_2 - x_1}. \quad (10.155)$$

The resulting element equation (10.154) can be expressed in a set of linear equations in matrix form:

$$V^{(m)} = [a_1, a_2]\begin{bmatrix} V_1 \\ V_2 \end{bmatrix} = [a]\{V_c^{(m)}\}. \tag{10.156}$$

The energy function can be written as

$$F(V) = \sum_{m=1}^{N} F^{(m)}\left(V^{(m)}\right) \tag{10.157}$$

where N is the number of elements with the domain $a \leq x \leq b$.
Now, substituting equation (10.156) into (10.153) can give

$$F^{(m)}(V^{(m)}) = \int_{x_1}^{x_2} \left(\frac{1}{2}\left(\left[\frac{da_1}{dx} \frac{da_2}{dx}\right]\begin{bmatrix} V_1 \\ V_2 \end{bmatrix}\right)^2 - \frac{\rho_v}{\varepsilon}\left([a_1 a_2]\begin{bmatrix} V_1 \\ V_2 \end{bmatrix}\right)^2 \right) dx. \tag{10.158}$$

By minimizing the $F^{(m)}(V^{(m)})$ with respect to the nodal values of V, we obtain the following equations for an element (m)

$$\frac{\partial F^{(m)}}{\partial V_1} = \int_{x_1}^{x_2} \left(\frac{da_1}{dx}\left[\frac{da_1}{dx} \frac{da_2}{dx}\right]\begin{bmatrix} V_1 \\ V_2 \end{bmatrix} - \frac{\rho_v}{\varepsilon} a_1 \right) dx = 0 \tag{10.159}$$

and

$$\frac{\partial F^{(m)}}{\partial V_2} = \int_{x_1}^{x_2} \left(\frac{da_2}{dx}\left[\frac{da_1}{dx} \frac{da_2}{dx}\right]\begin{bmatrix} V_1 \\ V_2 \end{bmatrix} - \frac{\rho_v}{\varepsilon} a_2 \right) dx = 0. \tag{10.160}$$

These equations can be expressed in matrix form as

$$[C^{(m)}]\{V_c^{(m)}\} = \{\Psi_c^{(m)}\} \tag{10.161}$$

where

$$[C^{(m)}] = \int_{x_1}^{x_2} \begin{bmatrix} \frac{da_1^{(m)}}{dx}\frac{da_1^{(m)}}{dx} & \frac{da_1^{(m)}}{dx}\frac{da_2^{(m)}}{dx} \\ \frac{da_2^{(m)}}{dx}\frac{da_1^{(m)}}{dx} & \frac{da_2^{(m)}}{dx}\frac{da_2^{(m)}}{dx} \end{bmatrix} dx \tag{10.162}$$

$$\{V_c^{(m)}\} = \begin{bmatrix} V_1 \\ V_2 \end{bmatrix} \qquad (10.163)$$

$$\{\Psi_c^{(m)}\} = \int_{x_1}^{x_2} \begin{bmatrix} -\dfrac{\rho_v}{\varepsilon} a_1 \\ -\dfrac{\rho_v}{\varepsilon} a_2 \end{bmatrix} dx \qquad (10.164)$$

where the elements of $\{\psi_c^{(m)}\}$ are the nodal forcing functions. The equations in (10.161) can give the characteristics of the Poisson's equation in 1D. Indeed, in spite of the type of element we choose to formulate the Poisson's equation in 1D, the element equations will have the form as equation (10.161). For the solution of the Poisson's equation in 1D, it is essential to derive the equations for all the elements in the assemblage and then to assemble these algebraic equations.

(2) Weighted Residuals Method

In the variational approach for 1D Poisson equation with boundary condition, we derive the element matrices $[C^{(m)}]$ and $\{\psi_c^{(m)}\}$ for a linear variation of potential $V(x)$ over element (m) with two nodes. Now, we will use Galerkin's method with weighting functions $W_i = a_i$ to derive the element matrices. We approximate the unknown exact solution $V_m(x)$ by

$$V_m(x) = \sum_{i=1}^{N} a_i(x) V_{mi} \qquad (10.165)$$

where
N is number of nodes (here $N = 2$), V_{mi} is the unknown nodal values, $i = 1, 2$.

Note that, we do not consider the fixed boundary conditions at the element level, but these are included after the assembly process as in the previous method. Now, by applying Galerkin's method we get:

$$\int_{x_1}^{x_2} \left(\frac{d^2 V_m}{dx^2} + \frac{\rho_v}{\varepsilon} \right) a_i(x)\, dx = 0, \qquad i = 1, 2 \qquad (10.166)$$

where
x_1 and x_2 are the coordinates of the end nodes of the line element.
By using integration by parts to the term with the derivatives of $V_m(x)$, that is,

$$a_i \frac{dV_m}{dx}\bigg|_{x_1}^{x_2} - \int_{x_1}^{x_2} \frac{dV_m}{dx} \frac{da_i}{dx} dx + \int_{x_1}^{x_2} \frac{\rho_v}{\varepsilon} a_i(x)\, dx = 0, \quad i = 1, 2 \qquad (10.167)$$

Taking the derivative of equation (2.29) as

$$\frac{dV_m}{dx} = \sum_{i=1}^{N} \frac{da_i}{dx} V_{mi} = \left[\frac{da_i}{dx}\right]\{V_c^{(m)}\} \qquad (10.168)$$

where
$\{V_c^{(m)}\}$ is the column vector of nodal unknowns for the element m.
Thus, equation (2.30) becomes

$$\int_{x_1}^{x_2} \left[\frac{da_i}{dx}\right] \frac{da_i}{dx} dx \{V_c^{(m)}\} = a_i \frac{dV_m}{dx}\bigg|_{x_1}^{x_2} + \int_{x_1}^{x_2} \frac{p_v}{\varepsilon} a_i(x)\, dx, \qquad i=1,2. \qquad (10.169)$$

Furthermore, the first term on the right-hand side of equation (10.169) represents natural boundary conditions for the element m. We obtain these as

$$i=1, \quad a_i \frac{dV_m}{dx}\bigg|_{x_1}^{x_2} = a_1(x_2)\frac{dV_m}{dx}(x_2) - a_1(x_1)\frac{dV_m}{dx}(x_1) = -\frac{dV_m}{dx}(x_1) \qquad (10.170)$$

because $a_1(x_2)=0$, $a_1(x_1)=1$,
and

$$i=2, \quad a_i \frac{dV_m}{dx}\bigg|_{x_1}^{x_2} = a_2(x_2)\frac{dV_m}{dx}(x_2) - a_2(x_1)\frac{dV_m}{dx}(x_1) = \frac{dV_m}{dx}(x_2) \qquad (10.171)$$

because $a_2(x_2)=1$, $a_2(x_2)=0$.

We use the end-point values of a_i shown in Figure 10.4. Thus, the element equations are presented as

$$\begin{bmatrix} C_{11}^{(m)} & C_{12}^{(m)} \\ C_{21}^{(m)} & C_{22}^{(m)} \end{bmatrix} \begin{bmatrix} V_1^{(m)} \\ V_2^{(m)} \end{bmatrix} = \begin{bmatrix} -\dfrac{dV^{(m)}}{dx}(x_1) \\ \dfrac{dV^{(m)}}{dx}(x_2) \end{bmatrix} + \begin{bmatrix} \Psi_1^{(m)} \\ \Psi_2^{(m)} \end{bmatrix} \qquad (10.172)$$

where

$$[C^{(m)}] = \int_{x_1}^{x_2} \begin{bmatrix} \dfrac{da_1}{dx}\dfrac{da_1}{dx} & \dfrac{da_1}{dx}\dfrac{da_2}{dx} \\ \dfrac{da_2}{dx}\dfrac{da_1}{dx} & \dfrac{da_2}{dx}\dfrac{da_2}{dx} \end{bmatrix} dx \quad \text{and} \quad \{\Psi_c^{(m)}\} = \int_{x_1}^{x_2} \begin{bmatrix} \dfrac{p_v}{\varepsilon} a_1 \\ \dfrac{p_v}{\varepsilon} a_2 \end{bmatrix} dx. \qquad (10.173)$$

The extension to an element with N nodes follows the same steps, but with $i = 1, 2,\ldots, N$. In addition, the matrices for an element with N nodes contain terms similar to the equation (10.172), but with additional rows and columns to account for N element equations.

10.13.1.3 Natural Coordinates in One-Dimension

We use natural (length) coordinates in deriving interpolation functions that can be used to evaluate the integrals in the element equations. In addition, we use the natural coordinate system in describing the location of a point inside an element in terms of the coordinates associated with the nodes of the element. Let η_i be the natural coordinates, where $i = 1, 2,\ldots, N$; N is the number of external nodes of the element. Knowing that, natural coordinates are functions of the global Cartesian coordinate system in which the element is defined, the one coordinate is associated with node i and has unit value there.

Figure 10.5 shows a line element with natural coordinates η_1, η_2 and location point x_l.

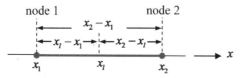

Figure 10.5. **Example of two-node line element in one-dimensional (1D) with global coordinate x_l.**

The global coordinate x_l can be expressed as

$$x_l = \eta_1 x_1 + \eta_2 x_2. \tag{10.174}$$

We can interpret natural (length) coordinates η_1 and η_2 as weighting functions relating the coordinates of the end modes to the coordinate of any interior point. As we know that,

$$\eta_1 + \eta_2 = 1 \tag{10.175}$$

although, the weighting functions are not independent. Let us consider $x_l = x$ and solving for η_1 and η_2 from equations (10.174) and (10.175), we get

$$\eta_1(x) = \frac{x_2 - x}{x_2 - x_1}, \quad \eta_2(x) = \frac{x - x_1}{x_2 - x_1}. \tag{10.176}$$

The linear interpolation used for the potential distribution variable $V(x)$ in the previous section, which can be written as

$$V(x) = V_1 \eta_1 + V_2 \eta_2. \tag{10.178}$$

By differential of $V(x)$ using the chain rule, we get

$$\frac{dV}{dx} = \frac{\partial V}{\partial \eta_1}\frac{\partial \eta_1}{\partial x} + \frac{\partial V}{\partial \eta_2}\frac{\partial \eta_2}{\partial x} \qquad (10.179)$$

where

$$\frac{\partial \eta_1}{\partial x} = \frac{-1}{x_2 - x_1}, \frac{\partial \eta_2}{\partial x} = \frac{1}{x_2 - x_1}. \qquad (10.180)$$

Now, taking the integration of length coordinates over the length of an element, that is,

$$\int_{x_1}^{x_2} \eta_1^i \eta_2^j dx = \frac{i!\,j!(x_2 - x_1)}{(1+j+1)!} \qquad (10.181)$$

where i and j are integer exponents.

10.13.2 Two-Dimensional Elements

10.13.2.1 Applications of FEM to Electrostatic Problems

It's often known that FEM is a numerical method used to find the approximate solutions either for partial differential equations or integral equations. These equations are most involved in electromagnetic problems. We illustrate the four steps above used to find the solution in FEM through three different types of differential equations, Laplace's equation, Poisson's equation, and wave equation.

10.13.2.1.1 Solution of Laplace's Equation $\nabla^2 V = 0$ with FEM

To find the potential distribution, $V(x, y)$, for the two-dimensional (2-D) solution region, as shown in Figure 10.6. We illustrate the following steps to get the solution of Laplace equation, $\nabla^2 V = 0$.

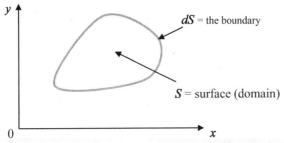

Figure 10.6. The solution region of the problem showing domain for the 2-D boundary value.

First step, using finite element discretization to find the potential distribution for the two-dimensional solution, $V(x, y)$ as shown in Figure 10.7, where the solution region is subdivided into seven nonoverlapping finite elements of triangles. It is always preferable in computation to have the same type of elements through the solution region which in our case is the triangle.

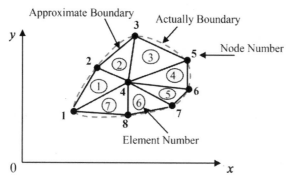

Figure 10.7. The finite element discretization of the solution.

We look for an approximation solution for the potential $V_m(x, y)$ within an element m and then interrelate the potential distribution in various elements such that the potential is continuous across interelement boundaries. We can express the approximation solution for the whole region as

$$V(x,y) \approx \sum_{m=1}^{N} V_m(x,y), \quad (10.182)$$

where N is the number of triangle elements into which the solution region is divided.

The most common form of approximation for $V_m(x, y)$ within an element is polynomial approximation for a triangle element, that is,

$$V_m(x,y) = a + bx + cy, \quad (10.183)$$

where the constants a, b and c are to be determined. The potential $V_m(x, y)$ in general is nonzero within element m, but zero outside m. Furthermore, our assumption of linear variation of potential within the triangle element as in equation (2.46) is the same as assuming that the electric field is uniform within the element, that is to say,

$$\mathbf{E}_m = -\nabla V_m = -(b\mathbf{a}_x + c\mathbf{a}_y). \quad (10.184)$$

Second step, developing equations for the element. Let us choose a typical triangle element shown in Figure 10.8.

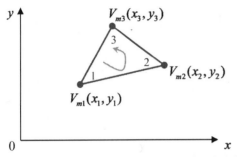

Figure 10.8. Typical triangle element; local node numbering 1-2-3 must proceed counter-clockwise as indicated by the arrow.

The potential $V_{m1}(x_1, y_1)$, $V_{m2}(x_2, y_2)$, and $V_{m3}(x_3, y_3)$ at nodes 1, 2, and 3, respectively, are obtained using equation (10.183), namely

$$\begin{bmatrix} V_{m1} \\ V_{m2} \\ V_{m3} \end{bmatrix} = \begin{bmatrix} 1 & x_1 & y_1 \\ 1 & x_2 & y_2 \\ 1 & x_3 & y_3 \end{bmatrix} \begin{bmatrix} a \\ b \\ c \end{bmatrix}. \qquad (10.185)$$

The coefficients a, b, and c are determined from equation (10.185) as

$$\begin{bmatrix} a \\ b \\ c \end{bmatrix} = \begin{bmatrix} 1 & x_1 & y_1 \\ 1 & x_2 & y_2 \\ 1 & x_3 & y_3 \end{bmatrix}^{-1} \begin{bmatrix} V_{m1} \\ V_{m2} \\ V_{m3} \end{bmatrix}. \qquad (10.186)$$

Therefore, equation (10.183) can be rewritten by substituting for a, b, and c, i.e.,

$$V_m = \begin{bmatrix} 1 & x & y \end{bmatrix} \begin{bmatrix} a \\ b \\ c \end{bmatrix} = \begin{bmatrix} 1 & x & y \end{bmatrix} \begin{bmatrix} 1 & x_1 & y_1 \\ 1 & x_2 & y_2 \\ 1 & x_3 & y_3 \end{bmatrix}^{-1} \begin{bmatrix} V_{m1} \\ V_{m2} \\ V_{m3} \end{bmatrix}. \qquad (10.187)$$

Equation (10.187) can be written as

$$V_m(x, y) = \sum_{i=1}^{3} a_i(x, y) V_{mi}, \qquad (10.188)$$

where $a_i(x, y)$ is given by

$$a_i(x, y) = \frac{1}{2A}(a_i + b_i x + c_i y), \text{ where } i = 1, 2, 3 \qquad (10.189)$$

and a_i, b_i, and c_i are given by

$$a_i = x_j y_k - x_k y_j \qquad (10.190)$$

$$b_i = y_j - y_k \qquad (10.191)$$

$$c_i = x_k - x_j \qquad (10.192)$$

where i, j, and k are cyclical, that is, $(i = 1, j = 2, k = 3)$, $(i = 2, j = 3, k = 1)$, and $(i = 3, j = 1, k = 2)$.

Note that, by substituting equations (10.190), (10.191), and (10.192) into equation (10.187) gives

$$V_m(x, y) = \begin{bmatrix} 1 & x & y \end{bmatrix} \frac{1}{2A} \begin{bmatrix} (x_2 y_3 - x_3 y_2) & (x_3 y_1 - x_1 y_3) & (x_1 y_2 - x_2 y_1) \\ (y_2 - y_3) & (y_3 - y_1) & (y_1 - y_2) \\ (x_3 - x_2) & (x_1 - x_3) & (x_2 - x_1) \end{bmatrix} \begin{bmatrix} V_{m1} \\ V_{m2} \\ V_{m3} \end{bmatrix}. \qquad (10.193)$$

Also, using equations (10.190), (10.191), and (10.192) into equation (10.188), gives

$$a_1(x, y) = \frac{1}{2A}[(x_2 y_3 - x_3 y_2) + (y_2 - y_3)x + (x_3 - x_2)y], \qquad (10.194a)$$

$$a_2(x, y) = \frac{1}{2A}[(x_3 y_1 - x_1 y_3) + (y_3 - y_1)x + (x_1 - x_3)y], \qquad (10.194b)$$

$$a_3(x, y) = \frac{1}{2A}[(x_1 y_2 - x_2 y_1) + (y_1 - y_2)x + (x_2 - x_1)y], \qquad (10.194c)$$

and A is given by

$$A = \frac{1}{2} \begin{vmatrix} 1 & x_1 & y_1 \\ 1 & x_2 & y_2 \\ 1 & x_3 & y_3 \end{vmatrix} = \frac{1}{2}[(x_1 y_2 - x_2 y_1) + (x_3 y_1 - x_1 y_3) + (x_2 y_3 - x_3 y_2)]$$

or

$$A = \frac{1}{2}[(x_2 - x_1)(y_3 - y_1) - (x_3 - x_1)(y_2 - y_1)] \qquad (10.195)$$

where A is the area of the element m.

The value of A is positive if the nodes are numbered counterclockwise, starting from any nodes, as shown by the arrow in Figure 10.5.

Furthermore, equation (10.188) gives the potential at any point (x, y) within the element provided that the potentials at the vertices are known. In addition, $a_i(x, y)$ are linear interpolation functions. They are called the element shape functions and they have the following properties:

$$a_i(x, y) = \begin{cases} 1, & i = j \\ 0, & i \neq j \end{cases}, \quad (10.196a)$$

$$\sum_{i=1}^{3} a_i(x, y) = 1. \quad (10.196b)$$

The shape of functions $a_1(x, y)$, $a_2(x, y)$, and $a_3(x, y)$, for example, are illustrated in Figure 10.9.

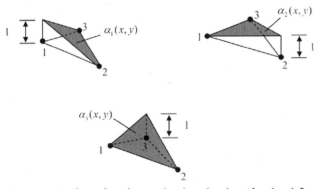

Figure 10.9. Shape functions $a_1(x, y)$, $a_2(x, y)$, and $a_3(x, y)$ for a triangle element.

The functional, W_m, corresponding to Laplace's equation, which physically is the energy per unit length associated with element m, is given by

$$W_m = \frac{1}{2}\int \varepsilon |\mathbf{E}_m|^2 dS = \frac{1}{2}\int \varepsilon |\nabla V_m|^2 dS. \quad (10.197)$$

But from equation (10.188),

$$\nabla V_m = \sum_{i=1}^{3} V_{mi} \nabla a_i. \quad (10.198)$$

By substituting equation (10.198) into equation (10.197), gives

$$W_m = \frac{1}{2}\sum_{i=1}^{3}\sum_{j=1}^{3} \varepsilon V_{mi}\left[\int \nabla a_i \cdot \nabla a_j dS\right] V_{mj}. \quad (10.199)$$

If we define the term in brackets as

$$C_{ij}^{(m)} = \int \nabla a_i \cdot \nabla a_j \, dS, \qquad (10.200)$$

now, we can write equation (10.199) in matrix form as

$$W_m = \frac{1}{2}\varepsilon [V_m]^t \left[C^{(m)}\right]\{V_m\}, \qquad (10.201)$$

where the superscript t denotes the transpose of the matrix,

$$[V_m]^t = \begin{bmatrix} V_{m1} & V_{m2} & V_{m3} \end{bmatrix}, \qquad (10.202a)$$

$$\{V_m\} = \begin{bmatrix} V_{m1} \\ V_{m2} \\ V_{m3} \end{bmatrix}, \qquad (10.202b)$$

and

$$\left[C^{(m)}\right] = \begin{bmatrix} C_{11}^{(m)} & C_{12}^{(m)} & C_{13}^{(m)} \\ C_{21}^{(m)} & C_{22}^{(m)} & C_{23}^{(m)} \\ C_{31}^{(m)} & C_{32}^{(m)} & C_{33}^{(m)} \end{bmatrix}. \qquad (10.202c)$$

The matrix $[C^{(m)}]$ is usually called the *element coefficient matrix* (or *stiffness matrix in structural analysis*). The element $C_{ij}^{(m)}$ of the coefficient matrix may be regarded as the coupling between nodes i and j; its value is obtained from equations (10.194) and (10.200). For instance,

$$\begin{aligned} C_{12}^{(m)} &= \int \nabla a_1 \cdot \nabla a_2 \\ &= \frac{1}{4A^2}[(y_2 - y_3)(y_3 - y_1) + (x_3 - x_2)(x_1 - x_3)]\int dS \qquad (10.203a) \\ &= \frac{1}{4A}[(y_2 - y_3)(y_3 - y_1) + (x_3 - x_2)(x_1 - x_3)]. \end{aligned}$$

Similarly,

$$C_{13}^{(m)} = \frac{1}{4A}[(y_2 - y_3)(y_1 - y_2) + (x_3 - x_2)(x_2 - x_1)], \qquad (10.203b)$$

$$C_{23}^{(m)} = \frac{1}{4A}[(y_3 - y_1)(y_1 - y_2) + (x_1 - x_3)(x_2 - x_1)], \qquad (10.203c)$$

$$C_{11}^{(m)} = \frac{1}{4A}\left[(y_2 - y_3)^2 + (x_3 - x_2)^2\right], \qquad (10.203\text{d})$$

$$C_{22}^{(m)} = \frac{1}{4A}\left[(y_3 - y_1)^2 + (x_1 - x_3)^2\right], \qquad (10.203\text{e})$$

$$C_{33}^{(m)} = \frac{1}{4A}\left[(y_1 - y_2)^2 + (x_2 - x_1)^2\right]. \qquad (10.203\text{f})$$

Additionally,

$$C_{23}^{(m)} = C_{32}^{(m)}, C_{13}^{(m)} = C_{31}^{(m)}, C_{12}^{(m)} = C_{21}^{(m)}. \qquad (10.204)$$

Now, for the third step, after having considered a typical element, the next step is to assemble all such elements in the solution region. The energy associated with the assemblage of elements assuming that the whole solution region is homogeneous so that ε is constant, i.e.,

$$W = \sum_{m=1}^{N} W_m = \frac{1}{2}\varepsilon[V]^t[C]\{V\}, \qquad (10.205)$$

where

$$\{V\} = \begin{bmatrix} V_1 \\ V_2 \\ V_3 \\ \cdot \\ \cdot \\ \cdot \\ V_n \end{bmatrix}, \qquad (10.206)$$

where

n is the number of nodes, N is the number of elements, and $[C]$ is called the overall or *global coefficient matrix*, which is the assemblage of individual element coefficient matrices.

For an inhomogeneous solution region such as that shown in Figure 10.10, the region is discretized with triangle elements such that each finite element is homogeneous. In this case, equation (10.197) still holds, however, equation (10.205) does not apply since ε ($\varepsilon = \varepsilon_r \varepsilon_0$) or simply ε_r varies from element to element. To apply equation (10.205), we need to replace ε by ε_0 and multiply the integrand in equation (10.200) by ε_r.

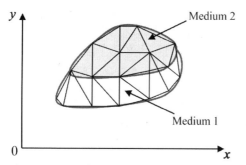

Figure 10.10. Discretization of an inhomogeneous solution region with triangle elements.

We use an example to illustrate the process by which individual element coefficient matrices are assembled to obtain the global coefficient matrix. In this example, we consider the finite element mesh consisting of three finite elements as shown in Figure 10.11. Observe the numberings of the mesh.

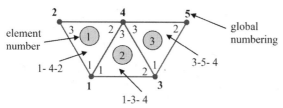

Figure 10.11. Assembly of three elements; $i - j - k$ corresponding to local numbering (1-2-3) of the element in Figure 10.5.

The numbering of nodes 1, 2, 3, 4, and 5 is called *global* numbering. The numbering $i - j - k$ is called *local* numbering, and it corresponds with 1-2-3 of the element in Figure 10.5, the local numbering must be in counterclockwise sequence starting from any node of the element. For element 1, we could choose 2-1-4 instead of 1-4-2 to correspond with 1–2–3 of the element to Figure 10.8. Thus, the numbering in Figure 10.8 is not unique. But whichever numbering is used, the global coefficient matrix remains the same. Assuming the particular numbering in Figure 10.11, the global coefficient matrix is expected to have the form

$$[C] = \begin{bmatrix} C_{11} & C_{12} & C_{13} & C_{14} & C_{15} \\ C_{21} & C_{22} & C_{23} & C_{24} & C_{25} \\ C_{31} & C_{32} & C_{33} & C_{34} & C_{35} \\ C_{41} & C_{42} & C_{43} & C_{44} & C_{45} \\ C_{51} & C_{52} & C_{53} & C_{54} & C_{55} \end{bmatrix}, \quad (10.207)$$

which is a 5 × 5 matrix since five nodes ($n = 5$) are involved. As we know, C_{ij} is the coupling between nodes i and j. The C_{ij} can be obtained by using the fact that the potential distribution must be continuous across interelement boundaries. The contribution to the i, j position in $[C]$ comes from all elements containing nodes i and j. For instance, in Figure 10.11, elements 1 and 2 have node 1 in common; therefore

$$C_{11} = C_{11}^{(1)} + C_{11}^{(2)}. \qquad (10.208a)$$

Node 2 belongs to element 1 only; therefore

$$C_{22} = C_{33}^{(1)}. \qquad (10.208b)$$

Node 4 belongs to element 1, 2, and 3; accordingly

$$C_{44} = C_{22}^{(1)} + C_{33}^{(2)} + C_{33}^{(3)}. \qquad (10.208c)$$

Nodes 1 and 4 belong simultaneously to element 1 and 2; as a result

$$C_{14} = C_{41} = C_{12}^{(1)} + C_{13}^{(2)}. \qquad (10.208d)$$

Since there is no coupling (or direct link) between nodes 2 and 3; hence

$$C_{23} = C_{32} = 0. \qquad (10.208e)$$

By continuing in this approach, we can obtain all the terms in the global coefficient matrix by inspection of Figure 10.11 as

$$[C] = \begin{bmatrix} C_{11}^{(1)} + C_{11}^{(2)} & C_{13}^{(1)} & C_{12}^{(2)} & C_{12}^{(1)} + C_{13}^{(2)} & 0 \\ C_{31}^{(1)} & C_{33}^{(1)} & 0 & C_{32}^{(1)} & 0 \\ C_{21}^{(2)} & 0 & C_{22}^{(2)} + C_{11}^{(3)} & C_{23}^{(2)} + C_{13}^{(3)} & C_{12}^{(3)} \\ C_{21}^{(1)} + C_{23}^{(2)} & C_{23}^{(1)} & C_{32}^{(2)} + C_{31}^{(3)} & C_{22}^{(1)} + C_{33}^{(2)} + C_{33}^{(3)} & C_{32}^{(3)} \\ 0 & 0 & C_{21}^{(3)} & C_{23}^{(3)} & C_{22}^{(3)} \end{bmatrix}. \qquad (10.209)$$

Note that element coefficient matrices overlap at nodes shared by elements and that are 27 terms (9 for each of the 3 elements) in the global coefficient matrix $[C]$. Also note the following properties of the matrix $[C]$:

1. It is symmetric ($C_{ij} = C_{ji}$) just as the element coefficient matrix.
2. Since $C_{ij} = 0$ if no coupling exists between nodes i and j, it is expected that for a large number of elements $[C]$ becomes sparse. Matrix $[C]$ is

also banded if the nodes are carefully numbered. It can be shown using equation (10.203), i.e.,

$$\sum_{i=1}^{3} C_{ij}^{(m)} = 0 = \sum_{j=1}^{3} C_{ij}^{(m)}. \qquad (10.210)$$

3. It is singular. Although this is not obvious, it can be shown using the finite element coefficient matrix of equation (10.202c).

Finally, *fourth step*, by solving the resulting equations. It can be shown that, from variational calculus, it is known that Laplace's (or Poisson's) equation is satisfied when the total energy in the solution region is minimum. Therefore, we require that the partial derivatives of W with respect to each nodal value of the potential be zero; that is,

$$\frac{\partial W}{\partial V_1} = \frac{\partial W}{\partial V_2} = \cdots = \frac{\partial W}{\partial V_n} = 0,$$

or

$$\frac{\partial W}{\partial V_k} = 0, \quad k = 1, 2, \ldots, n. \qquad (10.211)$$

For instance, to get $\dfrac{\partial W}{\partial V_1} = 0$ for the finite element mesh of Figure 2.9, we substitute equation (10.207) into equation (10.205) and take the partial derivative of W with respect to V_1. We obtain

$$0 = \frac{\partial W}{\partial V_1} = 2V_1 C_{11} + V_2 C_{12} + V_3 C_{13} + V_4 C_{14} + V_5 C_{15} + V_2 C_{21} + V_3 C_{31} + V_4 C_{41} + V_5 C_{51},$$

or

$$0 = V_1 C_{11} + V_2 C_{12} + V_3 C_{13} + V_4 C_{14} + V_5 C_{15}. \qquad (10.212)$$

In general case, $\dfrac{\partial W}{\partial V_k} = 0$ leads to

$$0 = \sum_{i=1}^{n} V_i C_{ik}, \qquad (10.213)$$

where n is the number of nodes in the mesh. By writing equation (10.213) for all nodes $k = 1, 2, 3, \ldots, n$, we obtain a set of simultaneous equations from which the solution of the transpose matrix for the potential distribution, $[V]^t = [V_1\ V_2\ \ldots\ V_n]$, can be found. This can be done in two ways:

(1) Iteration Method

Suppose node 1 in Figure 10.11, for example, is a free node. The potential at node 1 can be obtained from equation (10.212) as

$$V_1 = -\frac{1}{C_{11}} \sum_{i=2}^{5} V_i C_{1i}. \tag{10.214}$$

Thus, in general case, the potential at a free node k in a mesh with n nodes is obtained from equation (10.213) as

$$V_k = -\frac{1}{C_{kk}} \sum_{i=1, j \neq k}^{n} V_i C_{ik}. \tag{10.215}$$

Since $C_{ki} = 0$ is not directly connected to node i, only nodes that are directly linked to node k contribute to V_k in equation (10.215). Note equation (10.215) can be applied iteratively to all the free nodes. The iteration process begins by setting the potentials of fixed nodes (where the potentials are prescribed or known) to their prescribed values and the potentials at the free nodes (where the potentials are known) equal to zero or to the average potential

$$V_{ave} = \frac{1}{2}(V_{min} + V_{max}), \tag{10.216}$$

where V_{min} and V_{max} are the minimum and maximum values of the prescribed potentials at the fixed nodes, V, respectively. With these initial values, the potentials at the free nodes are calculated using equation (10.215). At the end of the first iteration, when the new values have been calculated for all the free nodes, they become the old values for the second iteration. Indeed, the procedure is repeated until the change between subsequent iteration is negligible enough.

(2) Band Matrix Method

If all free nodes are numbered first and the fixed nodes last, equation (10.205) can be written such that

$$W = \frac{1}{2}\varepsilon \begin{bmatrix} V_f & V_p \end{bmatrix} \begin{bmatrix} C_{ff} & C_{fp} \\ C_{pf} & C_{pp} \end{bmatrix} \begin{bmatrix} V_f \\ V_p \end{bmatrix}, \tag{10.217}$$

where subscripts f and p, refer to nodes with free and fixed (or prescribed) potentials, respectively. Since V_p is constant (it consists of known, fixed values), we only differentiate with respect to V_f so that applying equations (10.211) to (10.217) which yields to

$$\begin{bmatrix} C_{ff} & C_{fp} \end{bmatrix} \begin{bmatrix} V_f \\ V_p \end{bmatrix} = 0$$

or

$$\begin{bmatrix} C_{ff} \end{bmatrix}\begin{bmatrix} V_f \end{bmatrix} = -\begin{bmatrix} C_{fp} \end{bmatrix}\begin{bmatrix} V_p \end{bmatrix}. \tag{10.218}$$

This equation can be written as

$$[A][V] = [B] \tag{10.219a}$$

or

$$[V] = [A]^{-1}[B], \tag{10.219b}$$

where $[V] = [V_f]$, $[A] = [C_{ff}]$, $[B] = -[C_{fp}][V_p]$. Since, in general, nonsingular, the potential at the free nodes can be found using equation (10.219). Note we can solve for $[V]$ in equation (10.219a) using Gaussian elimination technique. Also, we can solve for $[V]$ in equation (10.219b) using matrix inversion if the size of the matrix to be inverted is not large.

In fact, it is sometimes necessary to impose the Neumann condition $\left(\frac{\partial V}{\partial n} = 0\right)$ as a boundary condition or at the line of symmetry when we take advantage of the symmetry of the problem. Indeed, suppose that for concreteness, a solution region is symmetric along the y–axis as in Figure 10.12. We impose condition $\left(\frac{\partial V}{\partial x} = 0\right)$ along the y–axis by making

$$V_1 = V_2, \quad V_4 = V_5, \quad V_7 = V_8. \tag{10.220}$$

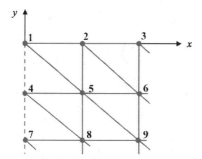

Figure 10.12. A solution region that is symmetric along the y–axis.

With this in mind that as from equation (10.197) onward; the solution has been restricted to a two-dimensional problem involving Laplace's equation, $\nabla^2 V = 0$.

10.13.2.1.2 Solution of Passion's Equation $\nabla^2 V = -\dfrac{\rho_v}{\varepsilon}$ with FEM

In this section, we solve the two-dimensional (2D) Poisson's equations

$$\nabla^2 V = -\frac{\rho_v}{\varepsilon}, \qquad (10.221)$$

using the same essential four steps as in previous section with FEM, we focus here on the source term and only the major differences.

The deriving element governing equations step. We divide the solution region into triangles, and then we approximate the potential distribution $V_m(x, y)$ and the source term ρ_{vm} over each triangle element by linear combinations of the local interpolation polynomial a_i, namely,

$$V_m = \sum_{i=1}^{3} V_{mi} a_i(x, y), \qquad (10.222)$$

$$\text{and} \quad \rho_{vm} = \sum_{i=1}^{3} \rho_{mi} a_i(x, y) \qquad (10.223)$$

where
V_{mi} is the values of V at vertex i of element m; ρ_{mi} is the values of ρ_v at vertex i of element m. The values of ρ_{mi} are known since $\rho_v(x, y)$ is prescribed, while the values of V_{mi} are to be determined.

An energy functional which associated Euler equation with equation (10.221) is

$$F(V_m) = \frac{1}{2} \int_S \left[\varepsilon |\nabla V_m|^2 - 2\rho_{vm} V_m \right] dS, \qquad (10.224)$$

where
$F(V_m)$ is the total energy per length within element m; $\dfrac{1}{2}\varepsilon|\nabla V_m|^2$ is the energy density in the electrostatic system and it is equal to $\dfrac{1}{2}\mathbf{D}\cdot\mathbf{E}$; $\rho_{vm} V_m dS$ is the work done in moving the charge $\rho_{vm} dS$ to its location at potential V_m.

Now, by substituting equations (10.222) and (10.223) into equation (10.224) we get

$$F(V_m) = \frac{1}{2} \sum_{i=1}^{3} \sum_{j=1}^{3} \varepsilon V_{mi} \left[\int \nabla a_i \cdot \nabla a_j dS \right] V_{mj} - \sum_{i=1}^{3} \sum_{j=1}^{3} V_{mi} \left[\int a_i a_j dS \right] \rho_{mj}. \qquad (10.225)$$

Equation (10.225) can be applied to every element in the solution region.

Also, it can be written in matrix form as

$$F(V_m) = \frac{1}{2}\varepsilon[V_m]^t[C^{(m)}][V_m] - [V_m]^t[T^{(m)}][\rho_m] \qquad (10.226)$$

where

$$C_{ij}^{(m)} = \int \nabla a_i \cdot \nabla a_j dS \qquad (10.227)$$

we know that equation (10.226) is already defined in equation (10.203) and

$$T_{ij}^{(m)} = \int a_i a_j dS. \qquad (10.228)$$

Also, $T_{ij}^{(m)}$ can be written as

$$T_{ij}^{(m)} = \begin{cases} A/6, & i = j \\ A/12, & i \neq j \end{cases} \qquad (10.229)$$

where A is the area of the triangle element.

We can obtain the discredited functional for the whole solution region, with N elements and n nodes, as the sum of the functional for the individual elements, that is, from equation (10.229),

$$F(V) = \sum_{m=1}^{N} F(V_m) = \frac{1}{2}\varepsilon[V]^t[C][V] - [V]^t[T][\rho] \qquad (10.230)$$

where
t is the transposition symbol. In equation (10.230), the column matrix $[V]$ consists of the values of V_{mi}, while the column matrix $[\rho]$ contains n values of the source function, ρ_v, at the nodes. The functional in equation (10.230) is now minimized by differentiating with respect to V_{mi} and setting the result equal to zero.

Now, we work on the *solving the resulting equations step*. We can solve the resulting equations by using either the iteration method or the band matrix method.

(1) Iteration Methods

By considering the solution region in Figure 10.8 which has five nodes, $n = 5$ and from the equation (10.230), we can get the energy functional as

$$F = \frac{1}{2}\varepsilon [V_1 \ V_2 \ \cdots \ V_5] \begin{bmatrix} C_{11} & C_{12} & \cdot & \cdot & C_{15} \\ C_{21} & C_{22} & \cdot & \cdot & C_{25} \\ \cdot & & & & \cdot \\ \cdot & & & & \cdot \\ \cdot & & & & \cdot \\ C_{51} & C_{52} & \cdot & \cdot & C_{55} \end{bmatrix} \begin{bmatrix} V_1 \\ V_2 \\ \cdot \\ \cdot \\ \cdot \\ V_5 \end{bmatrix}$$

$$- [V_1 \ V_2 \ \cdots \ V_5] \begin{bmatrix} T_{11} & T_{12} & \cdot & \cdot & T_{15} \\ T_{21} & T_{22} & \cdot & \cdot & T_{25} \\ \cdot & & & & \cdot \\ \cdot & & & & \cdot \\ \cdot & & & & \cdot \\ T_{51} & T_{52} & \cdot & \cdot & T_{55} \end{bmatrix} \begin{bmatrix} \rho_1 \\ \rho_2 \\ \cdot \\ \cdot \\ \cdot \\ \rho_5 \end{bmatrix} \qquad (10.231)$$

The energy can be minimized by applying

$$\frac{\partial F}{\partial V_k} = 0, \quad k = 1, 2, \cdots n. \qquad (10.232)$$

For example, from equation (10.231), we get $\frac{\partial F}{\partial V_1} = 0$, as

$$\frac{\partial F}{\partial V_1} = \varepsilon [V_1 C_{11} + V_2 C_{21} + \cdots + V_5 C_{51}] - [T_{11}\rho_1 + T_{21}\rho_2 + \cdots + T_{51}\rho_5] = 0$$

or

$$V_1 = -\frac{1}{C_{11}} \sum_{i=2}^{5} V_i C_{i1} + \frac{1}{\varepsilon C_{11}} \sum_{i=1}^{5} T_{i1}\rho_i. \qquad (10.233)$$

Therefore, in general, for a mesh with n nodes

$$V_k = -\frac{1}{C_{kk}} \sum_{i=1, i \neq k}^{n} V_i C_{ki} + \frac{1}{\varepsilon C_{kk}} \sum_{i=1}^{n} T_{ki}\rho_i \qquad (10.234)$$

where
node k is assumed to be a free node.

By fixing the potential at the prescribed nodes and setting the potential at the free nodes initially equal to zero, we apply equation (10.234) iteratively to all free nodes until convergence is reached.

(2) Band Matrix Method

In this method, we let the free nodes be numbered first and the prescribed nodes last. In doing this, equation (10.230) can be written as

$$F(V) = \frac{1}{2}\varepsilon \begin{bmatrix} V_f & V_p \end{bmatrix} \begin{bmatrix} C_{ff} & C_{fp} \\ C_{pf} & C_{pp} \end{bmatrix} \begin{bmatrix} V_f \\ V_p \end{bmatrix} - \begin{bmatrix} V_f & V_p \end{bmatrix} \begin{bmatrix} T_{ff} & T_{fp} \\ T_{pf} & T_{pp} \end{bmatrix} \begin{bmatrix} \rho_f \\ \rho_p \end{bmatrix} \quad (10.235)$$

where
subscript f is the free node; subscript p is the prescribed node;
ρ_f is the submatrix containing the values of ρ at free node; ρ_p is the submatrix containing the values of ρ at the prescribed node.

Minimizing $F(V)$ with respect to V_f, namely,

$$\frac{\partial F}{\partial V_f} = 0$$

gives

$$\varepsilon(C_{ff}V_f + C_{pf}V_p) - (T_{ff}\rho_f + T_{fp}\rho_p) = 0$$

or

$$[C_{ff}][V_f] = -[C_{fp}][V_p] + \frac{1}{\varepsilon}[T_{ff}][\rho_f] + \frac{1}{\varepsilon}[T_{fp}][\rho_p]. \quad (10.236)$$

Indeed, equation (10.236) can be written

$$[A][V] = [B] \quad (10.237)$$

where
$[A] = [C_{ff}]$, $[V] = [V_f]$, and $[B]$ is the right-hand side of equation (10.236). Equation (10.237) can be solved to determine $[V]$ either by matrix inversion or Gaussian elimination technique.

10.13.2.1.3 Solution of Wave's Equation $\nabla^2 \Phi + k^2 \Phi = g$ with FEM

A typical wave equation is the inhomogeneous scalar Helmholtz's equation

$$\nabla^2 \Phi + k^2 \Phi = g \tag{10.238}$$

where

Φ is the potential (for waveguide problem, $\Phi = H_z$ for *TE* mode or E_z for *TM* mode) to be determined, g is the source function, and $k = \omega\sqrt{\mu\varepsilon}$ is the wave number of the medium. The following three distinct special cases of equation (10.238) should be noted:

1. $k = 0$ and $g = 0$; Laplace's equation;
2. $k = 0$; Poisson's equation; and
3. k is an unknown, $g = 0$; homogeneous, scalar Helmholtz's equation.

It is known that the variational solution to the operator equation

$$L\Phi = g \tag{10.239}$$

is obtained by examining the functional

$$I(\Phi) = <L, \Phi> - 2<\Phi, g> \tag{10.240}$$

where L is an operator (differential, integral, or integro-differential), g is the unknown excitation or source, and Φ is the unknown function to be determined (here is the potential).

Therefore, the solution of equation (10.238) is equivalent to satisfying the boundary conditions and minimizing the functional

$$I(\Phi) = \frac{1}{2}\iint \left[|\nabla\Phi|^2 - k^2\Phi^2 + 2\Phi g\right] dS. \tag{10.241}$$

Note that, if other than the natural boundary conditions (i.e., Dirichlet of homogenous Neumann conditions) must be satisfied, appropriate terms must be added to the functional. The potential Φ and the source function g can be expressed now in terms of the shape functions a_i over a triangle element as

$$\Phi_m(x, y) = \sum_{i=1}^{3} a_i \Phi_{mi} \tag{10.242}$$

where
ϕ_{mi} is the value of Φ at the nodal point i of element m.

And

$$g_m(x,y) = \sum_{i=1}^{3} a_i g_{mi} \qquad (10.243)$$

g_{mi} is the value of g at the nodal point i of element m.
Substituting equations (10.242) and (10.243) into equation (10.241) gives

$$I(\Phi_m) = \frac{1}{2}\sum_{i=1}^{3}\sum_{j=1}^{3}\Phi_{mi}\Phi_{mj}\iint \nabla a_i \cdot \nabla a_j \, dS - \frac{k^2}{2}\sum_{i=1}^{3}\sum_{j=1}^{3}\Phi_{mi}\Phi_{mj}\iint a_i a_j \, dS$$

$$+ \sum_{i=1}^{3}\sum_{j=1}^{3}\Phi_{mi}g_{mj}\iint a_i a_j \, dS \qquad (10.244)$$

$$= \frac{1}{2}[\Phi_m]^t\left[C^{(m)}\right][\Phi_m] - \frac{k^2}{2}[\Phi_m]^t\left[T^{(m)}\right][\Phi_m] + [\Phi_m]^t\left[T^{(m)}\right][G_m]$$

where
$[\Phi_m] = [\Phi_{m1},\Phi_{m2},\Phi_{m3}]^t$, $[G_m] = [g_{m1},g_{m2},g_{m3}]^t$, and $\left[C^{(m)}\right]$ and $\left[T^{(m)}\right]$ are defined in equations (10.158) and (10.185), respectively.

The equation (10.244) is for a single element, but it can be applied for all N elements in the solution region. Therefore,

$$I(\Phi) = \sum_{m=1}^{N} I(\Phi_m). \qquad (10.245)$$

From equations (10.244) and (10.245), $I(\Phi)$ can be expressed in matrix form as

$$I(\Phi) = \frac{1}{2}[\Phi]^t [C][\Phi] - \frac{k^2}{2}[\Phi]^t [T][\Phi] + [\Phi]^t [T][G] \qquad (10.246)$$

where

$$[\Phi] = [\Phi_1, \ \Phi_2, \ \ldots \ , \Phi_N]^t, \qquad (10.247a)$$

$$[G] = [g_1, \ g_2, \ \ldots \ , g_N]^t \qquad (10.247b)$$

$[C]$, and $[T]$ are global matrices consisting of local matrices $[C^{(m)}]$ and $[T^{(m)}]$, respectively.

Now, if free nodes are numbered first and the prescribed nodes last, and considering the source function $g = 0$, we can write equation (2.109) as

$$I = \frac{1}{2}\begin{bmatrix}\Phi_f & \Phi_p\end{bmatrix}\begin{bmatrix}C_{ff} & C_{fp} \\ C_{pf} & C_{pp}\end{bmatrix}\begin{bmatrix}\Phi_f \\ \Phi_p\end{bmatrix} - \frac{k^2}{2}\begin{bmatrix}\Phi_f & \Phi_p\end{bmatrix}\begin{bmatrix}T_{ff} & T_{fp} \\ T_{pf} & T_{pp}\end{bmatrix}\begin{bmatrix}\Phi_f \\ \Phi_p\end{bmatrix}. \quad (10.248)$$

By setting $\dfrac{\partial I}{\partial \phi_f} = 0$, gives

$$\begin{bmatrix}C_{ff} & C_{fp}\end{bmatrix}\begin{bmatrix}\Phi_f \\ \Phi_p\end{bmatrix} - k^2\begin{bmatrix}T_{ff} & T_{fp}\end{bmatrix}\begin{bmatrix}\Phi_f \\ \Phi_p\end{bmatrix} = 0. \quad (10.249)$$

For *TM* modes, $\Phi_p = 0$ and hence

$$\left[C_{ff} - k^2 T_{ff}\right]\Phi_f = 0. \quad (10.250)$$

Premultiplying equation (10.250) by T_{ff}^{-1}, gives

$$\left[T_{ff}^{-1} C_{ff} - k^2 I\right]\Phi_f = 0. \quad (10.251)$$

By letting

$$T_{ff}^{-1} C_{ff} = A, \; k^2 = \beta, \; \Phi_f = X, \quad (10.252a)$$

and *U* is a unit matrix,
we can obtain the standard eigenvalue problem

$$(A - \beta U)X = 0. \quad (10.252b)$$

any standard procedure may be used to obtain some or all of the eigenvalues $\beta_1, \beta_2, \ldots, \beta_{nf}$ and eigenvectors $X_1, X_2, \ldots, X_{nf}$, where *nf* is the number of free nodes. The eigenvalues are always real since *C* and *T* are symmetric.

The solution of the algebraic eigenvalue problems in equation (10.252) furnishes eigenvalues and eigenvectors, which form good approximations to the eigenvalues and eigenfunctions of the Helmholtz problem, i.e., the cutoff wavelengths and field distribution patterns of the various modes possible in a given waveguide.

The solution of the problem of equation (10.238) is summarized in equation (10.251), and can be viewed as the finite element solution of homogeneous waveguides. The idea can be extended to handle inhomogeneous waveguide problems. However, in applying FEM to inhomogeneous problems, a serious difficulty is the appearance of spurious, nonphysical solutions. There are several techniques that have been proposed to overcome the difficulty.

10.14 AUTOMATIC MESH GENERATION

It is a fact that, one of the major difficulties encountered in the finite element analysis of continuum problems is the tedious and time-consuming effort required in data preparation. Indeed, efficient finite element programs must have node and element generating schemes, referred to collectively as *mesh generators*. Automatic mesh generation minimizes the input data required to specify a problem. In fact, it not only reduces the time involved in data preparation, it eliminates human errors that are introduced when data preparation is preformed manually. Furthermore, combining the automatic mesh generation program with computer graphics is particularly valuable since the output can be monitored visually.

A number of mesh generation algorithms of varying degrees of automation have been proposed. In this section, we focus on two types of domains, rectangular domains and arbitrary domains.

10.14.1 Rectangular Domains

Since some applications of FEM to EM problems involve simple rectangular domains, we consider the generation of simple meshes. Now, let us consider a rectangular solution region of a size $a \times b$ as shown in Figure 10.13. The goal here is to divide the region into rectangular elements, each of which is later divided into two triangular elements.

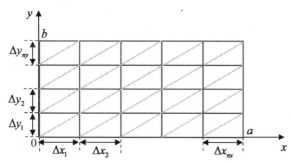

Figure 10.13. Discretization of a rectangular region into a nonuniform mesh.

Suppose n_x and n_y are the number of divisions in x and y directions, the total number of elements and nodes are, respectively, given by

$$n_m = 2n_x n_y$$
$$n_d = (n_x + 1)(n_y + 1). \qquad (10.253)$$

As a result, it is easy to figure out from Figure 10.13 a systematic way of numbering the elements and nodes. Indeed, to obtain the global coordinates (x, y) for each node, we need an array containing Δx_i, $i = 1, 2,..., n_x$ and Δy_j, $j = 1, 2,..., n_y$, which are, respectively, the distances between nodes in the x and y directions. If the order of node numbering is from left to right along horizontal rows and from bottom to top along the vertical rows, then the first node is the origin $(0, 0)$. The next node is obtained as $x \rightarrow x + \Delta x_1$ while $y = 0$ remains unchanged. The following node $x \rightarrow x + \Delta x_2$, $y = 0$, and so on until Δx_i are exhausted. We start the second next horizontal row by starting with $x = 0$, $y \rightarrow y + \Delta y_1$ and increasing x until Δx_i are exhausted. We repeat the process until the last node $(n_x + 1)(n_y + 1)$ is reached, i.e., when Δx_i and Δy_j are exhausted simultaneously.

The procedure presented here allows for generating uniform and nonuniform meshes. A mesh is uniform if all Δx_i are equal and all Δy_j are equal; it is nonuniform otherwise. A nonuniform mesh is preferred if it is known in advance that the parameter of interest varies rapidly in some parts of the solution domain. This allows a concentration of relatively small elements in the regions where the parameter changes rapidly, particularly since these regions are often of greatest interest in the solution. Additionally, without the preknowledge of the rapid change in the unknown parameter, a uniform mesh can be used. In that case, we set

$$\Delta x_1 = \Delta x_2 = \ldots = h_x$$
$$\Delta y_1 = \Delta y_2 = \ldots = h_y \tag{10.224}$$

where
$h_x = a/n_x$ and $h_y = a/n_y$.

In some cases, we also need a list of prescribed nodes. If we assume that all boundary points have prescribed potentials, the number n_p of prescribed nodes is given by

$$n_p = 2(n_x + n_y). \tag{10.255}$$

A simple way to obtain the list of boundary points is to enumerate points on the bottom, right, top, and left of the rectangular region in that order.

10.14.2 Arbitrary Domains

The basic steps involved in a mesh generation are as follows:

A. subdivide solution region into few quadrilateral blocks,
B. separately subdivide each block into elements,
C. connect individual blocks.

A. Definition of Blocks

The solution region is subdivided into quadrilateral blocks. Subdomains with different constitutive parameters $(\sigma, \mu, \varepsilon)$ must be represented by separate blocks. As input data, we specify block topologies and the coordinates at eight points describing each block. Each block is represented by an eight-node quadratic isoparametric element. With natural coordinate system (ζ, η), the x and y coordinates are represented as

$$x(\zeta, \eta) = \sum_{i=1}^{8} a_i(\zeta, \eta) x_i \tag{10.256}$$

$$y(\zeta, \eta) = \sum_{i=1}^{8} a_i(\zeta, \eta) y_i \tag{10.257}$$

where $a_i(\zeta, \eta)$ is a shape function associated with node i, and (x_i, y_i) are the coordinates of node i defining the boundary of the quadrilateral block as shown in Figure 10.14.

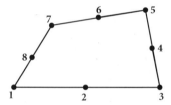

Figure 10.14. Typical quadrilateral block.

The shape functions are expressed in terms of the quadratic or parabolic isoparametric elements shown in Figure 10.15.

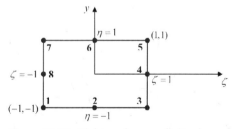

Figure 10.15. Eight-node serendipity element.

They are given by:

$$a_i = \frac{1}{4}(1 + \zeta\zeta_i)(1 + \eta\eta_i)(\zeta\zeta_i + \eta\eta_i + 1), i = 1,3,5,7. \tag{10.258}$$

For corner nodes,

$$a_i = \frac{1}{2}\zeta_i^2(1+\zeta\zeta_i)(1-\eta^2) + \frac{1}{2}\eta_i^2(1+\eta\eta_i+1)(1-\zeta^2), i=2,4,6,8. \quad (10.259)$$

For midside nodes, note the following properties of the shape functions:

1. They satisfy the conditions.

$$\sum_{i=1}^{n} a_i(\zeta,\eta) = 1 \quad (10.260a)$$

$$a_i(\zeta_j,\eta_j) = \begin{cases} 1, & i=j \\ 0, & i\neq j \end{cases} \quad (10.260b)$$

2. They become quadratic along element edges ($\zeta = \pm 1, \eta = \pm 1$).

B. Subdivision of Each Block

Furthermore, for each block, we specify N DIV X and N DIV Y, the number of element subdivisions to be made in the ζ and η directions, respectively. In addition, we specify the weighting factors $(W_\zeta)_i$ and $(W_\eta)_i$ allowing for graded mesh within a block. It is essential to know that, in specifying N DIV X, N DIV Y, $(W_\zeta)_i$ and $(W_\eta)_i$ care must be taken to ensure that the subdivision along block interfaces (for adjacent blocks) are compatible. We initialize ζ and η to a value of -1 so that the natural coordinates are incremented according to

$$\zeta_i = \zeta_i + \frac{2(W_\zeta)_i}{W_\zeta^T \times F} \quad (10.261)$$

$$\eta_i = \eta_i + \frac{2(W_\eta)_i}{W_\eta^T \times F} \quad (10.262)$$

where

$$W_\zeta^T = \sum_{j=1}^{N\ DIV\ X} (W_\zeta)_j \quad (10.263a)$$

$$W_\eta^T = \sum_{j=1}^{N\ DIV\ X} (W_\eta)_j \quad (10.263b)$$

and

$$F = \begin{cases} 1, & \text{for linear elements} \\ 2, & \text{for quadratic elements} \end{cases}.$$

Now, there are three elements types permitted: (1) linear four-node quadrilateral elements, (2) linear three-node triangle elements, and (3) quadratic eight-node isoparametric elements.

C. Connection of Individual Blocks

After subdividing each block and numbering its nodal points separately, it is necessary to connect the blocks and have each node numbered uniquely. This is accomplished by comparing the coordinates of all nodal points and assigning the same number to all nodes having identical coordinates. In other words, we compare the coordinates of node 1 with all other nodes, and then node 2 with other nodes, etc., until all repeated nodes are eliminated.

10.15 HIGHER ORDER ELEMENTS

Finite elements use higher order elements. The shape function or interpolation polynomial of the order two or more is called higher *order element*. To emphasize, the accuracy of a finite element solution can be improved by using finer mesh or using higher order elements or both. Desai and Abel studied mesh refinement versus higher order elements in [44]. Generally, fewer higher order elements are needed to achieve the same degree of accuracy in the final results. Moreover, the higher order elements are particularly useful when the gradient of the field variable is expected to vary rapidly.

10.15.1 Pascal Triangle

High order triangular elements can be systematically developed with the aid of the so-called Pascal triangle given in Figure 10.16. The family of finite elements generated in this matter with distribution of nodes illustrated in Figure 10.17. Note that in higher order elements, some secondary (side and/or interior) nodes are introduced in addition to the primary (corner) nodes so as to produce exactly the right number of nodes required to define the shape function of that order.

$$
\begin{array}{ccccccc}
& & & a_1 & & & & \text{Constant term, } n = 0 \\
& & a_2 x & & a_3 y & & & \text{Linear term, } n = 1 \\
& a_4 x^2 & & a_5 xy & & a_6 y^2 & & \text{Quadratic term, } n = 2 \\
a_7 x^3 & & a_8 x^2 y & & a_9 xy^2 & & a_{10} y^3 & \text{Cubic term, } n = 3
\end{array}
$$

Figure 10.16. The Pascal triangle (2D).

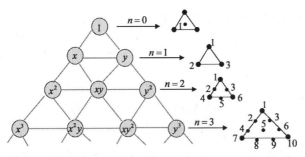

Figure 10.17. Pascal triangle (2D) and the associated polynomial basis functions degree $n = 1$ to 3.

Indeed, the Pascal triangle contains terms of the basic functions of various degrees in variable x and y. An arbitrary function $\Phi_i(x, y)$ can be approximated in an element in terms of a complete nth order polynomial as

$$\Phi_i(x, y) = \sum_{i=1}^{r} a_i \Phi_i \tag{10.264}$$

where

$$r = \frac{1}{2}(n+1)(n+2). \tag{10.265}$$

r is the number of terms in complete polynomials (also the number of nodes in the triangle). For example, for the third order ($n = 3$) or cubic (ten-node) triangular elements,

$$\Phi_m(x, y) = a_1 + a_2 x + a_3 y + a_4 x^2 + a_5 xy + a_6 y^2 + a_7 x^3 \\ + a_8 x^2 y + a_9 xy^2 + a_{10} y^3. \tag{10.266}$$

Equation (10.266) has ten coefficients, and hence the element must have ten nodes. It is also complete through the third order terms. A systematic derivation of the interpolation function a for the higher order elements involves the use of the local coordinates.

10.15.2 Local Coordinates

Now, the triangular local coordinates (η_1, η_2, η_3) are related to Cartesian coordinates (x, y) as

$$x = \eta_1 x_1 + \eta_2 x_2 + \eta_3 x_3 \tag{10.267a}$$

$$y = \eta_1 y_1 + \eta_2 y_2 + \eta_3 y_3. \tag{10.267b}$$

The local coordinates are dimensionless with values ranging from 0 to 1. Furthermore, by definition, η_i at any point within the triangle is the ratio of the perpendicular distance from the point to the side opposite to vertex i to the length of the altitude drawn from vertex i. Therefore, from Figure 10.18 the value of η_i at P, for example, is given by the ratio of the perpendicular distance d from the side opposite vertex 1 to the altitude h of that side, namely,

$$\eta_1 = \frac{d}{h}. \tag{10.268}$$

Alternatively, from Figure 10.15, η_i at P can be defined as

$$\eta_i = \frac{A_i}{A} \tag{10.269}$$

so that

$$\eta_1 + \eta_2 + \eta_3 = 1 \tag{10.270}$$

Since $A_1 + A_2 + A_3 = A$. The local coordinates η_i in equation (10.269) are also called *area coordinates*. The variation of (η_1, η_2, η_3) inside an element is shown in Figure 10.19.

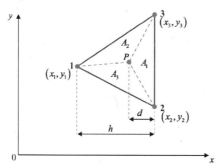

Figure 10.18. Definition of local coordinates.

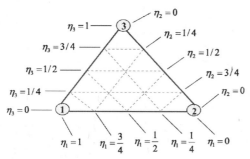

Figure 10.19. Variation of local coordinates.

Although the coordinates η_1, η_2, and η_3 are used to define a point P, only two are independent since they must satisfy equation (10.270). The inverted form of equations (10.267) and (10.268) is

$$\eta_i = \frac{1}{2A}(c_i + b_i x + a_i y) \tag{10.271}$$

where

$$a_i = x_k - x_j,$$
$$b_i = y_j - y_k,$$
$$c_i = x_j y_k - x_k y_j$$

$$A = \text{area of the triangle} = \frac{1}{2}(b_1 a_2 - b_2 a_1), \tag{10.272}$$

and (i, j, k) is an even permutation of $(1, 2, 3)$. The differentiation and integration in local coordinates are carried out using [47]:

$$\frac{\partial f}{\partial \eta_1} = a_2 \frac{\partial f}{\partial x} - b_2 \frac{\partial f}{\partial y} \tag{10.273a}$$

$$\frac{\partial f}{\partial \eta_2} = -a_1 \frac{\partial f}{\partial x} + b_1 \frac{\partial f}{\partial y} \tag{10.273b}$$

$$\frac{\partial f}{\partial x} = \frac{1}{2A}\left(b_1 \frac{\partial f}{\partial \eta_1} + b_2 \frac{\partial f}{\partial \eta_2}\right) \tag{10.273c}$$

$$\frac{\partial f}{\partial y} = \frac{1}{2A}\left(a_1 \frac{\partial f}{\partial \eta_1} + a_2 \frac{\partial f}{\partial \eta_2}\right) \tag{10.273d}$$

$$\iint f\, dS = 2A \int_0^1 \left(\int_0^{1-\eta_2} f(\eta_1, \eta_2)\, d\eta_1 \right) d\eta_2 \tag{10.273e}$$

$$\iint \eta_1^i \eta_2^j \eta_3^k\, dS = \left(\frac{i!\,j!\,k!}{(i+j+k+2)!}\right) \times 2A \tag{10.273f}$$

$$dS = 2A\, d\eta_1 d\eta_2 \tag{10.273g}$$

10.15.3 Shape Functions

Now, we may express the shape function for higher order elements in terms of local coordinates. Indeed, sometimes, it is convenient to label each point in the finite elements in Figure 10.17 with three integers, i, j, and k from which its local coordinates (η_1, η_2, η_3) can be found or vice versa. For instance, at each point P_{ijk}

$$(\eta_1, \eta_2, \eta_3) = \left(\frac{i}{n}, \frac{j}{n}, \frac{k}{n}\right). \tag{10.274}$$

Thus, if a value of Φ, say Φ_{ijk}, is prescribed at each point P_{ijk}, equation (10.264) can be expressed as

$$\Phi(\eta_1, \eta_2, \eta_3) = \sum_{i=1}^{r} \sum_{j=1}^{r-i} a_{ijk}(\eta_1, \eta_2, \eta_3) \Phi_{ijk} \tag{10.275}$$

where

$$a_l = a_{ijk} = p_i(\eta_1) p_j(\eta_2) p_k(\eta_3), \quad l = 1, 2, \ldots \tag{10.276}$$

$$p_e(\eta) = \begin{cases} \dfrac{1}{e!} \displaystyle\prod_{t=0}^{e-1} (n\eta - t), & e > 0 \\ 1, & e = 0 \end{cases} \tag{10.277}$$

and $e \in (i, j, k)$. Further, $p_e(\eta)$ may also be written as

$$p_e(\eta) = \frac{(n\eta - e + 1)}{e} \times p_{e-1}(\eta), \quad e > 0 \tag{10.278}$$

where $p_0(\eta) = 1$.

The relationships between the subscript $q \in \{1, 2, 3\}$ on η_q, $l \in \{1, 2, \ldots, r\}$ on a_l, and $e \in \{i, j, k\}$ on p_e and P_{ijk} in equations (10.276) to (10.278) are illustrated in Figure 10.20 for n ranging from 1 to 3. Furthermore, point P_{ijk} will be written as P_n for conciseness.

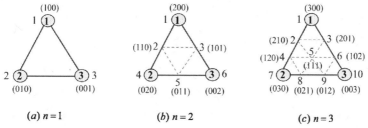

Figure 10.20. Distribution of nodes over triangles for n = 1 to 3. The triangles are in standard position.

Notice from equations (10.277) or (10.278) that

$$p_0(\eta) = 1$$
$$p_1(\eta) = n\eta$$
$$p_2(\eta) = \frac{1}{2}(n\eta - 1)n\eta \qquad (10.279)$$
$$p_3(\eta) = \frac{1}{6}(n\eta - 2)(n\eta - 1)n\eta, \text{ etc.}$$

Indeed, by substituting equation (10.279) into equation (10.276) gives the shape functions a_l for nodes $l = 1, 2, \ldots, r$, as shown in Table 10.1 for $n = 1$ to 3. In addition, observe that each a_l takes the value of 1 at node l and value 0 at all other nodes in the triangle. It can be verified by using equation (10.274) in conjunction with Figure 10.20.

Table 10.1. Polynomial Basic Functions $a_l (\eta_1, \eta_2, \eta_3)$ for First, Second, and Third

$n = 1$	$n = 2$	$n = 3$
$a_1 = \eta_1$	$a_1 = \eta_1(2\eta_1 - 1)$	$a_1 = \frac{1}{2}\eta_1(3\eta_1 - 2)(3\eta_1 - 1)$
$a_2 = \eta_2$	$a_2 = 4\eta_1\eta_2$	$a_2 = \frac{9}{2}\eta_1(3\eta_1 - 1)\eta_2$
$a_3 = \eta_3$	$a_3 = 4\eta_1\eta_3$	$a_3 = \frac{9}{2}\eta_1(3\eta_1 - 1)\eta_3$
	$a_4 = \eta_2(2\eta_2 - 1)$	$a_4 = \frac{9}{2}\eta_1(3\eta_2 - 1)\eta_2$
	$a_5 = 4\eta_2\eta_3$	$a_5 = 27\eta_1\eta_2\eta_3$
	$a_6 = \eta_3(2\eta_3 - 1)$	$a_6 = \frac{9}{2}\eta_1(3\eta_3 - 1)\eta_3$
		$a_7 = \frac{1}{2}\eta_2(3\eta_2 - 2)(3\eta_2 - 1)$
		$a_8 = \frac{9}{2}\eta_2(3\eta_2 - 1)\eta_3$
		$a_9 = \frac{9}{2}\eta_2(3\eta_3 - 1)\eta_3$
		$a_{10} = \frac{1}{2}\eta_3(3\eta_3 - 2)(3\eta_3 - 1)$

10.15.4 Fundamental Matrices

The fundamental matrices $[T]$ and $[Q]$ for triangle elements can be derived using the shape functions in Table 10.1. The matrix T is defined as

$$T_{ij} = \iint \alpha_i \alpha_j dS. \qquad (10.280)$$

From Table 10.1, we substitute α_l in equation (10.280) and apply equations (10.273f) and (10.273g) to obtain elements of T. For example, for $n = 1$,

$$T_{ij} = 2A \int_0^1 \int_0^{1-\eta_2} \eta_i \eta_j \, d\eta_1 d\eta_2. \qquad (10.281)$$

Furthermore, when $i \neq j$, T_{ij} can be written as

$$T_{ij} = \frac{2A(1!)(1!)(0!)}{4!} = \frac{A}{12}, \qquad (10.282a)$$

but, when $i = j$,

$$T_{ij} = \frac{2A(2!)}{4!} = \frac{A}{6}. \qquad (10.282b)$$

Thus,

$$T = \frac{A}{12} \begin{bmatrix} 2 & 1 & 1 \\ 1 & 2 & 1 \\ 1 & 1 & 2 \end{bmatrix}. \qquad (10.283)$$

Now, by following the same procedure, higher order T matrices can be obtained. The T matrices for orders up to $n = 3$ are tabulated in Table 10.2 where the factor A, the area of the element, has been expressed. The actual matrix elements are obtained from Table 2.2 by multiplying the tabulated numbers by A and dividing by the indicated common denominator. Indeed, the following properties of the T matrix are worth knowing:

1. T is symmetric with positive elements;
2. elements of T all add up to the area of the triangle, that is, $\sum_i^r \sum_j^r T_{ij} = A$, since by definition $\sum_{l=1}^r \alpha_l = 1$ at any point within the element;
3. elements for which the two triple subscripts from similar permutations are equal, that is, $T_{ijk,peq} = T_{ikj,peq} = T_{kij,epq} = T_{kji,eqp} = T_{jki,qep} = T_{jik,qpe}$; this should be obvious from equations (2.280) and (10.276).

As a result, the above properties are not only useful in checking the matrix; they have proved useful in saving computer time and storage.

Table 10.2. **Table of *T* Matrices for *n* = 1 to 3**

$n = 1$ Common denominator = 12

$$T = \begin{bmatrix} 2 & 1 & 1 \\ 1 & 2 & 1 \\ 1 & 1 & 2 \end{bmatrix}$$

$n = 2$ Common denominator = 180

$$T = \begin{bmatrix} 6 & 0 & 0 & -1 & -4 & -1 \\ 0 & 32 & 16 & 0 & 16 & -4 \\ 0 & 16 & 32 & -4 & 16 & 0 \\ -1 & 0 & -4 & 6 & 0 & -1 \\ -4 & 16 & 16 & 0 & 32 & 0 \\ -1 & -4 & 0 & -1 & 0 & 6 \end{bmatrix}$$

$n = 3$ Common denominator = 6720

$$T = \begin{bmatrix} 76 & 18 & 18 & 0 & 36 & 0 & 11 & 27 & 27 & 11 \\ 18 & 540 & 270 & -189 & 162 & -135 & 0 & -135 & -54 & 27 \\ 18 & 270 & 540 & -135 & 162 & -189 & 27 & -54 & -135 & 0 \\ 0 & -189 & -135 & 540 & 162 & -54 & 18 & 270 & -135 & 27 \\ 36 & 162 & 162 & 162 & 1944 & 162 & 36 & 162 & 162 & 36 \\ 0 & -135 & -189 & -54 & 162 & 540 & 27 & -135 & 270 & 18 \\ 11 & 0 & 27 & 18 & 36 & 27 & 76 & 18 & 0 & 11 \\ 27 & -135 & -54 & 270 & 162 & -135 & 18 & 540 & -189 & 0 \\ 27 & -54 & -135 & -135 & 162 & 270 & 0 & -189 & 540 & 18 \\ 11 & 27 & 0 & 27 & 36 & 18 & 11 & 0 & 18 & 76 \end{bmatrix}$$

In equation (10.227), elements of $[C]$ matrix are defined by

$$C_{ij} = \iint \left(\frac{\partial a_i}{\partial x} \frac{\partial a_j}{\partial x} + \frac{\partial a_i}{\partial y} \frac{\partial a_j}{\partial y} \right) dS. \tag{10.284}$$

By applying equations (10.273a) to (10.273d) to equation (2.147), it can be shown that:

$$C_{ij} = \frac{1}{2A} \sum_{q=1}^{3} \cot \theta_q \iint \left(\frac{\partial a_i}{\partial \eta_{q+1}} - \frac{\partial a_i}{\partial \eta_{q-1}} \right) \left(\frac{\partial a_j}{\partial \eta_{q+1}} - \frac{\partial a_j}{\partial \eta_{q-1}} \right) dS$$

or

$$C_{ij} = \sum_{q=1}^{3} Q_{ij}^{(q)} \cot \theta_q \tag{10.285}$$

where θ_q is the include angle of vertex $q \in \{1, 2, 3\}$ of the triangle and

$$Q_{ij}^{(q)} = \iint \left(\frac{\partial a_i}{\partial \eta_{q+1}} - \frac{\partial a_i}{\partial \eta_{q-1}} \right) \left(\frac{\partial a_j}{\partial \eta_{q+1}} - \frac{\partial a_j}{\partial \eta_{q-1}} \right) d\eta_1 d\eta_2. \quad (10.286)$$

It is clear that matrix C depends on the triangle shape, whereas the matrices $Q^{(q)}$ do not. The $Q^{(1)}$ matrices for $n = 1$ to 3 are tabulated in Table 10.3.
The following properties of Q matrices should be noted as:

1. they are symmetric;
2. the row and column sums of any Q matrix are zero, that is, $\sum_{i=1}^{r} Q_{ij}^{(q)} = 0 = \sum_{j=1}^{r} Q_{ij}^{(q)}$ so that the C matrix is singular.

$Q^{(2)}$ and $Q^{(3)}$ are easily obtained from $Q^{(1)}$ by row and column permutations so that the matrix C for any triangular element is constructed easily if $Q^{(1)}$ is known.

Table 10.3. **Table of Q Matrices for $n = 1$ to 3**

$n = 1$ Common denominator $= 2$

$$Q = \begin{bmatrix} 0 & 0 & 0 \\ 0 & 1 & -1 \\ 0 & -1 & 1 \end{bmatrix}$$

$n = 2$ Common denominator $= 6$

$$Q = \begin{bmatrix} 0 & 0 & 0 & 0 & 0 & 0 \\ 0 & 8 & -8 & 0 & 0 & 0 \\ 0 & -8 & 8 & 0 & 0 & 0 \\ 0 & 0 & 0 & 3 & -4 & 1 \\ 0 & 0 & 0 & -4 & 8 & -4 \\ 0 & 0 & 0 & 1 & -4 & 3 \end{bmatrix}$$

$n = 3$ Common denominator $= 80$

$$Q = \begin{bmatrix} 0 & 0 & 0 & 0 & 0 & 0 & 0 & 0 & 0 & 0 \\ 0 & 135 & -135 & -27 & 0 & 27 & 3 & 0 & 0 & -3 \\ 0 & -135 & 135 & 27 & 0 & -27 & -3 & 0 & 0 & 3 \\ 0 & -27 & 27 & 135 & -162 & 27 & 3 & 0 & 0 & -3 \\ 0 & 0 & 0 & -162 & 324 & -162 & 0 & 0 & 0 & 0 \\ 0 & 27 & -27 & 27 & -162 & 135 & -3 & 0 & 0 & 3 \\ 0 & 3 & -3 & 3 & 0 & -3 & 34 & -54 & 27 & -7 \\ 0 & 0 & 0 & 0 & 0 & 0 & -54 & 135 & -108 & 27 \\ 0 & 0 & 0 & 0 & 0 & 0 & 27 & -108 & 135 & -54 \\ 0 & -3 & 3 & -3 & 0 & 3 & -7 & 27 & -54 & 34 \end{bmatrix}$$

For example, for $n = 1$, the rotation matrix is basically derived from Figure 10.21 as

$$R = \begin{bmatrix} 0 & 0 & 1 \\ 1 & 0 & 0 \\ 0 & 1 & 0 \end{bmatrix} \qquad (10.287)$$

where $R_{ij} = 1$ node i is replaced by node j after one counterclockwise rotation, or $R_{ij} = 0$ otherwise.

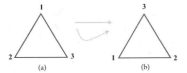

Figure 10.21. One counterclockwise rotation of the triangle in (a) gives the triangle in (b).

Moreover, Table 10.4 presents the R matrices for $n = 1$ to 3. Note that each row or column of R has only one nonzero element since R is essentially a unit matrix with rearranged elements.

Table 10.4. Table of R Matrices for $n = 1$ to 3

$n = 1$,
$$R = \begin{bmatrix} 0 & 0 & 1 \\ 1 & 0 & 0 \\ 0 & 1 & 0 \end{bmatrix}$$

$n = 2$,
$$R = \begin{bmatrix} 0 & 0 & 0 & 0 & 0 & 1 \\ 0 & 0 & 1 & 0 & 0 & 0 \\ 0 & 0 & 0 & 0 & 1 & 0 \\ 1 & 0 & 0 & 0 & 0 & 0 \\ 0 & 1 & 0 & 0 & 0 & 0 \\ 0 & 0 & 0 & 1 & 0 & 0 \end{bmatrix}$$

$n = 1$,
$$R = \begin{bmatrix} 0 & 0 & 0 & 0 & 0 & 0 & 0 & 0 & 0 & 1 \\ 0 & 0 & 0 & 0 & 0 & 1 & 0 & 0 & 0 & 0 \\ 0 & 0 & 0 & 0 & 0 & 0 & 0 & 0 & 1 & 0 \\ 0 & 0 & 1 & 0 & 0 & 0 & 0 & 0 & 0 & 0 \\ 0 & 0 & 0 & 0 & 1 & 0 & 0 & 0 & 0 & 0 \\ 0 & 0 & 0 & 0 & 0 & 0 & 0 & 1 & 0 & 0 \\ 1 & 0 & 0 & 0 & 0 & 0 & 0 & 0 & 0 & 0 \\ 0 & 1 & 0 & 0 & 0 & 0 & 0 & 0 & 0 & 0 \\ 0 & 0 & 0 & 1 & 0 & 0 & 0 & 0 & 0 & 0 \\ 0 & 0 & 0 & 0 & 0 & 0 & 1 & 0 & 0 & 0 \end{bmatrix}$$

Now, once the R is known, we can obtain $Q^{(2)}$ and $Q^{(3)}$ as

$$Q^{(2)} = RQ^{(1)}R^t \qquad (10.288a)$$

$$Q^{(3)} = RQ^{(2)}R^t \qquad (10.288b)$$

where R^t is the transpose of R.

10.16 THREE-DIMENSIONAL ELEMENT

In this section, we will discuss the finite element analysis of Helmholtz's equation in three dimensions, i.e.,

$$\nabla^2 \Phi + k^2 \Phi = g. \qquad (10.289)$$

First, we divide the solution region into tetrahedral or hexahedral (rectangular prism) elements as in Figure 10.22.

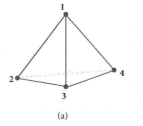
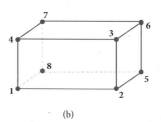

Figure 10.22. Three-dimensional elements: (a) Four-node or linear-order tetrahedral, (b) eight-node or linear-order hexahedral.

Now, assuming a four-node tetrahedral element, the function Φ is represented within element by

$$\Phi_m = a + bx + cy + dz. \qquad (10.290)$$

The same applies to the function g. Since equation (10.290) must be satisfied at the four nodes of the tetrahedral elements,

$$\Phi_{mi} = a + bx_i + cy_i + dz_i, \quad i = 1, 2, 3, 4. \qquad (10.291)$$

Therefore, we have four simultaneous equations with the potentials V_{m1}, V_{m2}, V_{m3}, and V_{m4} at nodes 1, 2, 3, and 4, respectively, i.e.,

$$\begin{bmatrix} V_{m1} \\ V_{m2} \\ V_{m3} \\ V_{m4} \end{bmatrix} = \begin{bmatrix} 1 & x_1 & y_1 & z_1 \\ 1 & x_2 & y_2 & z_2 \\ 1 & x_3 & y_3 & z_3 \\ 1 & x_4 & y_4 & z_4 \end{bmatrix} \begin{bmatrix} a \\ b \\ c \\ d \end{bmatrix}. \qquad (10.292)$$

The coefficients a, b, c, and d are determined from equation (10.291)

$$\begin{bmatrix} a \\ b \\ c \\ d \end{bmatrix} = \begin{bmatrix} 1 & x_1 & y_1 & z_1 \\ 1 & x_2 & y_2 & z_2 \\ 1 & x_3 & y_3 & z_3 \\ 1 & x_4 & y_4 & z_4 \end{bmatrix}^{-1} \begin{bmatrix} V_{m1} \\ V_{m2} \\ V_{m3} \\ V_{m4} \end{bmatrix}. \qquad (10.293)$$

The determinant of the system of equations is

$$\det = \begin{vmatrix} 1 & x_1 & y_1 & z_1 \\ 1 & x_2 & y_2 & z_2 \\ 1 & x_3 & y_3 & z_3 \\ 1 & x_4 & y_4 & z_4 \end{vmatrix} = 6v, \qquad (10.294)$$

where v is the volume of the tetrahedron. By finding a, b, c, and d, we can express Φ_m as,

$$\Phi_m = \sum_{i=1}^{4} a_i(x,y) \Phi_{mi} \qquad (10.295)$$

where

$$a_1 = \frac{1}{6v} \begin{vmatrix} 1 & x & y & z \\ 1 & x_2 & y_2 & z_2 \\ 1 & x_3 & y_3 & z_3 \\ 1 & x_4 & y_4 & z_4 \end{vmatrix}, \qquad (10.296a)$$

$$a_2 = \frac{1}{6v} \begin{vmatrix} 1 & x_1 & y_1 & z_1 \\ 1 & x & y & z \\ 1 & x_3 & y_3 & z_3 \\ 1 & x_4 & y_4 & z_4 \end{vmatrix}, \qquad (10.296b)$$

$$a_3 = \frac{1}{6v} \begin{vmatrix} 1 & x_1 & y_1 & z_1 \\ 1 & x_2 & y_2 & z_2 \\ 1 & x & y & z \\ 1 & x_4 & y_4 & z_4 \end{vmatrix}, \qquad (10.296c)$$

$$a_4 = \frac{1}{6v} \begin{vmatrix} 1 & x_1 & y_1 & z_1 \\ 1 & x_2 & y_2 & z_2 \\ 1 & x_3 & y_3 & z_3 \\ 1 & x & y & z \end{vmatrix}. \qquad (10.296d)$$

Indeed, for higher order approximation, the matrices for as become large in size and we resort to local coordinates. The existence of integration equations for local coordinates can simplify the evaluation of the fundamental matrices T and Q.

Now, for the tetrahedral element, the local coordinates are η_1, η_2, η_3, and η_4, each perpendicular to a side. They are defined at a given point as the ratio of the distance from that point to the appropriate apex to the perpendicular distance from the side to the opposite apex. In addition, they can be interpreted as volume ratios, that is, at a point P

$$\eta_i = \frac{v_i}{v} \tag{10.297}$$

where v_i is the volume bound by P and face i. It is evident that

$$\sum_{i=1}^{4} \eta_i = 1. \tag{10.298}$$

Note that, the following properties are useful in evaluating integration involving local coordinates:

$$dv = 6v\, d\eta_1 d\eta_2 d\eta_3, \tag{10.299a}$$

$$\iiint f dv = 6v \int_0^1 \left(\int_0^{1-\eta_3} \left(\int_0^{1-\eta_2-\eta_3} f\, d\eta_1 \right) d\eta_2 \right) d\eta_3, \tag{10.299b}$$

$$\iiint \eta_1^i \eta_2^j \eta_3^k \eta_4^l \, dv = \frac{i!\,j!\,k!\,l!}{(i+j+k+l+3)!} \times 6v. \tag{10.299c}$$

In terms of the local coordinates, an arbitrary function $\Phi(x, y)$ can be approximated within an element in terms of a complete nth order polynomial as

$$\Phi_m(x, y) = \sum_{i=1}^{r} a_i(x, y) \Phi_{mi} \tag{10.300}$$

where $r = \frac{1}{6}(n+1)(n+2)(n+3)$ is the number of nodes in the tetrahedron or number of terms in the polynomial. The terms in a complete three-dimensional polynomial may be arrayed for polynomial basic functions degree $n = 1$ to 3 as shown in Figure 10.23.

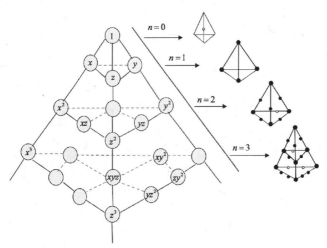

Figure 10.23. Pascal tetrahedral (3D) and the associated polynomial basic functions degree $n = 1$ to 3.

Each point in the tetrahedral element is represented by four integers, i, j, k, and l which can be used to determine the local coordinates $(\eta_1, \eta_2, \eta_3, \eta_4)$. That is at point P_{ijkl},

$$(\eta_1, \eta_2, \eta_3, \eta_4) = \left(\frac{i}{n}, \frac{j}{n}, \frac{k}{n}, \frac{l}{n}\right). \tag{10.301}$$

Thus, at each node,

$$a_q = a_{ijkl} = p_i(\eta_1) p_j(\eta_2) p_k(\eta_3) p_l(\eta_4) \tag{10.302}$$

where $q = 1, 2, \ldots, r$ and p_e is defined in equation (10.277) or (10.278). Figure 10.22 illustrates the relationship between the node numbers q and $ijkl$ for the second order tetrahedron ($n = 2$). The shape functions obtained by substituting equation (10.277) into (10.293) are presented in Table 2.5 for $n = 3$.

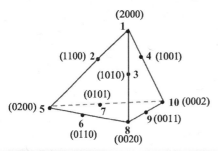

Figure 10.24. Numbering scheme for second-order tetrahedral.

Table 10.5. Polynomial Basic Functions $a_q(\eta_1, \eta_2, \eta_3)$ for $n = 1$ to 3

$n=1$	$n=2$	$n=3$
$a_1 = \eta_1$	$a_1 = \eta_1(2\eta_1 - 1)$	$a_1 = \dfrac{1}{2}\eta_1(3\eta_1 - 2)(3\eta_1 - 1)$
$a_2 = \eta_2$	$a_2 = 4\eta_1\eta_2$	$a_2 = \dfrac{9}{2}\eta_1(3\eta_1 - 1)\eta_2$
$a_3 = \eta_3$	$a_3 = 4\eta_1\eta_3$	$a_3 = \dfrac{9}{2}\eta_1(3\eta_1 - 1)\eta_3$
$a_4 = \eta_4$	$a_4 = 4\eta_1\eta_4$	$a_4 = \dfrac{9}{2}\eta_1(3\eta_1 - 1)\eta_4$
	$a_5 = \eta_2(2\eta_2 - 1)$	$a_5 = \dfrac{9}{2}\eta_1(3\eta_3 - 1)\eta_2$
	$a_6 = 4\eta_2\eta_3$	$a_6 = 27\eta_1\eta_2\eta_3$
	$a_7 = 4\eta_2\eta_4$	$a_7 = 27\eta_1\eta_2\eta_4$
	$a_8 = \eta_2(2\eta_3 - 1)$	$a_8 = \dfrac{9}{2}\eta_1(3\eta_3 - 1)\eta_3$
	$a_9 = 4\eta_3\eta_4$	$a_9 = 27\eta_1\eta_3\eta_4$
	$a_{10} = \eta_4(2\eta_4 - 1)$	$a_{10} = \dfrac{9}{2}\eta_1(3\eta_4 - 1)\eta_4$
		$a_{11} = \dfrac{1}{2}\eta_2(3\eta_2 - 1)(3\eta_2 - 2)$
		$a_{12} = \dfrac{9}{2}\eta_2(3\eta_2 - 1)\eta_3$
		$a_{13} = \dfrac{9}{2}\eta_2(3\eta_2 - 1)\eta_4$
		$a_{14} = \dfrac{9}{2}\eta_2(3\eta_3 - 1)\eta_3$
		$a_{15} = 27\eta_2\eta_3\eta_4$
		$a_{16} = \dfrac{9}{2}\eta_2(3\eta_3 - 1)\eta_3$
		$a_{17} = \dfrac{1}{2}\eta_3(3\eta_3 - 1)(3\eta_3 - 2)$
		$a_{18} = \dfrac{9}{2}\eta_3(3\eta_3 - 1)\eta_4$
		$a_{19} = \dfrac{9}{2}\eta_3(3\eta_4 - 1)\eta_4$
		$a_{20} = \dfrac{1}{2}\eta_4(3\eta_4 - 1)(3\eta_4 - 2)$

The fundamental matrices $[T]$ and $[Q]$ are involved triple integration. For Helmholtz equation, for example, equation (10.250) applies, namely,

$$\left[C_{ff} - k^2 T_{ff} \right] \Phi_f = 0 \qquad (10.303)$$

except that

$$C_{ij}^{(m)} = \int_v \nabla a_i \cdot \nabla a_j \, dv = \iint_v \left(\frac{\partial a_i}{\partial x} \frac{\partial a_j}{\partial x} + \frac{\partial a_i}{\partial y} \frac{\partial a_j}{\partial y} + \frac{\partial a_i}{\partial z} \frac{\partial a_j}{\partial z} \right) dv, \qquad (10.304)$$

$$T_{ij}^{(m)} = \int_v a_i a_j \, dv = 6v \iiint a_i a_j \, d\eta_1 d\eta_2 d\eta_3. \qquad (10.305)$$

10.17 FINITE ELEMENT METHODS FOR EXTERNAL PROBLEMS

We can apply the finite element to exterior or unbounded problems such as open-type transmission lines (e.g., microstrip). They pose certain difficulties. In this section, we will consider three common approaches: first, the infinite element method; second, the boundary element method; and third, the absorbing boundary condition.

10.17.1 Infinite Element Method

Let us consider the solution region shown in Figure 10.25. We can divide the entire domain into a near field region, which is bounded, and a far field region, which is unbounded. The near field region is divided into finite triangular elements as usual, while the far field region is divided into infinite elements. Knowing that, each infinite element shares two nodes with a finite element. We mainly will be focusing on the infinite elements.

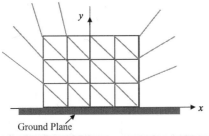

Figure 10.25. Division of solution region into finite and infinite elements.

Now, consider the infinite element in Figure 10.26 with nodes 1 and 2 and radial sides intersecting at point (x_0, y_0).

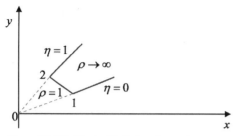

Figure 10.26. Typical infinite element.

We can relate triangular polar coordinates (ρ, η) to the global Cartesian coordinates (x, y) as:

$$x = x_0 + \rho((x_1 - x_0) + \eta(x_2 - x_1))$$
$$y = y_0 + \rho((y_1 - y_0) + \eta(y_2 - y_1))$$
(10.306)

where $1 \leq \rho < \infty$, $0 \leq \eta \leq 1$. The potential distribution within the element is approximated by a linear variation as

$$V = \sum_{i=1}^{2} a_i V_i \qquad (10.307)$$

or

$$V = \frac{1}{\rho}(V_1(1-\eta) + V_2 \eta)$$

where V_1 and V_2 are potentials at nodes 1 and 2 of the infinite elements, a_1 and a_2 are the interpolation or shape functions, that is,

$$a_1 = \frac{1-\eta}{\rho}, \quad a_2 = \frac{\eta}{\rho}. \qquad (10.308)$$

Moreover, the infinite element is compatible with the ordinary first order finite element and satisfies the boundary condition at infinity. Indeed, with the shape functions in equation (10.308), we can obtain the $[C^{(m)}]$ and $[T^{(m)}]$ matrices. We obtain solution for the exterior problem by using a standard finite element program with the $[C^{(m)}]$ and $[T^{(m)}]$ matrices of the infinite elements added to the $[C]$ and $[T]$ matrices of the near field region.

10.17.2 Boundary Element Method

The boundary element method is a finite element approach for handling exterior problems. It involves obtaining the integral equation formulation of the boundary value problem, and solving this by a discretization procedure similar to that used in regular finite element analysis. But, since the boundary element method is based on the boundary integral equivalent to the governing differential equation, only the surface of the problem domain needs to be modeled. Moreover, for the dimension of 2D problems, the boundary elements are taken to be straight line segments, whereas for 3D problems, they are taken as triangular elements.

10.17.3 Absorbing Boundary Conditions

To apply the finite element approach to open region problems, an artificial boundary is introduced in order to bound the region and limit the number of unknowns to a manageable size. It can be expected that, as the boundary approaches infinity, the approximate solution tends to the exact one. However, the closer the boundary to the modeled object, the less computer memory is required. To avoid the error caused by this truncation, an *absorbing boundary condition* (ABC) can be imposed on the artificial boundary S, as typically portrayed in Figure 10.27.

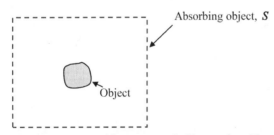

Figure 10.27. An object surrounded by an absorbing boundary.

Indeed, the ABC minimizes the nonphysical reflections from the boundary. The major challenge of these ABCs is to bring the truncation boundary as close as possible to the object without sacrificing accuracy and to absorb the outgoing waves with little or no reflection.

$$\Phi(r,\theta,\phi) = \frac{e^{-jkr}}{kr} \sum_{i=0}^{\infty} \frac{F_i(\theta,\phi)}{(kr)^i}. \tag{10.309}$$

Furthermore, the sequence of BGT operators can be obtained by the recursion relation, i.e.,

$$B_1 = \left(\frac{\partial}{\partial r} + jk + \frac{1}{r}\right)$$

$$B_m = \left(\frac{\partial}{\partial r} + jk + \frac{2m-1}{r}\right) B_{m-1}, \quad m = 2, 3, \dots \quad (10.310)$$

Now, since Φ satisfies the higher-order radiation condition

$$B_m \Phi = O\left(\frac{1}{(r)^{2m+1}}\right). \quad (10.311)$$

By imposing the mth-order boundary condition

$$B_m \Phi = 0, \quad \text{on } S \quad (10.312)$$

will compel the solution Φ to match the first $2m$ terms of the expansion in equation (10.309). Equation (10.312) along with other appropriate equations is solved for Φ using the FEM.

10.18 MODELING AND SIMULATION OF SHIELDED MICROSTRIP LINES WITH COMSOL MULTIPHYSICS

In today's fast-paced research and development culture, simulation power gives you the competitive edge. COMSOL Multiphysics delivers the ideal tool to build simulations that accurately replicate the important characteristics of your designs. Its unparalleled ability to include all relevant physical effects that exist in the real world is known as multiphysics. This approach delivers results—tangible results that save precious development time and spark innovation. COMSOL Multiphysics brings you this remarkable technology in an easy-to-use, intuitive interface, making it accessible to all engineers including designers, analysts, and researchers.

Today, electromagnetic propagation on multiple parallel transmission lines has been a very attractive area in computational electromagnetics. Multiple parallel transmission lines have been successfully applied and used by designers in compact packaging, semiconductor device, high speed interconnecting buses, monolithic integrated circuits, and other applications. Microstrip lines are the most commonly used in all planar circuits despite the frequencies ranges of the applied signals. Microstrip lines are the most commonly used transmission lines at high frequencies. Quasi-static analysis of microstrip lines involves evaluating them as parallel plates transmission lines, supporting a pure "TEM" mode. Development in microwave circuits using rectangular coaxial lines as a transmission medium has been improving over the past decades. Reid and Webster used rectangular coaxial transmission lines to fabricate a 60 GHz branch line coupler. The finite difference time domain method has been used for analyzing a satellite beamforming network consisting of rectangular coaxial line.

Advances in microwave solid-state devices have stimulated interest in the integration of microwave circuits. Today, microstrip transmission lines have attracted great attention and interest in microwave integrated circuit applications. This creates the need for accurate modeling and simulation of microstrip transmission lines. Due to the difficulties associated with analytical methods for calculating the capacitance of shielded microstrip transmission lines, other methods have been applied. Such methods include finite difference technique, extrapolation, point-matching method, boundary element method, spectral-space domain method, finite element method, conformal mapping method, transverse modal analysis, and mode-matching method.

In this book, we consider systems of rectangular coaxial lines as well as single-strip, double-strip, three-strip, six-strip, and eight-strip (multiconductor) shielded microstrip lines. Using COMSOL, a finite element package, we performed the simulation of these systems of microstrip lines. We compared the results with other methods and found them to be in good agreement.

The rectangular coaxial line consists of a two-conductor transmission system along which TEM wave propagates. The characteristic impedance of such a lossless line is given by

$$Z = \sqrt{\frac{L}{C}} = \frac{1}{cC} \tag{10.313}$$

where

Z = characteristic impedance of the line
L = inductance per unit length of the line
C = capacitance per unit length of the line
$c = 3 \times 10^8$ m/s (the speed of light in vacuum).

As shown in Figure 10.28, a rectangular coaxial line consists of inner and outer rectangular conductors with a dielectric material separating them.

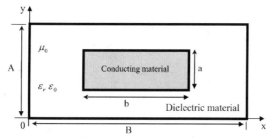

Figure 10.28. Cross-section of the rectangular coaxial line.

Using COMSOL for each type of the rectangular lines involves taking the following steps:

1. Develop the geometry of the inner and outer conductors, such as shown in Figure 10.28(a).

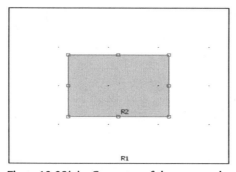

Figure 10.28(a). Geometry of the rectangular coaxial line model.

2. Select both conductors/rectangle and take the difference.
3. We select the relative permittivity as 1 for the difference in Step 2. For the boundary, we select the outer conductor as ground and inner conductor as port.
4. We generate the finite element mesh as in Figure 10.29.

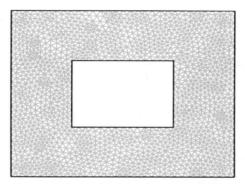

Figure 10.29. Mesh of the rectangular coaxial line.

5. We solve the model and obtain the potential shown in Figure 10.30.

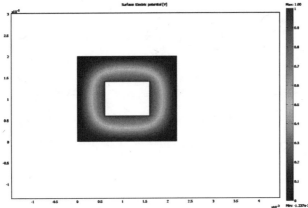

Figure 10.30. 2D for the potential distribution of the rectangular coaxial line.

6. As post-processing, we select Point Evaluation and choose capacitance element 11 to find the capacitance per unit length of the line.

We now consider the following three models.

10.18.1 Rectangular Cross-Section Transmission Line

For COMSOL, we use the following values.
Dielectric material:

$$\varepsilon_r = 1, \mu_r = 1, \sigma = 0 \text{ S/m (air)}$$

Conducting material:

$$\varepsilon_r = 1, \mu_r = 1, \sigma = 5.8 \times 10^7 \text{ S/m (copper)}$$

where

ε_0 = permittivity of free space = $\dfrac{1}{36\pi} \times 10^{-9} = 8.854 \times 10^{-12}$ F/m
ε_r = dielectric constant
μ_r = relative permeability
μ_0 = permeability of free space = $4\pi \times 10^{-7} = 1.257 \times 10^{-6}$ H/m
σ = conductivity of the conductor
a = width of the inner conductor = 1 mm
b = height of the inner conductor = 0.8 mm
A = width of the outer conductor = 2.2 mm
B = height of the outer conductor = 2 mm

From the COMSOL model, we obtained the capacitance per unit length (based on the dimensions given above) as 72.94 pF/m. Using the finite difference (FD) method, we obtained the capacitance per unit length of the line as 71.51 pF/m. Table 10.6 shows the comparison of the characteristic impedance using equation (10.313) of several models. It is evident from the table that the results are very close.

Table 10.6. Comparison of Characteristic Impedance Values of Rectangular Coaxial Line

Name	Z_0
Zheng	45.789
Chen	45.759
Costamagna and Fanni	45.767
Lau	45.778
Finite difference (FD)	46.612
COMSOL	45.70

10.18.2 Square Cross-Section Transmission Line

This is only a special case of the rectangular line. We used the same values for the dielectric and conducting materials. We used the following dimensions for the line.

a = width of the inner conductor = 2 mm
b = height of the inner conductor = 2 mm
A = width of the outer conductor = 4 mm
B = height of the outer conductor = 4 mm

From the COMSOL model, we obtained the capacitance per unit length as 90.696 pF/m. Using the finite difference (FD) method, we obtained the capacitance per unit length of the line as 90.714 pF/m. Table 10.7 presents the comparison of the characteristic impedance of several models. It is evident from the table that the results are in good agreement.

Table 10.7. Comparison of Characteristic Impedance Values of Square Coaxial Line

Name	Z_0
Zheng	36.79
Lau	36.81
Cockcroft	36.80
Bowan	36.81
Green	36.58

(Continued)

Table 10.7. Comparison of Characteristic Impedance Values of Square Coaxial Line (Continued)

Name	Z_0
Ivanov and Djankov	36.97
Costamagna and Fanni	36.81
Riblet	36.80
Finite difference (FD)	36.75
COMSOL	36.75

10.18.3 Rectangular Line with Diamondwise Structure

The geometry of the cross-section of this line is shown in Figure 10.31. The same dielectric and conducting materials used for the rectangular line are used for this line.

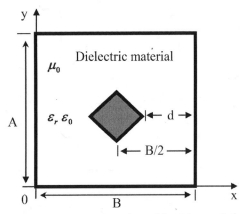

Figure 10.31. Cross-section of the Diamondwise (or Rhombus) structure with 45° offset angle.

The following values are used for the COMSOL model of the line.

d = 1 mm
A = width of the outer conductor = 4 mm
B = height of the outer conductor = 4 mm

For the COMSOL model, we obtained the capacitance per unit line as 57.393 pF/m.

Table 10.8 displays the comparison of the characteristic impedance of several models. It is evident from the table that the results are in good agreement.

Table 10.8. Comparison of Characteristic Impedance Values of Diamondwise Structure

Name	Z_0
Zheng et al.	56.742
Bowan	56.745
Riblet	56.745
COMSOL	58.079

10.18.4 A Single-Strip Shielded Transmission Line

Figure 10.32 presents the cross-section of a single-strip shielded transmission line.

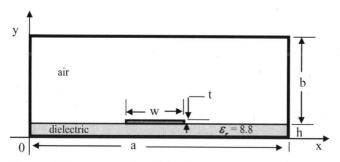

Figure 10.32. Cross-section of the Single-strip Shielded Transmission Line.

The following parameters are used in modeling the line. The characteristic impedance of such a lossless line is given by

$$Z = \frac{1}{c\sqrt{CC_o}} \quad (10.314)$$

where

Z = characteristic impedance of the line
C_o = capacitance per unit length of the line when the substrate is replaced with air
C = capacitance per unit length of the line when the substrate is in place
$c = 3 \times 10^8$ m/s (the speed of light in vacuum).

For COMSOL, the simulation was done twice on Figure 10.32 (to find C_o and C) using the following values.
Air:

$$\varepsilon_r = 1, \mu_r = 1, \sigma = 0 \text{ S/m}$$

Dielectric material:

$$\varepsilon_r = 8.8, \mu_r = 1, \sigma = 0 \text{ S/m}$$

Conducting material:

$$\varepsilon_r = 1, \mu_r = 1, \sigma = 5.8 \times 10^7 \text{ S/m (copper)}$$

w = width of the inner conductor = 1 mm
t = height of the inner conductor = 0.1×10^{-4} m
h = height of dielectric material = 1 mm
a = width of the outer conductor = 19 mm
b = height of the air-filled region = 9 mm

Using COMSOL for modeling and simulation of the lines involves taking the following steps:

1. Develop the geometry of the line, such as shown in Figure 10.33.

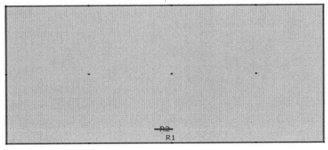

Figure 10.33. Geometry of a Single-strip Shielded Transmission Line at Air.

2. We take the difference between the conductor and dielectric material.
3. We select the relative permittivity as 1 for the difference in Step 2.
4. For the boundary, we select the outer conductor as ground and inner conductor as port.
5. We generate the finite element mesh, and then we solve the model and obtain the potential.
6. As post-processing, we select Point Evaluation and choose capacitance element 11 to find the capacitance per unit length of the line.
7. We add a dielectric region under the inner conductor with relative permittivity as 8.8, as in Figure 10.33. Then we take the same steps from 3 to 6 to generate the mesh as in Figure 10.34 and the potential distribution as in Figure 10.35.

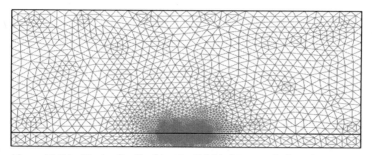

Figure 10.34. Mesh of a Single-strip Shielded Transmission Line.

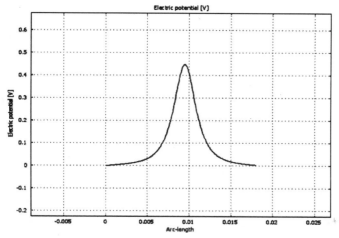

Figure 10.35. The potential distribution along y = 0.002.

Table 10.9 shows the comparison between our method using COMSOL and other methods. It is evident that the results are very close.

Table 10.9. Comparison of Capacitance Values for a Single-strip Shielded Transmission Line

Methods	C_0 (pF/m)	C (pF/m)
Finite difference method	26.79	1405.2
Extrapolation	26.88	1393.6
Analytical derivation	27.00	1400.9
COMSOL	26.87	1574.0

10.19 MULTISTRIP TRANSMISSION LINES

Recently, with the advent of integrated circuit technology, the coupled microstrip transmission lines consisting of multiple conductors embedded in a multilayer dielectric medium have led to a new class of microwave networks. Multiconductor transmission lines have been utilized as filters in the microwave region which make it interesting in various circuit components. For coupled multiconductor microstrip lines, it is convenient to write:

$$Q_i = \sum_{j=1}^{m} C_{sij} V_j \quad (i = 1, 2, \ldots, m) \tag{10.315}$$

where Q_i is the charge per unit length, V_j is the voltage of jth conductor with reference to the ground plane, C_{sij} is the short circuit capacitance between ith conductor and jth conductor. The short circuit capacitances can be obtained either from measurement or from numerical computation. From the short circuit capacitances, we obtain

$$C_{ii} = \sum_{j=1}^{m} C_{sij} \tag{10.316}$$

where C_{ii} is the capacitance per unit length between the ith conductor and the ground plane. Also,

$$C_{ij} = -C_{sij}, \quad j \neq i \tag{10.317}$$

where C_{ij} is the coupling capacitance per unit length between the ith conductor and jth conductor. The coupling capacitances are illustrated in Figure 10.36.

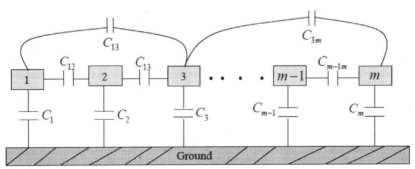

Figure 10.36. The Per-unit Length Capacitances of a General *m*-conductor Transmission Line.

For m-strip line, the per-unit-length capacitance matrix is given by

$$C = \begin{bmatrix} C_{11} & -C_{12} & \cdots & -C_{1m} \\ -C_{21} & C_{22} & \cdots & -C_{2m} \\ \vdots & \vdots & & \vdots \\ -C_{m1} & -C_{m2} & \cdots & C_{mm} \end{bmatrix}. \quad (10.318)$$

Also, we can determine the characteristic impedance matrix for m-strip line by using

$$Z_o = \begin{bmatrix} Z_{11} & Z_{12} & \cdots & Z_{1m} \\ Z_{21} & Z_{22} & \cdots & Z_{2m} \\ \vdots & \vdots & & \vdots \\ Z_{m1} & Z_{m2} & \cdots & Z_{mm} \end{bmatrix} \quad (10.319)$$

where Z_o is the characteristic impedance per unit length.

Using COMSOL for modeling and simulation of the lines involves taking the following steps:

1. Develop the geometry of the line.
2. We take the difference between the conductor and dielectric material.
3. We select the relative permittivity as 1 for the difference in Step 2.
4. We add a dielectric region under the inner conductors with specified relative permittivity.
5. For the boundary, we select the outer conductor as ground and the inner conductors as ports.
6. We generate the finite element mesh, and then we solve the model.
7. As post-processing, we select Point Evaluation and choose capacitance elements to find the coupling capacitance per unit length of the line.

These steps were taken for the following four cases.

10.19.1 Double-strip Shielded Transmission Line

Figure 10.37 presents the cross-section of double-strip shielded transmission line, which consists of two inner conductors.

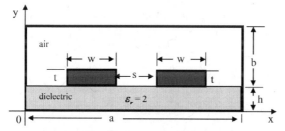

Figure 10.37. Cross-section of the Double-strip Shielded Transmission Line.

For COMSOL, the simulation was done twice on Figure 10.36 (one for C_o and other for C) using the following values.
Air:

$$\varepsilon_r = 1, \mu_r = 1, \sigma = 0 \text{ S/m}$$

Dielectric material:

$$\varepsilon_r = 2, \mu_r = 1, \sigma = 0 \text{ S/m}$$

Conducting material:

$$\varepsilon_r = 1, \mu_r = 1, \sigma = 5.8 \times 10^7 \text{ S/m (copper)}$$

For the geometry (see Figure 10.37), we followed the following values:

w = width of each of the inner conductors = 3 mm
t = height (or thickness) of the inner conductors = 1 mm
s = distance between the inner conductors = 2 mm
h = height of dielectric material = 1 mm
a = width of the outer conductor = 11 mm
b = height of the air-filled region = 2.7 mm

From the COMSOL model, the simulation was done twice, one for the case in which the line is air-filled (the dielectric was replaced by air) and the other case in which the dielectric is in place as shown in Figure 10.37. Figure 10.38 shows the finite element mesh while Figure 10.39 depicts the potential distribution for the dielectric case. The potential distribution for $y = 1$ mm is portrayed in Figure 10.10.

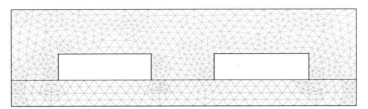

Figure 10.38. Mesh the of Double-strip Shielded Transmission Line.

Figure 10.39. Potential distribution.

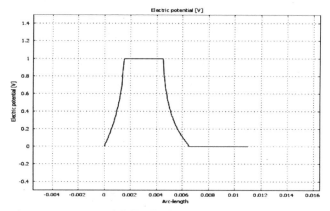

Figure 10.40. Potential distribution at y =1 mm.

We obtained the capacitances per unit length (C_o and C) by taking steps enumerated above for the single-strip transmission line. The results are shown in Table 10.9. Table 10.10 is for the case which the line is air-filled, i.e., the dielectric in Figure 10.37 is replaced by air. Table 10.11 is for the case in which the dielectric is in place. The results in Table 10.11 are compared with other methods and found to be close.

Table 10.10. Capacitance Values for Double-strip Air-filled Shielded Transmission Line

Methods	$C_{11} = C_{22}$ (pF/m)	$C_{12} = C_{21}$ (pF/m)
COMSOL	72.9	−4.591

Table 10.11. Comparison of Capacitance Values for Double-strip Shielded Transmission Line Shown in Figure 10.36

Methods	$C_{11} = C_{22}$ (pF/m)	$C_{12} = C_{21}$ (pF/m)
Spectral-space domain method	108.1	−4.571
Finite element method	109.1	−4.712
Point-matching method	108.8	−4.683
COMSOL	108.5	−4.618

10.19.2 Three-strip Line

Figure 10.40(a) shows the cross-section for three-strip transmission line. For COMSOL, the simulation was done twice on Figure 10.40 (one for C_o and other for C) using the following values:
Air:

$$\varepsilon_r = 1, \mu_r = 1, \sigma = 0 \text{ S/m}$$

Dielectric material:

$$\varepsilon_r = 8.6, \mu_r = 1, \sigma = 0 \text{ S/m}$$

Conducting material:

$$\varepsilon_r = 1, \mu_r = 1, \sigma = 5.8 \times 10^7 \text{ S/m (copper)}$$

For the geometry (see Figure 10.40(a)), we used the following values:

a = width of the outer conductor = 13 mm
b = height of the free space region (air) = 4 mm
h = height of the dielectric region = 2 mm
w = width of each inner strip = 2 mm
t = thickness of each inner strip = 0.01 mm
D = distance between the outer conductor and the first strip = 2.5 mm
s = distance between two consecutive strips = 1 mm

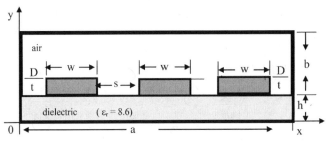

Figure 10.40(a). **Cross-section of the Three-strip Transmission Line.**

Figure 10.41 shows the finite element mesh, while Figure 10.42 illustrates the potential distribution along line $y = h$.

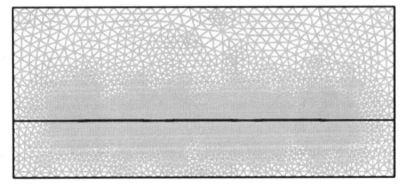

Figure 10.41. **Mesh for the Three-strip Transmission Line.**

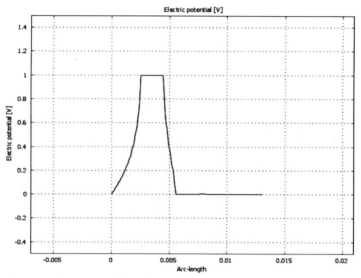

Figure 10.42. Potential distribution along the Air-dielectric interface ($y = h$) for the Three-strip Transmission Line.

Table 10.12 shows the finite element results for the three-strip line. Unfortunately, we could not find any work in the literature to compare our results.

Table 10.12. Capacitance Values (in pF/m) for Three-strip Shielded Microstrip Line

Methods	C_{11}	C_{21}	C_{31}
COMSOL	163.956	−27.505	−0.4301

10.19.3 Six-strip Line

Figure 10.43 shows the cross-section for six-strip transmission line. For COMSOL, the simulation was done twice on Figure 10.42 (one for C_o and other for C) using the following values:

Air:
$$\varepsilon_r = 1, \mu_r = 1, \sigma = 0 \text{ S/m}$$

Dielectric material:
$$\varepsilon_r = 6, \mu_r = 1, \sigma = 0 \text{ S/m}$$

Conducting material:
$$\varepsilon_r = 1, \mu_r = 1, \sigma = 5.8 \times 10^7 \text{ S/m (copper)}$$

For the geometry (see Figure 10.43), we used the following values:

a = width of the outer conductor = 15 mm
b = height of the free space region (air) = 2 mm
h = height of the dielectric region = 8 mm
w = width of each inner strip = 1 mm
t = thickness of each inner strip = 0.01 mm
D = distance between the outer conductor and the first strip = 2 mm
s = distance between two consecutive strips = 1 mm

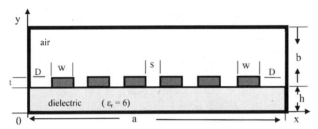

Figure 10.43. Cross-section of the Six-strip Transmission Line.

Figure 10.44 shows the finite element mesh, while Figure 10.45 depicts the potential distribution along line $y = h$.

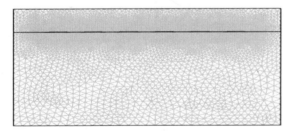

Figure 10.44. Mesh for the Six-strip Transmission Line.

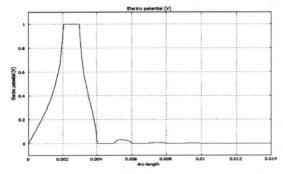

Figure 10.45. Potential distribution along the Air-dielectric Interface ($y = h$) for the Six-strip Transmission Line.

The capacitance values for six-strip shielded microstrip line are compared with other methods as shown in Table 10.13, where "iterative" refers to an iterative method and ABC refers to the asymptotic boundary condition. It is evident from the table that the finite element methods based closely agree. The finite element methods seem to be more accurate than the iterative and ABC techniques. (The negative capacitances are expected from equation (10.318).)

Table 10.13. Capacitance Values (in pF/m) for Six-strip Shielded Microstrip Line

Methods	C_{11}	C_{21}	C_{31}	C_{41}	C_{51}	C_{61}
Iterative	66.8	−27.9	−5.49	−2.08	−0.999	−0.704
Finite Element	84.8	−26.4	−3.71	−1.17	−0.456	−0.812
ABC	68.6	−31.5	−6.00	−2.25	−0.792	−0.602
COMSOL	80.4	−23.9	−3.61	−1.15	−0.451	−0.180

10.19.4 Eight-strip Line

Figure 10.46 shows the cross-section for eight-strip transmission line. For COMSOL, the simulation was done twice on Figure 10.45 (one for C_o and other for C) using the following values:

Air:
$$\varepsilon_r = 1, \mu_r = 1, \sigma = 0 \text{ S/m}$$

Dielectric material:
$$\varepsilon_r = 12.9, \mu_r = 1, \sigma = 0 \text{ S/m}$$

Conducting material:
$$\varepsilon_r = 1, \mu_r = 1, \sigma = 5.8 \times 10^7 \text{ S/m (copper)}$$

For the geometry (see Figure 10.46), we used the following values:

a = width of the outer conductor = 175 mm
b = height of the free space region (air) = 100 mm
h = height of the dielectric region = 16 mm
w = width of each inner strip = 1 mm
t = thickness of each inner strip = 0.01 mm
D = distance between the outer conductor and the first strip = 80 mm
s = distance between two consecutive strips = 1 mm

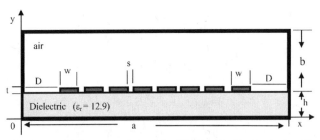

Figure 10.46. Cross-section of the Eight-strip Transmission Line.

Figure 10.47 shows the finite element mesh, while Figure 10.48 depicts the potential distribution along line $y = 20$ mm.

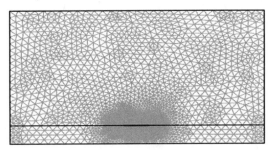

Figure 10.47. Mesh for the Eight-strip Transmission Line.

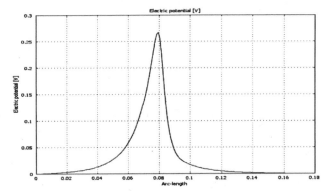

Figure 10.48. Potential distribution along the Air-dielectric interface ($y = 20$ mm) for the Eight-strip Transmission Line.

The capacitance values for eight-strip shielded microstrip line are compared with other methods as shown in Table 10.14, where other authors used the analytic approach and Fourier series expansion. It is evident from the table that the results from the finite element method (COMSOL) closely agree with the analytic approach.

Table 10.14. **Capacitance Values (in pF/m) for Eight-strip Shielded Microstrip Line**

Method	C_{11}	C_{21}	C_{31}	C_{41}	C_{51}	C_{61}	C_{71}	C_{81}
Analytic	127.776	−58.446	−13.024	−5.721	−3.104	−1.892	−1.282	−1.211
Fourier series	126.149	−57.066	−12.927	−5.684	−3.086	−1.875	−1.264	−1.185
COMSOL	128.204	−58.759	−13.064	−5.739	−3.1206	−1.902	−1.290	−1.226

10.20 SOLENOID ACTUATOR ANALYSIS WITH ANSYS

We use Ansys to do magnetic analysis (linear static) of a solenoid actuator. A solenoid actuator is to be analyzed as a 2D axisymmetric model as shown in Figure 10.49. For the given current, we determine the force on the armature.

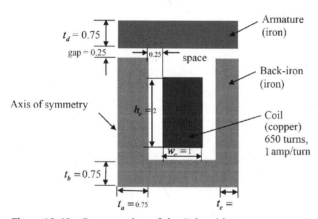

Figure 10.49. **Cross-section of the Solenoid Actuator.**

The dimensions of the solenoid actuator are in centimeters. The *armature* is the moving component of the actuator. The *back-iron* is the stationary iron component of the actuator that completes the magnetic circuit around the coil. The stranded, wound *coil* of 650 windings with 1 amp/turn supplies the predefined current. The current per winding is 1 amp. The *air-gap* is the thin rectangular region of air between the armature and the pole faces of the back-iron.

The magnetic flux produced by the coil current is assumed to be so small that no saturation of the iron occurs. This allows a single iteration linear analysis. The flux leakage out of the iron at the perimeter of the model is assumed to be negligible. This assumption is made simple to keep the model small. The model would normally be created with a layer of air surrounding the iron equal to or greater than the maximum radius of the iron.

The air gap is modeled so that a quadrilateral mesh is possible. A quadrilateral mesh allows for an uniform thickness of the air elements adjacent to the armature where the virtual work force calculation is performed. This is desirable for an accurate force calculation. The program requires the current to be input in the form of current density (current over the area of the coil). The assumption of no leakage at the perimeter of the model means that the flux will be acting parallel to this surface. This assumption is enforced by the "flux parallel" boundary condition placed around the model. This boundary condition is used for models in which the flux is contained in an iron circuit. Forces for the virtual work calculation are stored in an element table and then summed. The force is also calculated by the Maxwell Stress Tensor method and the two values are found to be relatively close. Table 10.15 summarizes the parameters of the model for the actuator geometry.

Table 10.15. **Parameters of the Model for the Actuator Geometry**

Parameter	Value
Number of turns in the coil; used in post-processing	$n = 650$
Current per turn	$I = 1.0$
Thickness of inner leg of magnetic circuit	$t_a = 0.75$
Thickness of lower leg of magnetic circuit	$t_b = 0.75$
Thickness of outer leg of magnetic circuit	$t_c = 0.50$
Armature thickness	$t_d = 0.75$
Width of coil	$w_c = 1$
Height of coil	$h_c = 2$
Air Gap	$gap = 0.25$
Space around coil	$space = 0.25$
w_s	$w_s = w_c + 2 *\ space$
h_s	$h_s = h_c + 0.75$
Total width of model	$w = t_a + w_s + t_c$
h_b	$h_b = t_b + h_s$
Total height of model	$h = h_b + gap + t_d$
Coil area	$acoil = w_c * h_c$
Current density of coil	$idens = n * i / acoil$

The below steps are a guideline in solving the above model.

1. **Input the geometry of the model**
   We use the information in the problem description to make Figure 10.50.

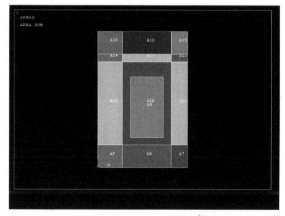

Figure 10.50. The 2D geometry of the Solenoid Actuator model.

2. **Define the materials**
   (a) Set preferences
   You will now set preferences in order to filter quantities that pertain to this discipline only.

   1. **Main Menu > Preferences**
   2. **Check** "Magnetic-Nodal" as in Figure 10.50(a)
   3. **OK**

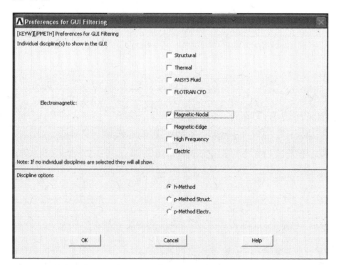

Figure 10.50(a). Preferences for GUI filtering.

(b) Specify material properties

Now specify the material properties for the magnetic permeability of air, back-iron, coil, and armature. For simplicity, all material properties are assumed to be linear. (Typically, iron is input as a nonlinear B-H curve.) Material 1 will be used for the air elements. Material 2 will be used for the back-iron elements. Material 3 will be used for the coil elements. Material 4 will be used for the armature elements.

1. **Main Menu > Preprocessor > Material Props > Material Models**
2. **Double-click** "Electromagnetics", then "Relative Permeability", then "Constant"
3. "MURX" = 1
4. **OK**
5. **Edit > Copy**
6. **OK** to copy Material Model Number 1 to become Material Model Number 2.
7. **Double-click** "Material Model Number 2", then "Permeability (Constant)"
8. "MURX" = 1000 as shown in Figure 10.51
9. **OK**
10. **Edit > Copy**
11. "from Material Number" = 1
12. "to Material Number" = 3
13. **OK**
14. **Edit > Copy**
15. "from Material Number" = 2
16. "to Material Number" = 4
17. **OK**
18. **Double-click** "Material Model Number 4", then "Permeability (Constant)"
19. "MURX" = 2000 as shown in Figure 10.52
20. **OK**
21. **Material > Exit**
22. **Utility Menu > List > Properties > All Materials**
23. Review the list of materials, then: as shown in Figure 10.53
    File > Close (Windows)

Engineering Electromagnetics Analysis

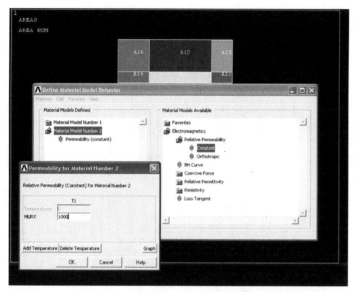

Figure 10.51. Definition of material model behavior for Number 1 and 2.

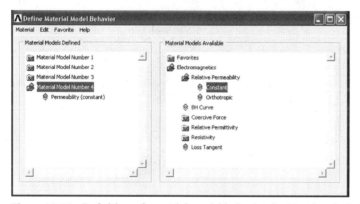

Figure 10.52. Definition of material model behavior for Number 1, 2, 3, and 4.

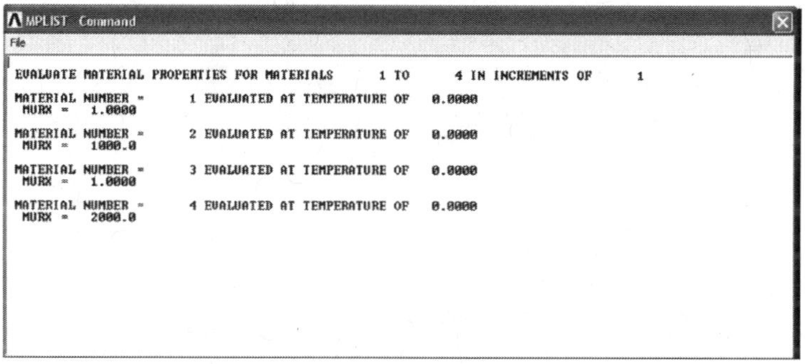

Figure 10.53. Review the list of materials of the model.

3. **Generating the mesh**
   (a) **Define element types and options**
   In this step, you will define element types and specify options associated with these element types.

   The higher-order element PLANE53 is normally preferred, but to keep the model size small, use the lower-order element PLANE13.

   1. **Main Menu > Preprocessor > Element Type > Add/Edit/Delete**
   2. **Add...**
   3. "Magnetic Vector" (left column)
   4. "Vect Quad 4nod13 (PLANE13" (right column)
   5. **OK**
   6. **Options...**
   7. (drop down) "Element behavior" = Axisymmetric, as shown in Figure 10.54
   8. **OK**
   9. **Close**

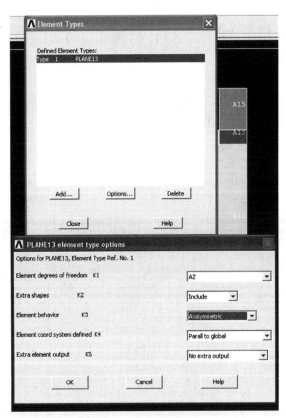

Figure 10.54. Element type PLANE 13.

(b) Assign material properties.

Now assign material properties to air gaps, iron, coil, and armature areas.

1. **Main Menu > Preprocessor > Meshing > MeshTool**
2. (drop down) "Element Attributes" = Areas; then [Set] as in Figure 10.55
3. Pick four areas of air gaps, A13, A14, A17, and A18 (the picking "hot spot" is at the area number label).

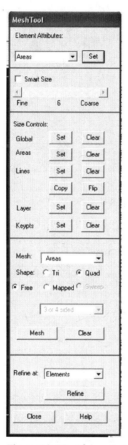

Figure 10.55. Element attribute for MeshTool.

4. **OK**
5. (drop down) "Material number" = 1
6. **Apply**
7. Pick the five back-iron areas, A7, A8, A9, A11, A12 as in Figure 10.56

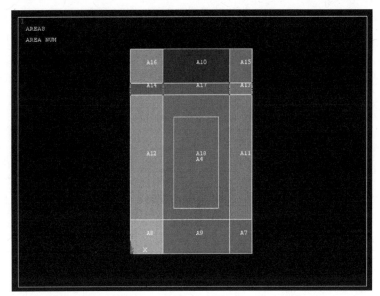

Figure 10.56. Five Back-iron areas, A7, A8, A9, A11, A12.

   8. **OK**
   9. (drop down) "Material number" = 2
10. **Apply**
11. Pick coil area, A4
12. **OK**
13. (drop down) "Material number" = 3
14. **Apply**
15. Pick armature area, A10, A15, A16
16. **OK**
17. (drop down) "Material number" = 4
18. **OK**
19. Toolbar: **SAVE_DB**

(c) Specify meshing-size controls on air gap
Adjust meshing size controls to get two element divisions through the air gap.

1. Main Menu > Preprocessor > Meshing > Size Cntrls > ManualSize > Lines > Picked Lines
2. Pick four vertical lines through air gap
3. **OK**
4. "No. of element divisions" = 2
5. **OK**

(d) Mesh the model using the MeshTool

1. "Size control global" = [Set]
2. "Element edge length" = 0.25
3. **OK** as in Figure 10.57
4. (drop down) "Mesh" = Areas
5. **Mesh**
6. **Pick All**
7. **Close**

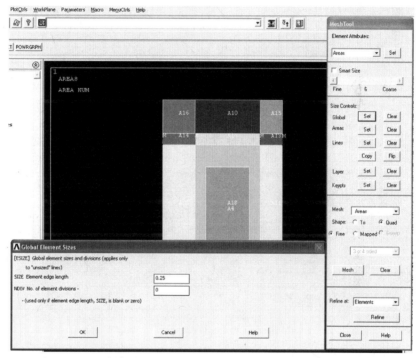

Figure 10.57. Global element sizes.

8. **Utility Menu > PlotCtrls > Numbering** as in Figure 10.58
9. (drop down) "Elem / attrib numbering" = Material numbers as in Figure 10.59
10. **OK** as in Figure 10.60

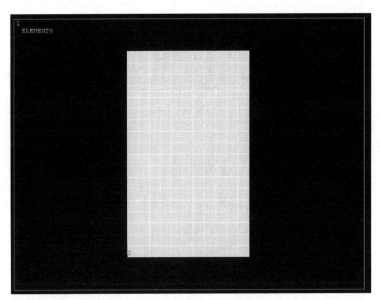

Figure 10.58. Numbering after PlotCtrls.

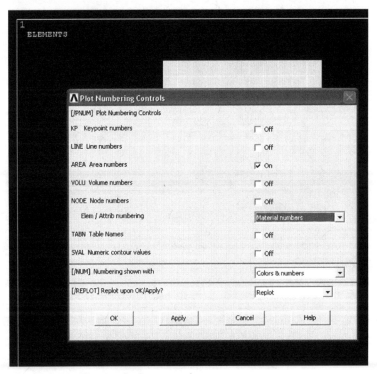

Figure 10.59. Plot numbering control.

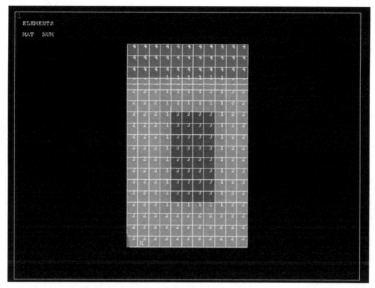

Figure 10.60. Numbering of the model.

(e) Scale model to meters for solution

For a magnetic analysis, a consistent set of units must be used. In this tutorial, MKS units are used, so you must scale the model from centimeters to meters.

1. **Main Menu > Preprocessor > Modeling > Operate > Scale > Areas**
2. **Pick All**
3. "RX,RY,RZ Scale Factors" = 0.01, 0.01, 1
4. (drop down) "Existing areas will be" = Moved
5. **OK** as in Figure 10.61
6. Toolbar: **SAVE_DB**

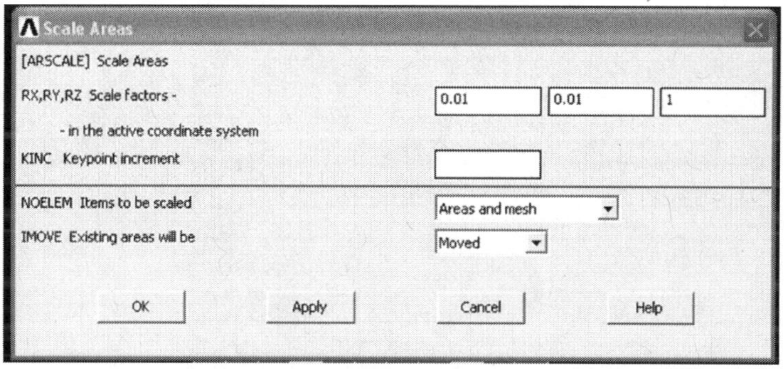

Figure 10.61. Scale area of the model.

4. **Apply Loads**
   (a) Define the armature as a component
   The armature can conveniently be defined as a component by selecting its elements.

   1. **Utility Menu > Select > Entities**
   2. (first drop down) "Elements"
   3. (second drop down) "By Attributes"
   4. "Min, Max, Inc" = 4
   5. **OK** as in Figure 10.62

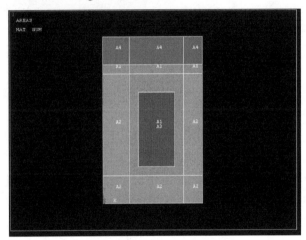

Figure 10.62. The entities of the model.

   6. **Utility Menu > Plot > Elements** as in Figure 10.63

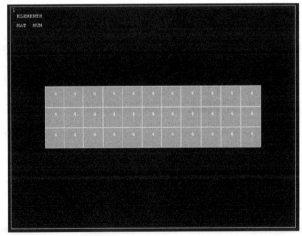

Figure 10.63. The Armature as a component of the model.

7. Utility Menu > Select > Comp/Assembly > Create Component
8. "Component name" = ARM
9. (drop down) "Component is made of" = Elements
10. OK

(b) Apply force boundary conditions to armature

1. Main Menu > Preprocessor > Loads > Define Loads > Apply > Magnetic > Flag > Comp. Force/Torq
2. (highlight) "Component name" = ARM
3. OK
4. Utility Menu > Select > Everything
5. Utility Menu > Plot > Elements as in Figure 10.64

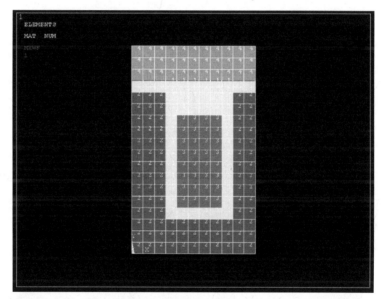

Figure 10.64. Plot of apply Force Boundary Conditions to Armature.

(c) Apply the current density

The current density is defined as the number of coil windings times the current, divided by the coil area. This equals (650)(1)/2, or 325. To account for scaling from centimeters to meters, the calculated value needs to be divided by .01**2.

1. Utility Menu > Plot > Areas
2. Main Menu > Preprocessor > Loads > Define Loads > Apply > Magnetic > Excitation > Curr Density > On Areas
3. Pick the coil area, which is the area in the center

4. OK
5. "Curr density value" = 325/.01**2
6. OK
Close any warning messages that appear.

(d) Obtain a flux parallel field solution
Apply a perimeter boundary condition to obtain a "flux parallel" field solution. This boundary condition assumes that the flux does not leak out of the iron at the perimeter of the model. Of course, at the centerline, this is true due to axisymmetry.

1. **Utility Menu > Plot > Lines**
2. **Main Menu > Preprocessor > Loads > Define Loads > Apply > Magnetic > Boundary > Vector Poten > Flux Par'l > On Lines**
3. Pick all lines around perimeter of model (14 lines) as in Figure 10.65

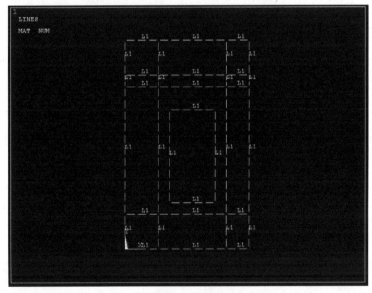

Figure 10.65. Plot of lines for Flux Parallel field of the model.

4. **OK**
5. Toolbar: **SAVE_DB**

5. **Obtain solution**
   (a) Solve

   1. **Main Menu > Solution > Solve > Electromagnet > Static Analysis > Opt & Solve**
   2. **OK** to initiate the solution
   3. **Close** the information window when solution is done

6. Review results
   (a) **Plot the flux lines in the model**
   Note that a certain amount of undesirable flux leakage occurs out of the back-iron.

   1. **Main Menu > General Postproc > Plot Results > Contour Plot > 2D Flux Lines**
   2. **OK** as in Figure 10.66

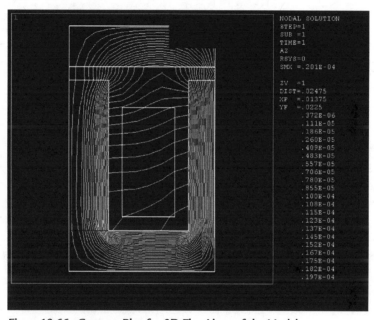

Figure 10.66. Contour Plot for 2D Flux Lines of the Model.

Your results may vary slightly from what is shown here due to variations in the mesh.

   (b) **Summarize magnetic forces**

   1. **Main Menu > General Postproc > Elec & Mag Calc > Component Based > Force**
   2. (highlight) "Component name(s)" = ARM
   3. **OK**
   4. Review the information, then choose:
      File > Close (Windows),
      or
      Close (X11/Motif) to close the window.

(c) Plot the flux density as vectors

1. **Main Menu > General Postproc > Plot Results > Vector Plot > Predefined**
2. "Flux & gradient" (left column)
3. "Mag flux dens B" (right column)
4. **OK** as in Figure 10.67

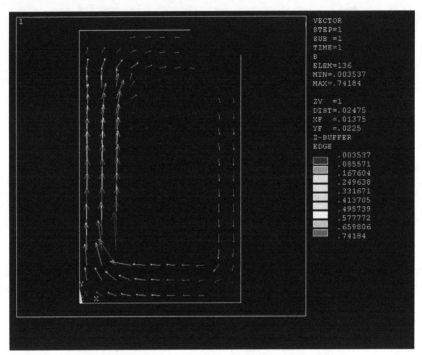

Figure 10.67. Plot the Flux Density as Vectors of the model.

(d) Plot the magnitude of the flux density

Plot the magnitude of the flux density without averaging the results across material discontinuities.

1. **Main Menu > General Postproc > Plot Results > Contour Plot > Nodal Solu**
2. Choose "Magnetic Flux Density," then "Magnetic flux density vector sum"
3. **OK** as in Figure 10.68

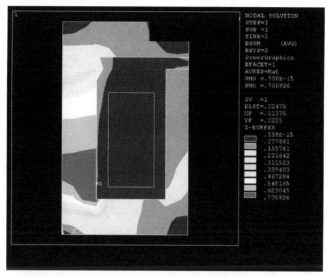

Figure 10.68. Contour Plot of the model.

Next, you will see how the flux density is distributed throughout the entire actuator. Up to this point, the analysis and all associated plots have used the 2D axisymmetric model, with the axis of symmetry aligned with the left vertical portion of the device. ANSYS will continue the analysis on the 2D finite element model, but will allow you to produce a three-quarter expanded plot representation of the flux density throughout the device, based on the defined axisymmetry. This function is purely graphical. No changes to the database will be made when you produce this expanded plot.

4. **Utility Menu > PlotCtrls > Style > Symmetry Expansion > 2D Axi-Symmetric**
5. (check) "3/4 expansion" as in Figure 10.69
6. **OK** as in Figure 10.70

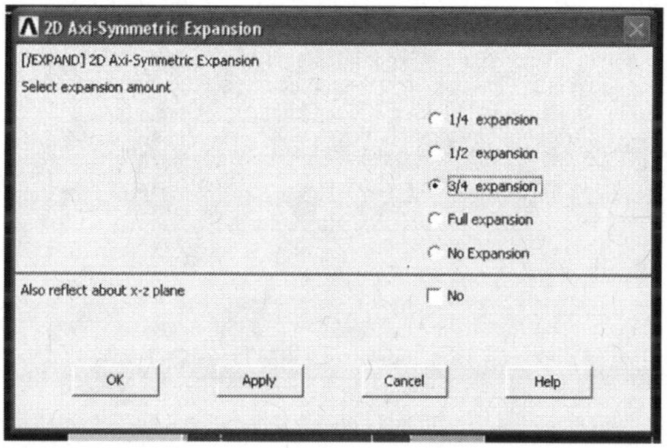

Figure 10.69. 2D Axi-symmetric expansion with Amount ¾.

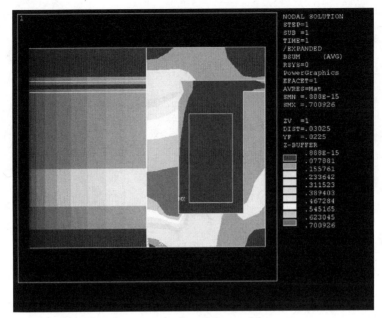

Figure 10.70. 2D Axi-symmetric plot of the model.

7. Utility Menu > PlotCtrls > Pan, Zoom, Rotate, as in Figure 10.71
8. Iso
9. Close

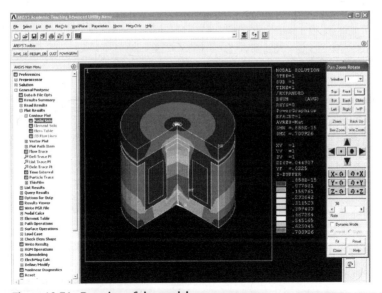

Figure 10.71. Rotation of the model.

(e) Exit the ANSYS program

1. Toolbar: **QUIT**
2. (check) "Quit - No Save!"
3. OK

PROBLEMS

1. Given $\mathbf{H} = He^{j(\omega t + 2\beta z)}\mathbf{a}_x$ in free space, known that, $\nabla \times \mathbf{H} = \dfrac{\partial \mathbf{D}}{\partial t}$, find **E**.
2. Calculate the skin depth, δ, for a copper conductor in 50 Hz field ($\sigma = 56 \times 10^6$ S/m).
3. EM problems and examples.
4. For the axisymmetric coaxial cable illustrated in Figure 10.71. Determine one dimension finite element general solution based on the following:

   a. Obtain and solve the governing differential solution for the coaxial cable, hint: $\dfrac{\varepsilon}{r}\dfrac{d}{dr}\left(r\dfrac{d\phi}{dr}\right) = -\rho.$

   b. Obtain the boundary conditions and continuity conditions, hint: $\phi_1\,(r = a) = \phi_a$, $\phi_2\,(r = c) = 0$, and the electric potential and the electric displacement are continuous at $r = b$.

   c. Formulate the equations of part (b) as a matrix equation that can be solved for the constants of integrations.

   d. Determine the shape functions for a general three-node quadratic element in terms of $x_1, x_2,$ and x_3.

   e. Determine the shape functions for a general three-node quadratic element when $x_1 = -L$, $x_2 = 0$, and $x_3 = L$.

   f. Find the local stiffness matrix for an element of length 2L with coordinates $(-L, 0, L)$.

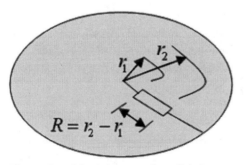

Figure 10.71(a). Axisymmetric radial element.

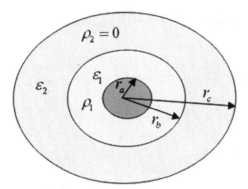

Figure 10.71(b). Coaxial cable.

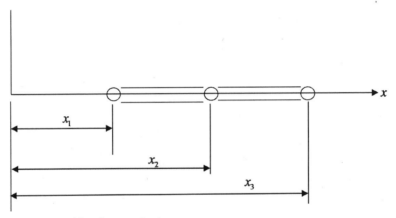

Figure 10.71(c). Three-node element.

5. Determine the variational function for two-dimensional axisymmetric heat conduction in r, z coordinate and formulate the corresponding local finite stiffness matrix using three-node triangular elements.

6. Use COMSOL in modeling of the four-conductor transmission lines with the following parameters as in Figure 10.72:

   ε_{r1} = dielectric constant of the dielectric material = 4.2
   ε_{r2} = dielectric constant of the free space = 1.0
   W = width of the dielectric material = 10 mm
   w = width of a single conductor line = 1 mm
   H_1 = distance of conductors 1 and 2 from the ground plane = 3 mm
   H_2 = distance of conductor 4 from the ground plane = 1 mm
   H_3 = distance of conductor 3 from the ground plane = 2 mm
   s = distance between the two coupled conductors = 1 mm
   t = thickness of the strips = 0.01 mm

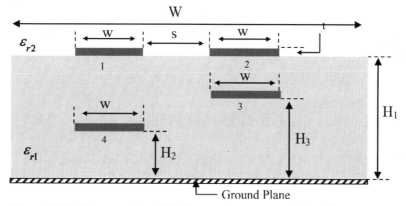

Figure 10.72. Cross-section of the four-conductor transmission lines.

The geometry is enclosed by a 10 × 10 mm shield. Find the capacitances per unit length, C_{11}, C_{12}, C_{13}, C_{14}, C_{22}, C_{23}, C_{24}, C_{33}, C_{34}, and C_{44}.

7. Use COMSOL in modeling of the shielded two vertically coupled striplines geometry is enclosed by a 3.4 × 1 mm shield with the following parameters as in Figure 10.73:

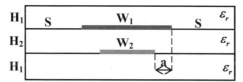

Figure 10.73. Cross-section of the two shielded vertically coupled striplines embedded in dielectric material.

ε_r = dielectric constant = 1 and 7.5
W_1 = width of the stripline 1 = 1.4 mm
W_2 = width of the stripline 2 = 1 mm
H_1 = height from stripline 1 and stripline 2 to the upper side and lower side of the shield respectively = 0.4 mm
H_2 = distance between the two striplines = 0.2 mm
S = distance between the stripline 1 and right/left side of the shield = 1 mm
a = $(W_1 - W_2)/2$ = 0.2 mm
t = thickness of the striplines = 0.01 mm
Find the capacitances per unit length, C_{11}, C_{21}, and C_{22}.

8. Use ANSYS Modeling of harmonic high-frequency electromagnetic of a coaxial waveguide as shown in Figure 10.74. The properties of the model is summarized as
   Material property:
   $$\mu_r = 1.0, \varepsilon_r = 1.0,$$
   Geometric property:
   $$r_i = 0.025 \text{ m}, r_0 = 0.075 \text{ m}, I = 0.375 \text{ m},$$
   Load used
   Port voltage = 1.0
   $\Omega = 0.8$ GHz

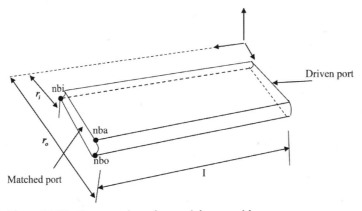

Figure 10.74. Cross-section of a coaxial waveguide.

Find $S11$, $S12$, Z_{Re}, Z_{im}, RL

9. Use ANSYS Modeling of electrostatic of a shielded microstrip transmission line consisting of a substrate, microstrip, and a shield. The strip is at potential V_1, and the shield is at a potential V_0. Find the capacitance of the transmission line as shown in Figure 10.75.
   The properties of the model is summarized as

   Material property:
   Air: $\varepsilon_r = 1$
   Substrate: $\varepsilon_r = 12$
   Geometric property:
   a = 10 cm
   b = 1 cm
   w = 2 cm
   Loading property:
   $V_0 = 1$ V
   $V_1 = 10$ V

Knowing that the electrostatic energy, W_e is defined as

$$W_e = \frac{1}{2}C(V_1 - V_0)^2.$$

Also, you need to type the following values in scalar parameters as:

$C = (w*2)/((V_1-V_0)**2)$ and $C = ((C*2)*1e12)$.

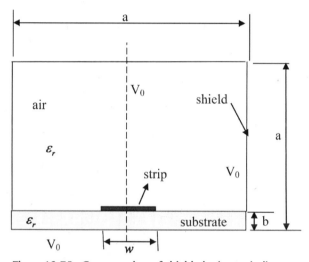

Figure 10.75. Cross-section of shielded microstrip line.

REFERENCES

1. M. N. O. Sadiku, "A Simple Introduction to Finite Element Analysis of Electromagnetic Problem," *IEEE Trans. Educ.*, vol. 32, no. 2, 1989, pp. 85-93.
2. P. Tong, "Exact Solution of Certain Problems by the Finite Element Method," *AIAA*, vol. 7, no. 1, 1969, pp. 179-180.
3. E. A. Thornton, P. Dechaumphai, and K. K. Tamma, "Exact Finite Elements for Conduction and Convection," *Proceedings of Second International Conference on Numerical Methods in Thermal Problems*, Venice, Italy, July 7-10, 1981, Swansea, Wales, Pineridge Press, 1981, pp. 1133-1144.
4. B. A. Finlayson and L. E. Scriven, "The method of weighted residuals-a review," *Appl. Mech. Rev.*, vol. 19, no. 9, 1966, pp. 735-748.
5. R. W. Klopfenstein and C. P. Wu, "Computer Solution of One-Dimensional Poisson's Equation," *IEEE Transactions on Electron Devices*, vol. 22, no. 6, 1976, pp. 329-333.

6. P. P. Silverster and R. L. Ferrari, *Finite Elements for Electrical Engineers*. Cambridge: Cambridge University Press, 3rd ed., 1996.
7. C. R. Paul and S. A. Nasar, *Introduction to Electromagnetic Fields*. New York: McGraw-Hill, 1982, pp. 465-472.
8. O. W. Anderson, "Laplacian Electrostatic Field Calculations by Finite Elements with Automatic Grid Generation," *IEEE Trans. Power App. Syst.*, vol. PAS-92, no. 5, 1973, pp. 1485-1492.
9. M. L. James et al., *Applied Numerical Methods for Digital Computation*. New York: Harper and Row, 1985, pp. 146-274.
10. R. E. Collin, *Field Theory of Guide Waves*. New York: McGraw-Hill, 1960, p. 128.
11. B. S. Garhow, *Matrix Eigensystem Routine-EISPACK Guide Extension*. Berlin: Springer-Verlag, 1977.
12. S. Ahmed and P. Daly, "Finite-Element Methods for Inhomogeneous Waveguides," *Proc. IEEE*, vol. 116, no. 10, Oct. 1969, pp. 1661-1664.
13. Z. J. Csendes and P. Silvester, "Numerical Solution of Dielectric Loaded Waveguides: I -Finite-Element Analysis," *IEEE Trans. Micro. Tech.*, vol. MTT-18, no. 12, Dec. 1970, pp. 1124-1131.
14. Z. J. Csendes and P. Silvester, "Numerical Solution of Dielectric Loaded Waveguides: II-Modal Approximation Technique," *IEEE Trans. Micro. Theo. Tech.*, vol. MTT-19, no. 6, 1971, pp. 504-509.
15. M. Hano, "Finite-Element Analysis of Dielectric-Loaded Waveguides," *IEEE Trans. Micro. Theo. Tech.*, vol. MTT-32, no. 10, 1984, pp. 1275-1279.
16. A. Konrad, "Vector Variational Formulation of Electromagnetic Fields in Anisotropic Media," *IEEE Trans. Micro. Theo. Tech.*, vol. MTT-24, Sept. 1976, pp. 553-559.
17. M. Koshiba, et al., "Improve Finite-Element Formulation in Terms of the Magnetic Field Vector for Dielectric Waveguides," *IEEE Trans. Micro. Theo. Tech.*, vol. MTT-33, no. 3, 1985, pp. 227-233.
18. M. Koshiba, et al., "Finite-Element Formulation in Terms of the Electric-Field Vector for Electromagnetic Waveguide Problems," *IEEE Trans. Micro. Theo. Tech.*, vol. MTT-33, no. 10, 1985, pp. 900-905.
19. K. Hayata, et al., "Vectorial Finite-Element Method Without Any Spurious Solutions for Dielectric Waveguiding Problems Using Transverse Magnetic-Field Component," *IEEE Trans. Micro. Theo. Tech.*, vol. MTT-34, no. 11, 1986, pp. 1120-1124.
20. K. Hayata, et al., "Novel Finite-Element Formulation Without Any Spurious Solutions for Dielectric Waveguides," *Elect. Lett.*, vol. 22, no. 6, 1986, pp. 295-296.
21. S. Dervain, "Finite Element Analysis of Inhomogeneous Waveguides," Master thesis, Department of Electrical and Computer Engineering, Florida Atlantic University, Boca Raton, 1987.
22. J. R. Winkler and J. B. Davies, "Elimination of Spurious Modes in Finite Element Analysis," *J. Comp. Phys.*, vol. 56, no. 1, 1984, pp. 1-14.
23. W. R. Buell and B. A. Bush, "Mesh Generation-A Survey," *J. Eng. Ind.*, Feb. 1973, pp. 332-338.

24. W. C. Thacker, "A Brief Review of Techniques for Generating Irregular Computational Grids," *Int. J. Num. Meth. Eng.*, vol. 15, 1980, pp. 1335-1341.
25. E. Hinton and D. R. J. Owen, *An Introduction to Finite Element Computations*. Swansea, U. K.: Pineridge, 1980, pp. 247, 260, 328-346.
26. J. N. Reddy, *An Introduction to the Finite Element Method*. New York: McGraw-Hill, 1984, pp. 340-345, 436.
27. M. N. O. Sadiku, et al., "A Further Introduction to Finite Element Analysis of Electromagnetic Problems," *IEEE Trans. Educ.*, vol. 34, no. 4, 1991, pp. 322-329.
28. M. Kono, "A Generalized Automatic Mesh Generation Scheme for Finite Element Method," *Inter. J. Num. Meth. Engr.*, vol. 15, 1980, pp. 713-731.
29. J. C. Cavendish, "Automatic Triangulation of Arbitary Planar Domains for the Finite Element Method," *Inter. J. Num. Meth. Engr.*, vol. 8, 1974, pp. 676-696.
30. A. O. Moscardini, et al., "AGTHOM-Automatic Generation of Triangular and Higher Order Meshes," *Inter. J. Num. Meth. Engr.*, vol. 19, 1983, pp. 1331-1353.
31. C. O. Frederick, et al., "Two-Dimensional Automatic Mesh Generation for Structured Analysis," *Inter. J. Num. Meth. Engr.*, vol. 2, no. 1, 1970, pp. 133-144.
32. E. A. Heighway, "A Mesh Generation for Automatically Subdividing Irregular Polygon into Quadrilaterals," *IEEE Trans. Mag.*, vol. MAG-19, no. 6, 1983, pp. 2535-2538.
33. C. Kleinstreuer and J. T. Holdeman, "A Triangular Finite Element Mesh Generator for Fluid Dynamic Systems of Arbitrary Geometry," *Inter. J. Num. Meth. Engr.*, vol. 15, 1980, pp. 1325-1334.
34. A. Bykat, "Automatic Generation of Triangular Grid I-Subdivision of a General Polygon into Convex Subregions. II-Triangulation of Convex Polygons," *Inter. J. Num. Meth. Engr.*, vol. 10, 1976, pp. 1329-1342.
35. N. V. Phai, "Automatic Mesh Generator with Tetrahedron Elements," *Inter. J. Num. Meth. Engr.*, vol. 18, 1982, pp. 273-289.
36. F. A. Akyuz, "Natural Coordinates Systems-an Automatic Input Data Generation Scheme for a Finite Element Method," *Nuclear Engr. Design*, vol. 11, 1970, pp. 195-207.
37. P. Girdinio, et al., "New Developments of Grid Optimization by the Grid Iteration Method," in Z. J. Csendes (ed.), *Computational Electromagnetism*. New York: North-Holland, 1986, pp. 3-12.
38. M. Yokoyama, "Automated Computer Simulation of Two-Dimensional Electrostatic Problems by Finite Element Method," *Inter. J. Num. Meth. Engr.*, vol. 21, 1985, pp. 2273-2287.
39. G. F. Carey, "A Mesh-Refinement Scheme for Finite Element Computations," *Comp. Meth. Appl. Mech. Engr.*, vol. 7, 1976, pp. 93-105.
40. K. Preiss, "Checking the Topological Consistency of a Finite Element Mesh," *Inter. J. Meth. Engr.*, vol. 14, 1979, pp. 1805-1812.
41. H. Kardestuncer (ed.), *Finite Element Handbook*. New York: McGraw-Hill, 1987, pp. 4.191-4.207.
42. W. C. Thacker, "A Brief Review of Techniques for Generating Irregular Computational Grids," *Inter. J. Num. Meth. Engr.*, vol. 15, 1980, pp. 1335-1341.

43. E. Hinton and D. R. J. Owen, *An Introduction to Finite Element Computations*. Swansea, UK: Pineridge Press, 1979, pp. 247, 328-346.
44. C. S. Desai and J. F. Abel, *Introduction to the Finite Element Method: A Numerical Approach for Engineering Analysis*. New York: Van Nostrand Reinhold, 1972.
45. M. V. K. Chari and P. P. Silvester (eds.), *Finite Elements for Electrical and Magnetic Field Problems*. Chichester: John Wiley, 1980, pp. 125-143.
46. P. Silvester, "Construction of Triangular Finite Element for Universal Matrices," *Inter. J. Num. Meth. Engr.*, vol. 12, 1978, pp. 237-244.
47. P. Silvester, "Higher-Order Polynomial Triangular Finite Elements for Potential Problems," *Inter. J. Engr. Sci.*, vol. 7, 1969, pp. 849-861.
48. G. O. Stone, "High-Order Finite Elements for Inhomogeneous Acoustic Guiding Structures," *IEEE Trans. Micro. Theory Tech.*, vol. MTT-21, no. 8, 1973, pp. 538-542.
49. A. Konrad, "High-Order Triangular Finite Elements for Electromagnetic Waves in Anisotropic Media," *IEEE Trans. Micro. Theory Tech.*, vol. MTT-25, no. 5, 1977, pp. 353-360.
50. P. Daly, "Finite Elements for Field Problems in Cylindrical Coordinates," *Inter. J. Num. Meth. Engr.*, vol. 6, 1073, pp. 169-178.
51. C. A. Brebbia and J. J. Connor, *Fundamentals of Finite Element Technique*. London: Butterworth, 1973, pp. 114-118, 150-163, 191.
52. M. N. O. Sadiku and L. Agba, "New Rules for Generating Finite Elements Fundamental Matrices," *Proc. IEEE Southeastcon*, 1989, pp. 797-801.
53. R. L. Ferrari and G. L. Maile, "Three-Dimensional Finite Element Method for Solving Electromagnetic Problems," *Elect. Lett.*, vol. 14, no. 15, 1978, p. 467.
54. M. dePourcq, "Field and Power-Density Calculation by Three-Dimensional Finite Elements," *IEEE Proc.*, vol. 130, no. 6, 1983, pp. 377-384.
55. M V. K. Chari, et al., "Finite Element Computation of Three-Dimensional Electrostatic and Magnetostatic Filed Problems," *IEEE Trans. Mag.*, vol. 19, no. 16, 1983, pp. 2321-2324.
56. O. A. Mohammed, et al., "Validity of Finite Element Formulation and Solution of Three Dimensional Magnetostatic Problems in Electrical Devices with Applications to Transformers and Reactors," *IEEE Trans. Pow. App. Syst.*, vol. 103, no. 7, 1984, pp. 1846-1853.
57. J. S. Savage and A. F. Peterson, "Higher-Order Vector Finite Elements for Tetrahedral Cells," *IEEE Trans. Micro. Theo. Theo. Tech.*, vol. 44, no. 6, 1996, pp. 874-879.
58. J. F. Lee and Z. J. Cendes, "Transfinite Elements: A Highly Efficient Procedure for Modeling Open Field Problems," *Jour. Appl. Phys.*, vol. 61, no. 8, 1987, pp. 3913-3915.
59. B. H. McDonald and A. Wexler, "Finite-Element Solution of Unbounded Field Problems," *IEEE Trans. Micro. Theo. Tech.* vol. 20, no. 12, 1977, pp. 1267-1270.
60. P. P. Silvester, et al., "Exterior Finite Elements for 2-Dimentional Field Problems with Open Boundaries," *Proc. IEEE*, vol. 124, no. 12, 1972, pp. 841-847.
61. S. Washisu, et al., "Extension of Finite-Element Method to Unbounded Field Problems," *Elect. Lett.*, vol. 15, no. 24, 1979, pp. 772-774.

62. P. P. Silvester and M. S. Hsieh, "Finite-Element Solution of 2-Dimentional Exterior-Field Problems," *Proc. IEEE*, vol. 118, no. 12, 1971, pp. 1743-1747.
63. Z. J. Csendes, "A Note on the Finite-Element Solution of Exterior-Field Problems," *IEEE Trans. Micro. Theo. Tech.*, vol. 24, no. 7, 1976, pp. 468-473.
64. T. Corzani, et al., "Numerical Analysis of Surface Wave Propagation Using Finite and Infinite Elements," *Alta Frequenza*, vol. 51, no. 3, 1982, pp. 127-133.
65. O. C. Zienkiewicz, et al., "Mapped Infinite Elements for Exterior Wave Problems," *Iner. J. Num. Meth. Engr.*, vol. 21, 1985.
66. F. Medina, "An Axisymmetric Infinite Element," *Int. J. Num. Meth Engr.*, vol. 17, 1981, pp. 1177-1185.
67. S. Pissanetzky, "A Simple Infinite Element," *Int. J. Comp. Math Elect. Engr.*, vol. 3, no. 2, 1984, pp. 107-114.
68. Z. Pantic and R. Mittra, "Quasi-TEM Analysis of Microwave Transmission Lines by the Finite-Element Method," *IEEE Trans. Micro. Theo. Tech.*, vol. 34, no. 11, 1986, pp. 1096-1103.
69. K. Hayata, et al., "Self-Consistent Finite/Infinite Element Scheme for Unbounded Guided Wave Problems," *IEEE Trans. Micro. Theo. Tech.*, vol. 36, no. 3, 1988, pp. 614-616.
70. P. Petre and L. Zombory, "Infinite Elements and Base Functions for Rotationally Symmetric Electromagnetic Waves," *IEEE Trans. Ant. Prog.*, vol. 36, no. 10, 1988, pp. 1490-1491.
71. Z. J. Csenes and J. F. Lee, "The Transfinite Element Method for Modeling MMIC Device," *IEEE Trans. Micro. Theo. Tech.*, vol. 36, no. 12, 1988, pp. 1639-1649.
72. K. H. Lee, et al., "A Hybrid Three-Dimensional Electromagnetic Modeling Scheme," *Geophys.*, vol. 46, no. 5, 1981, pp. 779-805.
73. S. J. Salon and J. M. Schneider, "A Hybrid Finite Element-Boundary Integral Formulation of Poisson's Equation," *IEEE Trans. Mag.*, vol. 17, no. 6, 1981, pp. 2574-2576.
74. S. J. Salon and J. Peng, "Hybrid Finite-Element Boundary-Element Solutions to Axisymmetric Scalar Potential Problems," in Z. J. Csendes (ed.), *Computational Electromagnetics*. New York: North-Holland/Elsevier, 1986, pp. 251-261.
75. J. M. Lin and V. V. Liepa, "Application of Hybrid Finite Element Method for Electromagnetic Scattering from Coated Cylinders," *IEEE Trans. Ant. Prop.*, vol. 36, no. 1, 1988, pp. 50-54.
76. J. M. Lin and V. V. Liepa, "A Note on Hybrid Finite Element Method for Solving Scattering Problems," *IEEE Trans. Ant. Prop.*, vol. 36, no. 10, 1988, pp. 1486-1490.
77. M. H. Lean and A. Wexler, "Accurate Field Computation with Boundary Element Method," *IEEE Trans. Mag.*, vol. 18, no. 2, 1982, pp. 331-335.
78. R. F. Harrington and T. K. Sarkar, "Boundary Elements and Method of Moments," in C. A. Brebbia, et al. (eds), *Boundary Elements*. Southampton: CML Publ., 1983, pp. 31-40.
79. M. A. Morgan, et al., "Finite Element-Boundary Integral Formulation for Electromagnetic Scattering," *Wave Motion*, vol. 6, no. 1, 1984, pp. 91-103.

80. S. Kagami and I. Fukai, "Application of Boundary-Element Method to Electromagnetic Field Problems," *IEEE Trans. Micro. Theo. Tech.*, vol. 32, no. 4, 1984, pp. 455-461.
81. Y. Tanaka, et al., "A Boundary-Element Analysis of TEM Cells in Three Dimensions," *IEEE Trans. Elect. Comp.*, vol. 28, no. 4, 1986, pp. 179-184.
82. N. Kishi and T. Okoshi, "Proposal for a Boundary-Integral Method Without Using Green's Function," *IEEE Trans. Micro. Theo. Tech.*, vol. 35, no. 10, 1987, pp. 887-892.
83. D. B. Ingham, et al., "Boundary Integral Equation Analysis of Transmission-Line Singularities," *IEEE Trans. Micro. Theo. Tech.*, vol. 29, no. 11, 1981, pp. 1240-1243.
84. S. Washiru, et al., "An Analysis of Unbounded Field Problems by Finite Element Method," *Electr. Comm. Japan*, vol. 64-B, no. 1, 1981, pp. 60-66.
85. T. Yamabuchi and Y. Kagawa, "Finite Element Approach to Unbounded Poisson and Helmholtz Problems Using-Type Infinite Element," *Electr. Comm. Japan*, Pt. I, vol. 68, no. 3, 1986, pp. 65-74.
86. K. L. Wu and J. Litva, "Boundary Element Method for Modeling MIC Devices," *Elect. Lett.*, vol. 26, no. 8, 1990, pp. 518-520.
87. P. K. Kythe, *An Introduction to Boundary Element Methods*. Boca Raton, FL: CRC Press, 1995, p. 2.
88. J. M. Jin et al., "Fictitious Absorber for Truncating Finite Element Meshes in Scattering," *IEEE Proc. H*, vol. 139, 1992, pp. 472-476.
89. R. Mittra and O. Ramahi, "Absorbing Bounding Conditions for Direct Solution of Partial Differential Equations Arising in Electromagnetic Scattering Problems," in M. A. Morgan (ed.), *Finite Element and Finite Difference Methods in Electromaganetics*. New York: Elsevier, 1990, pp. 133-173.
90. U. Pekel and R. Mittra, "Absorbing Boundary Conditions for Finite Element Mesh Truncation," in T. Itoh se al. (eds.), *Finite Element Software for Microwave Engineering*. New York: John Wiley & Sons, 1996, pp. 267-312.
91. U. Pekel and R. Mittra, "A Finite Element Method Frequency Domain Application of the Perfectly Matched Layer (PML) Concept," *Micro. Opt. Technol. Lett.*, vol. 9, pp. 117-122.
92. A. Boag and R. Mittra, "A numerical absorbing boundary condition for finite difference and finite element analysis of open periodic structures," *IEEE Trans. Micro. Theo. Tech.*, vol. 43, no. 1, 1995, pp. 150-154.
93. P. P. Silvester and G. Pelosi (eds.), *Finite Elements for Waves Electromagnetics: Methods and Techniques*. New York: IEEE Press, 1994, pp. 351-490.
94. A. M. Bayliss, M. Gunzburger, and E. Turkel, "Boundary conditions for the numerical solution of elliptic equation in exterior regions," *SIAM Jour. Appl. Math.*, vol. 42, 1982, pp. 430-451.
95. M. N. O. Sadiku, *Elements of Electromagnetics*. Fifth Edition. New York: Oxford University Press, 2010.
96. J. A. Kong, *Electromagnetic Wave Theory*. New York: John Wiley and Sons, Inc., 1986.

97. www.comsol.com
98. J. Jin, *The Finite Element Method in Electromagnetics*, Second Edition, New York: John Wiley & Sons Inc., 2002.
99. S. R. H. Hoole, Computer-Aided and Design of Electromagnetic Devices. New York: Elsevier, 1989.
100. U. S. Inan and R. A. Marshall, *Numerical Electromagnetics: The FDTD Method*, Cambridge, New York: Cambridge University Press, 2011.
101. C. W. Steele, *Numerical Computation of Electric and Magnetic Fields*, New York: Van Nostrand Reinhold Company, 1987.
102. J. Jin, *Theory and Computation of Electromagnetic Fields*, New York: John Wiley & Sons Inc., 2010.
103. P. P. Silvester and R. L. Ferrari, *Finite Elements for Electrical Engineering*. Second Edition, Cambridge, New York: Cambridge University Press, 1990.
104. D. K. Cheng, *Field and Wave Electromagnetic*. Second Edition. New York: John Wiley and Sons, Inc., 1992.
105. R. F. Harrington, *Time-Harmonic Electromagnetic Fields*, New York: McGraw-Hill, 1961.
106. M. L. Crawford, "Generation of Standard EM Fields Using TEM Transmission Cells," *IEEE Transactions on. Electromagnetic Compatibility*, vol. EMC-16, pp. 189-195, Nov. 1974.
107. J. R. Reid and R. T. Webster, "A 60 GHz Branch Line Coupler Fabricated Using Integrated Rectangular Coaxial Lines," *Microwave Symposium Digest, 2004 IEEE MTT-S International*, vol. 2, pp. 441-444, 6-11 June 2004.
108. S. Xu and P. Zhou, "FDTD Analysis for Satellite BFN Consisting of Rectangular Coaxial Lines," *Asia Pacific Microwave Conference*, pp. 877-880, 1997.
109. J. G. Fikioris, J. L. Tsalamengas, and G. J. Fikioris, "Exact Solutions for Shielded Printed Microstrip Lines by the Carleman-Vekua Method", *IEEE Transactions on Microwave Theory and Techniques*, vol. 37, no. 1, pp. 21-33, Jan. 1989.
110. S. Khoulji and M. Essaaidi, "Quasi-Static Analysis of Microstrip Lines with Variable-Thickness Substrates Considering Finite Metallization Thickness", *Microwave and Optic Technology Letters*, vol. 33, no. 1, pp. 19-22, April. 2002.
111. T. K. Seshadri, S. Mahapatra, and K. Rajaiah, "Corner Function Analysis of Microstrip Transmission Lines", *IEEE Transactions on Microwave Theory and Techniques*, vol. 28, no. 4, pp. 376-380, April. 1980.
112. S. V. Judd, I. Whiteley, R. J. Clowes, and D. C. Rickard, "An Analytical Method for Calculating Microstrip Transmission Line Parameters", *IEEE Transactions on Microwave Theory and Techniques*, vol. 18, no. 2, pp. 78-87, Feb. 1970.
113. N. H. Zhu, W. Qiu, E. Y. B. Pun, and P. S. Chung, "Quasi-Static Analysis of Shielded Microstrip Transmission Lines with Thick Electrodes", *IEEE Transactions on Microwave Theory and Techniques*, vol. 45, no. 2, pp. 288-290, Feb. 1997.
114. T. Chang and C. Tan, "Analysis of a Shielded Microstrip Line with Finite Metallization Thickness by the Boundary Element Method," *IEEE Transactions on Microwave Theory and Techniques*, vol. 38, no. 8, pp. 1130-1132, Aug. 1990.

115. G. G. Gentili and G. Macchiarella, "Quasi-Static Analysis of Shielded Planar Transmission Lines with Finite Metallization Thickness by a Mixed Spectral-Space Domain Method", *IEEE Transactions on Microwave Theory and Techniques*, vol. 42, no. 2, pp. 249-255, Feb. 1994.
116. A. Khebir, A. B. Kouki, and R. M. Mittra, "Higher Order Asymptotic Boundary Condition for Finite Element Modeling of Two-Dimensional Transmission Line Structures", *IEEE Transactions on Microwave Theory and Techniques*, vol. 38, no. 10, pp. 1433-1438, Oct. 1990.
117. G. W. Slade and K. J. Webb, "Computation of Characteristic Impedance for Multiple Microstrip Transmission Lines Using a Vector Finite Element Method", *IEEE Transactions on Microwave Theory and Techniques*, vol. 40, no. 1, pp. 34-40, Jan. 1992.
118. M. S. Alam, K. Hirayama, Y. Hayashi, and M. Koshiba, "Analysis of Shielded Microstrip Lines with Arbitrary Metallization Cross Section Using a Vector Finite Element Method", *IEEE Transactions on Microwave Theory and Techniques*, vol. 42, no. 11, pp. 2112-2117, Nov. 1994.
119. J. Svacina, "A New Method for Analysis of Shielded Microstrips", *Proceedings of Electrical Performance of Electronic Packaging*, pp. 111-114, 1993.
120. H. Y. Yee, and K. Wu, "Printed Circuit Transmission-Line Characteristic Impedance by Transverse Modal Analysis", *IEEE Transactions on Microwave Theory and Techniques*, vol. 34, no. 11, pp. 1157-1163, Nov. 1986.
121. I. P. Hong, N. Yoon, S. K. Park, and H. K. Park, "Investigation of Metal-Penetrating Depth in Shielded Microstrip Line", *Microwave and Optic Technology Letters*, vol. 19, no. 6, pp. 396-398, Dec. 1998.
122. http://www.comsol.com/
123. Q. Zheng, W. Lin, F. Xie, and M. Li, "Multipole Theory Analysis of a Rectangular Transmission Line Family," *Microwave and Optical Technology Letters*, vol. 18, no. 6, pp. 382-384, Aug. 1998.
124. T. S. Chen, "Determination of the Capacitance, Inductance, and Characteristic Impedance of Rectangular Lines," *IEEE Transactions on Microwave Theory and Techniques*, Volume 8, Issue 5, pp. 510–519, Sep. 1960.
125. E. Costamagna and A. Fanni, "Analysis of Rectangular Coaxial Structures by Numerical Inversion of the Schwarz-Christoffel Transformation," *IEEE Transactions on Magnets*, vol. 28, pp. 1454-1457, Mar. 1992.
126. K. H. Lau, "Loss Calculation for Rectangular Coaxial Lines," *IEE Proceedings*, vol. 135, Pt. H. no. 3, pp. 207-209, June 1988.
127. Personal computer program based on finite difference method.
128. J. D. Cockcroft, "The Effect of Curved Boundaries on the Distribution of Electrical Stress Round Conductors," *J. IEE*, vol. 66, pp. 385-409, Apr. 1926.
129. F. Bowan, "Notes on Two Dimensional Electric Field Problems," *Proc. London Mathematical Society.*, vol. 39, no. 211, pp. 205-215, 1935.
130. H. E. Green, "The Characteristic Impedance of Square Coaxial Line," *IEEE Transactions Microwave Theory and Techniques*, vol. MTT-11, pp. 554-555, Nov. 1963.

131. S. A. Ivanov and G. L. Djankov, "Determination of the Characteristic Impedance by a Step Current Density Approximation," *IEEE Transactions on Microwave Theory and Techniques.*, vol. MTT-32, pp. 450-452, Apr. 1984.
132. H. J. Riblet, "Expansion for the Capacitance of a Square in a Square with a Comparison," *IEEE Transactions on Microwave Theory and Techniques*, vol. 44, pp. 338-340, Feb. 1996.
133. M. S. Lin, "Measured Capacitance Coefficients of Multiconductor Microstrip Lines with Small Dimensions," *IEEE Transactions on Microwave Theory and Techniques*, vol. 13, no. 4, pp. 1050-1054, Dec. 1990.
134. F. Y. Chang, "Transient Analysis of Lossless Coupled Transmission Lines in a Nonhomogeneous Dielectric Media," *IEEE Transactions on Microwave Theory and Techniques*, vol. 18, no. 9, pp. 616-626, Aug. 1970.
135. P. N. Harms, C. H. Chan, and R. Mittra, "Modeling of Planar Transmission Line Structures for Digital Circuit Applications," *Arch. Eleck. Ubertragung.*, vol. 43, pp. 245-250, 1989.
136. A. Kherbir, A. B. Kouki, and R. Mittra, "Absorbing Boundary Condition for Quasi-TEM Analysis of Microwave Transmission Lines via the Finite Element Method," *J. Electromagnetic Waves and Applications*, vol. 4, no. 2, 1990.
137. A. Khebir, A. B. Kouki, and R. Mittra, "High Order Asymptotic Boundary Condition for the Finite Element Modeling of Two-Dimensional Transmission Line Structures," *IEEE Transactions on Microwave Theory and Techniques*, vol. 38, no. 10, pp. 1433-1438, Oct. 1990.
138. D. Homentcovschi, G. Ghione, C. Naldi, and R. Oprea, "Analytic Determination of the Capacitance Matrix of Planar or Cylindrical Multiconductor Lines," *IEEE Transactions on. Microwave Theory and Techniques*, pp. 363-373, Feb. 1995.
139. M. K. Amirhosseini, "Determination of Capacitance and Conductance Matrices of Lossy Shielded Coupled Microstrip Transmission Lines," *Progress In Electromagnetics Research*, PIER 50, pp. 267-278, 2005.
140. http://www.ansys.com/
141. S. M. Musa and M. N. O. Sadiku, "Modeling and Simulation of Shielded Microstrip Lines," *The Technology Interface*, Fall 2007.
142. J. A. Edminister, "Theory and Problems of Electromagnetics," Schaum's Outline Series, McGraw-Hill, 1979.
143. G. R. Buchanan, "Finite Element Analysis," Schaum's outline, McGraw-Hill, 1995.

ON THE DVD

The appendices are located on the companion disc in the back of the book:

Appendix A ANSYS   (pp. 473–502)
Appendix B MATLAB   (pp. 503–520)
Appendix C Color Figures   (pp. 521–554)

Index

A

ABC. *See* Absorbing boundary conditions
Absorbing boundary conditions (ABC), 422–423
Airfoil, potential flow
   post-processing, 301–302
   preprocessing, 297–300
   software results, 297
Analytical method, 48
   axial vibrations, 335
   cantilever beams, 202–203, 208–209
   constant cross-section area, 65–66, 70–72, 76–77, 82, 85–87
   engineering problem, 48
   natural frequency determination, 307
   one-dimensional heat conduction problems, 259
   simply supported beams, 176, 180–181, 187–188, 192–193
   stepped bar, 118, 122–123, 127–129
   stress analysis, rectangular plate with circular hole, 240–241, 242–243
   truss, 147–150
   varying cross-section area, 97–98, 103–104
ANSYS
   classic/traditional, 473–482
   design modeler, 483
   design optimization-design, 484
   engineering data, 486–501
   finite element modeler, 483
   geometry definition, 483
   graphical user interface (GUI), 473
   materials definition, 483
   mechanisms, 484
   reporting, 483
   software method, 485–486
   Solenoid actuator analysis, 441–459
   workbench, 483–501
   workbench user interface, 483
ANSYS program, 145
Anti-symmetric (Skew-symmetric) matrix, 5
Argument matrix, 19
Automatic mesh generation
   arbitrary domains, 402–405
   rectangular domains, 401–402
Axially loaded members
   constant cross-section area, 65–96
   stepped bar, 117–144
   three-node bar element, 64
   two-node bar element, 62–63
   varying cross-section area, 97–117
Axial vibrations
   analytical method, 335
   FEM by hand calculations, 336–337

555

frequency values, 338
post-processing, 344
preprocessing, 339–344
software results, 338
Axisymmetric radial element, 459

B

Banded matrix, 5
Beam
    beam element, 175
    cantilever beams, 202–223
    definition, 175
    simply supported beams, 176–202
Bio-Savart's law and field intensity, 365
Boundary conditions, 362–363
Boundary element method, 422

C

CAD model, 52
Cantilever beam, 347
Cantilever beams
    analytical method, 202–203, 208–209
    bending moment diagram, 218
    bending stress diagram, 219
    deflection values at nodes, 206, 213, 218, 224
    FEM by hand calculations, 203–206, 209–212, 214–217, 220–225
    maximum stress diagram, 219
    nodal force calculation, 210–212, 215–217, 221–223
    post-processing, 230–232
    preprocessing, 226–230
    reaction calculation, 212, 217, 223–224
    reaction values, 207, 213, 218, 225
    rotational deflection values at nodes, 206, 213, 218, 225
    shear force diagram, 219
    software results, 206–207, 212–214, 217–220
    total values, 218
Cantilever beam, transverse vibrations
    analytical solution, 317–318
    FEM by hand calculations, 318–320
    of software results, 320–322

Coaxial cable, 460
Coaxial waveguide, 462
Column matrix, 4
Compressible fluid, 291
Compressive forces, 59
Conduction boundary conditions, two-dimensional problem, 285–286
Constant cross-section area
    analytical method, 65–66, 70–72, 76–77, 82, 85–87
    deflection values as node, 68–69, 75, 80, 84, 92–93
    displacement calculation, 65–66, 71–71
    FEM by hand calculations, 66–68, 72–76, 77–79, 82–84, 87–91
    post-processing, 95–96
    preprocessing, 93–95
    reaction calculation, 66, 70, 79, 86–87
    reaction value, computer generated output, 69, 76
    software results, 68–70, 80–81, 84–85, 91–93
    stress calculation, 66, 67, 70–71, 74, 79, 86
    stress values, computer generated output, 69
    Young's modulus, 65
Constitutive relations, 357–360
Convection boundary conditions, two-dimensional problem, 285–286
Coulomb's law and field intensity, 364
Cramer's rule, 21–23

D

Diagonal matrix, 4
Dielectric loss, 369
Direct approach, 49
Displacement method, 52
Double-strip shielded transmission line, 433–435
Dynamic analysis
    bar, axial vibrations of, 335–344
    cantilever beam, transverse vibrations of, 316–322
    forcing function, bar subjected, 345–346

forcing function, fixed-fixed beam subjected to, 323–334
natural frequency determination, fixed-fixed beam for, 307–316
procedure of finite element analysis, 306

E

Eigenvalues, 34–36
Eigenvectors, 34–36
Eight-strip Line, 439–441
Electromagnetic analysis, 373–400
Electromagnetic energy and power flow, 365–369
Elementary row operations, 19
Element stiffness matrix, 146, 151
Element strains and stresses, 50
Elimination method, 50–51
Engineering electromagnetics analysis
    automatic mesh generation, 401–405
    Bio-Savart's law and field intensity, 365
    boundary conditions, 362–363
    constitutive relations, 357–360
    Coulomb's law and field intensity, 364
    electromagnetic analysis, 373–400
    electromagnetic energy and power flow, 365–369
    external problems, 420–423
    FEM by hand calculations (*See* Hand calculations, FEM by)
    higher order elements, 405–415
    Lorentz force law and continuity equation, 356–357
    loss in medium, 370
    Maxwell's equations and continuity equation, 351–356
    multistrip transmission lines, 432–441
    Poison's and Laplace's equations, 371
    potential equations, 360–361
    shielded microstrip lines, modeling and simulation of, 423–431
    skin depth, 370–371
    solenoid actuator analysis, with ANSYS, 441–459
    three-dimensional element, 415–420
    wave equations, 372–373
Engineering problem
    analytical method, 48
    experimental method, 48
    numerical method, 48
    post-processing, 52
    preprocessing, 52
    processing, 52
Equation of continuity, 352
Equations
    of continuity, 352
    Laplace's equation, 371, 382–392
    Lorentz force law and continuity equation, 356–357
    Maxwell's equations and continuity equation, 351–356, 352–356
    Poison's and Laplace's equations, 371
    Poisson's equation, 377–381
    potential equations, 360–361
    symbolic expressions, solving equations, 519–520
    wave equations, 372–373
Equilibrium, 61
Experimental method, 48

F

FEA. *See* Finite element method
Finite element method (FEA)
    dynamic problems, procedure of, 306
    fluid flow problems, procedure of, 292–293
    linear spring, direct method for, 55–56
    practical applications of, 51
    prescribing boundary conditions, 50–51
    procedure of, 48–50
    software package, 52
    solving engineering problems, 47–48
    structural problems, procedure of, 48–50
    for structure, 52–53
    thermal problems, procedure of, 258
    types of, 53–54

Fixed-fixed beam, 347
    forcing function, 323–334
    natural frequency determination, 307–316
Fluid flow analysis
    airfoil, potential flow, 296–304
    cylinder, potential flow, 293–295
    procedure of finite element analysis, 292–293
Force and elongation behaviors, 60
Force method, 52
Forcing function, bar
    displacement values, 346
    maximum displacement values, 346
    software results, 345–346
Forcing function, fixed-fixed beam
    maximum displacement values, 334
    preprocessing, 325–334
    software results, 322–324
Four-conductor transmission lines, 461

G

Gaussian elimination method, 19–20
Gauss's law, 371
Global stiffness matrix, 152
Gradient operator, 31

H

Hand calculations, FEM by
    axial vibrations, 336–337
    cantilever beams, 203–206, 209–212, 214–217, 220–225
    cantilever beam, transverse vibrations of, 318–320
    constant cross-section area, 66–68, 72–76, 77–79, 82–84, 87–91
    natural frequency determination, 308–310
    simply supported beams, 177–178, 181–184, 188–191, 193–195
    stepped bar, 118–121, 123–125, 130–132
    truss, 150–160, 162–165
    varying cross-section area, 98–103, 104–106, 108–111

Higher order elements
    fundamental matrices, 411–415
    local coordinates, 406–408
    pascal triangle, 405–406
    shape functions, 409–410
Homogeneous, 362
Hybrid method, 53

I

Identity (unit) matrix, 4–5
Incompressible fluid, 291
Infinite element method, 420–421
Inhomogeneous, 362
Integration of matrix, 15
Inverse of matrix, 24–26
Invertible matrix (nonsingular matrix), 25
Irrotational flow, 291–292

L

Laplace's equations, 371
Linear spring element, 55–56
Linear systems, direct methods for, 19
Lorentz force law and continuity equation, 356–357
Loss tangent, 370

M

Magnetostatic energy, 367
MATLAB, 36–43
MATLAB (Matrix Laboratory)
    calculations, 505–512
    differentiating symbolic expressions, 516
    integrating symbolic expressions, 517
    limits symbolic expressions, 517
    simplifying symbolic expressions, 515
    sums symbolic expressions, 518
    symbolic computation, 512–514
    symbolic expressions, solving equations, 519–520
    Taylor series symbolic expressions, 518
    windows, 503–505
Matrix
    addition of, 6–7
    Cramer's rule, 21–23

definition, 1–3
determinant of, 16–17
differentiation of, 14
eigenvalues and eigenvectors, 34–36
equality of, 15
Gaussian elimination method, 19–20
integration of, 15
inverse of, 24–26
linear systems, direct methods for, 19
MATLAB, 36–43
multiplication of, 8–9
multiplied by scalar, 8
operations, 38
rules of multiplications, 9–12
subtraction of, 6–7
trace of, 14
transpose of, 12–13
types of, 3–6
vector analysis, 27–34
Matrix Laboratory. *See* MATLAB
Matrix multiplications
by another matrix, 8–9
rules of, 9–12
by scalar, 8
Maximum bending moment, 232–233
Maxwell's equations and continuity equation
differential form, 352
divergence and Stokes theorems, 353
integral form, 353
quasi-statics case, 354
space case, source-free regions of, 354–355
statics case, 354
time-harmonic fields case, 355–356
Mixed method, 52
Multipoint constrains method, 51
Multistrip transmission lines, 432–441

N

Natural frequency determination
analytical method, 307
FEM by hand calculations, 308–310
post-processing, 316
preprocessing, 312–316
software results, 310–311
Nodal displacements, 50, 141, 142, 149, 156
Nodal lines, 374
Nodes, 47
Non-invertible matrix (singular matrix), 25
Nonsingular matrix, 25
Null matrix, 4
Numerical method, 48

O

One-dimensional elements, 53
FEM standard steps procedure, 374–377
natural coordinates, 381–382
Poisson's equation, 377–381
variational approach, 377–379
weighted residuals method, 379–381
One-dimensional heat conduction
problems, 257, 258–262
analytical method, 259
FEM by calculations, 260–261
post-processing, 268–272, 277–280, 285
preprocessing, 262–267, 272–277, 281–285
software results, 261–262, 280

P

Partitioned matrix (Super-matrix), 6
Penalty method, 51
Plot velocity distribution, 303
Poison's and Laplace's equations, 371
Poisson's equation, 377–381
Post-processing
axial vibrations, 344
cantilever beams, 230–232
constant cross-section area, 95–96
engineering problem, solving, 52
natural frequency determination, 316
simply supported beams, 200–202
stepped bar, 140–141
stress analysis, rectangular plate with circular hole, 251
truss, 169–170
varying cross-section area, 116–117

Potential equations, 360–361
Poynting's theorem, 368
Poynting vector, 367
Practical applications, 51
Preprocessing
    axial vibrations, 339–344
    cantilever beams, 226–230
    constant cross-section area, 93–95
    engineering problem, solving, 52
    natural frequency determination, 312–316
    simply supported beams, 197–200
    stepped bar, 134–139
    stress analysis, rectangular plate with circular hole, 246–251
    truss, 167–169
Prescribing boundary conditions
    elimination method, 50–51
    multipoint constrains method, 51
    penalty method, 51
Processing
    engineering problem, solving, 52
    varying cross-section area, 113–116

R

Rectangular cross-section transmission line, 426–427
Rectangular line, with diamondwise structure, 428–429
Rectangular matrix, 3
Rectangular plate with circular hole. *See* Stress analysis
Row matrix, 3

S

Scalar matrix, 4
Simply supported beams
    analytical method, 176, 180–181, 187–188, 192–193
    deflection values at nodes, 179, 185, 191, 196
    FEM by hand calculations, 177–178, 181–184, 188–191, 193–195
    nodal force calculation, 182–184
    post-processing, 200–202
    preprocessing, 197–200
    reaction calculation, 178, 191
    reaction values, 179, 185, 192
    slope values at nodes, 185, 191, 196
    software results, 179–180, 184–187, 191–192, 195–196
    total values, 185–186, 192
Single-strip shielded transmission line, 429–431
Singular matrix, 25
Six-strip line, 437–439
Skew-symmetric matrix, 5
Skin depth, 370–371
Software package, finite element analysis, 52
Software results
    airfoil, potential flow, 296
    axial vibrations, 338
    cantilever beams, 206–207, 212–214, 217–220
    cantilever beam, transverse vibrations of, 320–322
    constant cross-section area, 68–70, 80–81, 84–85, 91–93
    forcing function, bar, 345–346
    forcing function, fixed-fixed beam, 323–324
    natural frequency determination, 310–311
    one-dimensional heat conduction problems, 261–262, 280
    simply supported beams, 179–180, 184–187, 191–192, 195–196
    stepped bar, 125–127, 133–134
    stress analysis, rectangular plate with circular hole, 241–242, 243–245
    truss, 160–162, 165–167
    varying cross-section area, 106–108, 111–113
Solenoid actuator analysis, with ANSYS, 441–459
Square cross-section transmission line, 427–428

Square matrix, 3
Static equilibrium, 61
Stepped bar
    analytical method, 118, 122–123, 127–129
    deflection values at nodes, 120, 125–126, 133
    displacement calculation, 118, 119, 122
    FEM by hand calculations, 118–121, 123–125, 130–132
    nodal displacements, 117
    post-processing, 140–141
    preprocessing, 134–139
    reaction calculation, 118, 120, 125, 132
    reaction value, 121, 126
    reaction values, 134
    software results, 125–127, 133–134
    stress calculation, 118, 119, 123, 124, 132
    stress value at elements, 121, 126, 134
Stiffness matrix, 72, 99
Stokes theorems, 353
Strain, 60
Streamline, 292
Stress, 59
Stress analysis, rectangular plate with circular hole
    analytical method, 240–241, 242–243
    boundary of hole, 239
    center of plate, 240
    deflection pattern, 245
    element options, 247
    first principal stress distribution, 245
    one-quarter of plate, 239, 240
    plate thickness, 247
    post-processing, 251
    preprocessing, 246–251
    real constants, 247
    software results, 241–242, 243–245
    sub cases, 238–240
    tensile load at both edges, 238
    tensile load at one edge, 238
    validation of results, 246
    Von Mises stress distribution, 243
Stress calculation
    constant cross-section area, 66, 67, 70–71, 74, 79, 86
    stepped bar, 118, 119, 123, 124, 132
    truss, 153–154, 159–160, 165
    varying cross-section area, 98, 101, 104, 106, 110
Stress-strain behavior, 61
Structural problems
    direct approach, 49
    discretization of, 48
    element stiffness matrices and load vectors, 49
    element strains and stresses, 50
    nodal displacements, 50
    proper interpolation, 48
    variational approach, 49
    weighted residual approach, 49–50
Super-matrix, 6
Symbolic expressions, MATLAB
    differentiating symbolic expressions, 516
    integrating symbolic expressions, 517
    limits symbolic expressions, 517
    simplifying symbolic expressions, 515
    sums symbolic expressions, 518
    symbolic expressions, solving equations, 519
    taylor series symbolic expressions, 518
Symmetric matrix, 5

T

Temperature distribution, 286–289
Tensile forces, 59
Tension, 59
Thermal analysis
    one-dimensional heat conduction problems, 257, 258–262
    procedure of FEM, 258
    two-dimensional heat conduction problems, 258, 285–286
Three-dimensional element, 54, 415–420
Three-node bar element, 64
Three-node element, 460

Three-strip line, 435–437
Triangular matrix, 5–6
Truss
    analytical method, 147–150
    angle calculation, 150–153, 162–164
    definition, 145
    deflection value at nodes, 155, 160–161, 166
    2-D trusses, 145
    element stiffness matrix, 146
    element stress, 146
    FEM by hand calculation, 150–160, 162–165
    nodal displacements, 146
    post-processing, 169–170
    preprocessing, 167–169
    reaction calculation, 154
    software results, 160–162, 165–167
    stress calculation, 153–154, 159–160, 165
    stress values of elements, 155–156, 161–162, 166–167
Two-dimensional elements, 53–54
    band matrix method, 392–395, 397–400
    FEM to electrostatic problems, 382–400
    iteration method, 392, 396–397
    Laplace's equation, 382–392
Two-dimensional problem, 285–286
Two-node bar element, 62–63

V

Variational approach, 49
Varying cross-section area
    analytical method, 97–98, 103–104
    deflection values at node, 101, 107, 111–112
    displacement and stress, 97
    displacement calculation, 98, 104
    equivalent model, 99
    FEM by hand calculations, 98–103, 104–106, 108–111
    post-processing, 116–117
    processing, 113–116
    reaction calculation, 111
    reaction value, 102, 112
    software results, 106–108, 111–113
    stress values, 102, 107, 112
    Young's modulus, 97
Vector
    addition and subtraction, 28
    algebra, 28
    components of, 27
    $Del\,(\nabla)$ operator, 31–34
    equality, 28
    multiplication, 29–30
    multiplication of scalar, 28
    right-hand rule, 29
    unit, 27
Vector analysis, 27–34
Von Mises stress distribution, 243

W

Wave equations, 372–373
Weighted residual approach, 49

Y

Young's modulus
    constant cross-section area, 65
    varying cross-section area, 97